AF597678

Civil Aerospace Technologies – FITEC '98
(Farnborough International Technology Exploitation Conference)

Shareholders' Committee

J Bentley
IEE

A Little
RAeS

D Noble
IMechE

T Sidebottom (Chair)
SBAC

Management Committee

D Allerton
Cranfield University

D Atton
IMechE

C Bowyer
IEE

I Brackenbury
RAF

S Craven
IMechE

M Garrigan
SBAC

A Hamlett
IMechE

J Henderson
Real Time Engineering Limited

T Knibb (Chair)
British Aerospace

I Moir
Consultant

C Philbin
RAeS

Programme Committee

A Bradley
Rolls-Royce

J Calnan
Consultant

J Coldwell
British Aerospace Defence System

M Darkins
British Aerospace Military Aircraft and Aerostructures

I Ephgrave
Dowty Aerospace

I Moir (Chair)
Consultant

B Perry
RAeS Avionics and Systems Group

B Schofield
British Aerospace

I Strachan
RAeS Flight Simulation Group

R Taplin
British Aerospace Airbus

A Vincent
GKN-Westland Helicopters

Papers presented at a conference *Civil Aerospace* Technologies, FITEC'98
held at IEE, Savoy Place, London, UK, on 8–10 September 1998.

IMechE Conference Transactions

International Conference

Civil Aerospace Technologies FITEC '98

Organized by

IMechE Conference Transactions 1998–9

Published by Professional Engineering Publishing Limited for
The Institution of Mechanical Engineers, Bury St Edmunds and London, UK.

First Published 1998

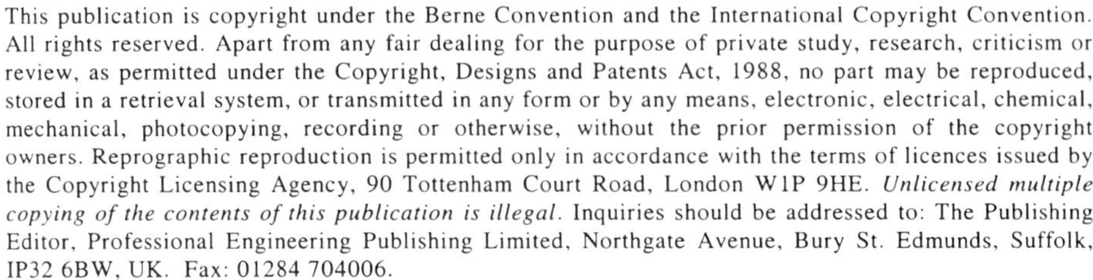

ISSN 1356–1448
ISBN 1 86058 168 4

A CIP catalogue record for this book is available from the British Library.

Printed by The Ipswich Book Company, Suffolk, UK.

Related Titles of Interest

Title	**Editor/Author**	**ISBN**
Designing Cost Effective Composites	IMechE Conference 1998–8	1 86058 148 X
Foreign Object Impact and Energy Absorbing Structure	IMechE Seminar 1998–9	1 86058 156 0
Aerospace Transmission Systems	IMechE Seminar 1998–6	1 86058 161 7
Gas Turbines – Reducing Time and Cost from Concept to Product	IMechE Seminar 1997–13	1 86058 126 9
The Manufacturing Challenge in Aerospace	IMechE Seminar 1997–3	1 86058 111 0

For the full range of titles published by Professional Engineering Publishing contact:

Sales Department
Professional Engineering Publishing Limited
Northgate Avenue
Bury St Edmunds
Suffolk
IP32 6BW
UK

Tel: 01284 724384
Fax: 01284 718692

Contents

Plenary Session

C545/074/98

Size – does it matter ?

R WHITFIELD
British Aerospace Airbus, Bristol, UK

One of the time honoured questions that has just as much relevance for the next Millennium in aerospace as it has had for the last in a number of other fields, size – does it matter?

Competing aphorisms include 'the bigger the better' and 'small is beautiful'. When looking at the relevance of size to a commercial company the key driver is the size of the natural market boundaries for competition, namely whether it is local, national, international, or truly global. Similarly, the scale of the investment and risk involved is a prime driver. But in general terms, size is not necessarily good. Financial statistics show for instance that smaller companies tend to have a higher growth rate than larger companies.

The merits of corporate size have varied over the years and in practice, there has been an element of fashion. In the UK for instance in the '50s and '60s there was long period of conglomerate growth. The merits of leverage were discovered in the late '70s and a bonanza of acquisitions in the '80s was followed by a period of "return to basics" and "back to core activities". Whilst there is a fair correlation between the rate of change of corporate size and the economic cycle, nevertheless fashion and other extraneous factors play a part.

Within Aerospace size does matter however and there has been an inexorable trend over the last 50 years towards consolidation. Over time, the number of new projects launched has gone down as the cost of developing a new product has increased. The scale of the investment required, the range of skills and the level of technology and risk are such that a small company cannot survive today as a prime aircraft manufacturer. Over and above the inexorable trend, there has, within our industry, been the impact of the end of the cold war with its consequential reduction in defence expenditure. Quick off the mark was the United States where the US leading manufacturers were actively invited by the Department of Defence to consolidate. As we know this they have done and with a vengeance leading to the establishment of the famous three gorilla companies. Consolidation in the US has now however largely run up against the buffers of Anti-Trust. Indeed the most recent consolidation move namely the proposed acquisition of Northrop Grumman by Lockheed Martin has effectively been stopped by the Anti-Trust Authorities.

Interestingly however those same authorities found the merger of the number 1 and number 3 players in the large civil aircraft market to be acceptable (albeit subject to certain conditions imposed by the European Anti-trust authorities). This leaves Boeing responsible for 87% of the world's civil aircraft fleet.

How has Europe responded to this headlong rush to consolidation in the United States? On the civil side, the merger of Boeing and Douglas has been a further spur towards the establishment of Airbus SCE. It was a coincidence that the Airbus SCE MOU signed in January last year was only one month after the announcement by Boeing and McDonnell Douglas of their proposed merger. Nevertheless, the consolidation of Airbus as a single company is a key step in ensuring Airbus's long term competitiveness and success in this newly created duopoly.

On the broader Aerospace and Defence front, the long period of talk has started to lead to some action. Recent transnational developments include BAe acquiring a stake in STE, a 49% stake in STN Atlas, a joint acquisition with DASA of Siemens Plessey and a 35% stake in Saab. Matra BAe has acquired a 30% stake in DASA LFK. Now in France the linking of Aerospatiale and Dassault and the recently proposed merger of Matra Defence and Space with Aerospatiale indicate important movement both in terms of structure and level of government shareholding. The key issue remains of course whether Europe will go ahead transnationally to form the European Aerospace and Defence Company and if so, when. The Trilateral Government declaration has been a timely demonstration of a Governmental desire for appropriate consolidation. Industry has made its proposal and there is a high degree of commonality in basic structural thinking. It remains to be seen however whether differences of opinion as to shareholder structure and Government shareholdings can be resolved.

It is interesting to contrast this inexorable trend towards greater size in aircraft manufacturing (subject only to global Anti-Trust issues) with other allied sectors such as airlines and Government, defence services.

In terms of airlines the ease of entry into the industry is in strong contrast to that of prime aircraft manufacturers. The basic unit of activity can be considered as the lease of one aircraft for a year in contrast to the full development and production of a 600 aircraft programme over twenty years. Economies of scale are significant however as manifested by the recent trend towards establishing global airline alliances.

In terms of Government there is little relevant structural change except in Europe. The whole concept of EADC is predicated on greater flexibility regarding issues of national security for individual European nations. This move is consistent with greater levels of military cooperation within Europe and indeed with the move towards greater European economic integration.

Europe is however presented with an interesting challenge regarding Airbus SCE. With the full weight of US government financial and political support now able to be focused, in civil terms, on their support for Boeing (a level of support which incidentally dwarfs that of Europe), it remains to be seen how European Governments respond, particularly in the light of the industrial move towards a single company being made by the Airbus partners.

Returning to the aircraft manufacturing industry, Boeing's recent financial problems merely go to undermine the fact that size in itself is no panacea. The key steps to be taken are to rationalise, to develop an effective culture of performance improvement and to retain flexibility and responsiveness to market change.

Rationalisation and the creation of one single identity for the new company is a process where, in the past, industry has not been particularly successful. Certainly when I joined British Aerospace in 1985 I learnt within weeks how to spot a BAC man or a Hawker Siddley man. You merely had to ask them their views on organisation structure. Although they had been part of this new British Aerospace Company for some 10 years, to a man an ex BAC man would argue the structural thinking of BAC whereas an ex Hawker Siddley man would argue for the organisational approach of Hawker Siddley. In my experience other mergers and consolidations at the time were little different. In the US the establishment of the gorillas has demonstrated the scale of rationalisation and cost reduction available within the US industry. There is inevitable duplication and inefficiency within newly consolidated structures and with the opportunity for performance improvement, Europe cannot afford to continue with an outdated cost structure. Having first achieved the appropriate size, Europe's first challenge is to carry out the consequential rationalisation and creation of one single company "culture". That single company culture then needs to be developed to ensure the optimum rate of performance improvement with clear leadership throughout the company, a clear set of values that underpin the whole business and an effective set of actions to develop the performance and culture with the new organisation.

A critical danger and a natural consequence of size is the risk of a lack of flexibility, adaptability and responsiveness. The transfer of shareholder value from IBM to Microsoft over the last 15 years would have been considered quite unimaginable 15 years ago. IBM at the time seemed quite unassailable. It was not. Similarly the Government services division of IBM in the UK demonstrated by winning the EH101 Merlin ASW prime contractorship that one no longer necessarily needed to be a helicopter manufacturer to play the leadership role in a helicopter contract. The development of systems integration and now systems of systems underlines the need within our industry for flexibility and rapid reaction to changing opportunities. In both the military and civil sectors one is seeing a movement in the demarcation line between the role of the customer and the role of the supplier. On the military side there is a rapid switch in terms of support away from military organisations and towards the private sector leading to the contemplation of power by the hour delivery to meet certain requirements. On the civil side one sees airlines increasingly considering moving away from the responsibility for the aircraft (in terms of ownership, maintenance and support) with the airline increasingly moving to be a marketing organisation oriented towards the customer service end of the business.

Civil aircraft manufacturers need to listen to their customers in terms of the type of service required as for example has Airbus in terms of the development of an A3XX super jumbo. Airbus has been highly proactive in seeking out airlines opinions on their requirements for large aircraft has similarly been responding to their replies. Here again its another case where it seems that size does matter. Whilst aircraft such as the A340-500 and 600 are ideally placed for developing the new long routes as part of an overall route fragmentation nevertheless we and our airline customers see significant demand in relation to traffic between the major hubs for aircraft larger than those available today. Airbus plans to respond to that demand by launching next year the A3XX.

Reduced environmental impact is another example of a trend which has increasing scale effects. Global climatic considerations and local demands for reduced environmental disturbance are demanding a response by the industry both in substance and in establishing the perception of what has and is being done.

In terms of substance, the technologies required to improve aircraft efficiency by further leaps and bounds is getting more and more expensive to investigate, develop and trial. The technology came "relatively" cheaply in the decades after jet airliners became common with quite large efficiency gains that meant customers would often "pay" for the improvements. An example might be the increase in combustion temperatures or the improvement in aerofoil cruise drag design. Future gains however will be smaller and more expensive to derive such as laminar flow technology, external fan engines or hydrogen fuelled aircraft. This means that it is only one or two organisations globally that will be able to invest to deliver these improvements and size will be of the essence to remain competitive.

I also mentioned the need to improve the image of the industry on environmental performance. A recent Friends of the Earth study supported by the European Commission on alternative forms of medium haul travel still used Boeing 727 data to represent aviation rather than today's generation of aircraft such as the A320. The B727 achieved sterling service in its day but on noise the modern aircraft 85dBA footprint is 90% less in area and usually kept within the airport perimeter, on fuel efficiency it is 30% better and yet in range performance, which increases the fuel takeoff weight, it is also 40% better. These facts are not generally attributed to the civil aircraft industry in analysis or the fact that modern high speed trains can be less efficient in CO2 generation per passenger km, produce noise disturbance all along their route and do this on huge government infrastructure and operating subsidy compared to air travel.

To sum up then, I think one has to accept that size does matter: the key issue is how effectively it is used. The US has taken the steps to achieve appropriate size; in Europe we must be sure to complete Airbus restructuring in 1999 and then complete the establishment of the European Aerospace and Defence Company with a distributed shareholder base. Consolidation however is merely the starting point. The challenges for the next Millennium are to

a) rationalise to ensure that the benefits of consolidation are rapidly realised
b) to focus on the development of a clear common performance oriented culture within the organisation
c) and thirdly to ensure that size never becomes a disadvantage and ensure that the new large organisations retain sufficient flexibility and responsiveness to cope with the ever evolving market place.

C545/058/98

Delivering value to the customer through technology

P C RUFFLES BSc, FEng, FIMechE, FRAeS, RDI
Rolls-Royce, Derby, UK

ABSTRACT

Airline customers demand aero-engines which operate in a safe and reliable manner while at the same time giving good economics to enable them to compete in the global air transport industry. Military customers require high performance aircraft weapons systems and are increasingly aware of the costs of procuring and maintaining such systems. Manufacturers aim to significantly improve their civil and military business success and market share by constantly striving to deliver better value to their customers. This is achieved by developing, manufacturing and supporting world competitive engines on time and to cost by demonstrating and transferring new technologies into marketable products in a timely and cost effective fashion. The advanced technologies designed into both civil and military propulsion systems are directly driven by customers' needs, with application specific and 'Dual Use' technology programmes enabling efficient investment in Research & Technology activities.

1 INTRODUCTION

The civil aerospace industry is both global and extremely competitive and has driven manufacturers to deliver increasing value to their airline customers over many years. The military market, on the other hand, is driven by the need to maintain an effective defence capability in contract arrangements where historically government has borne the majority of the risk. This environment has been less beneficial in delivering value to the defence customer than their civil counterparts. As we move forwards, it is important that the best practices of the civil sector are adapted to the military sector so that defence needs are satisfied at affordable costs in a political climate where defence now takes a lower priority relative to other national needs.

Delivering value to the customer goes beyond simply providing a product which meets the market need and is better than the previous generation. World competitive products are key to securing market share and giving a good return on investment. Technology is essential to delivering these product improvements, and must be incorporated effectively to agreed timescales and costs.

Technology is also able to improve business profitability by investing in the processes for design, manufacture and aftermarket support in order to reduce costs and shorten timescales. These benefits will also flow down to the customer.

These considerations are particularly pertinent to aero-engines which work at the limits of technology, which take longer to develop than the airframe, which are often in direct competition on the same aircraft and which have an aftermarket value typically up to double that of the original purchase price. It is therefore important to consider the whole life cycle of the engine at the outset of the design in order to arrive at the optimum product for a particular market requirement.

2 AERO-ENGINE MARKETS

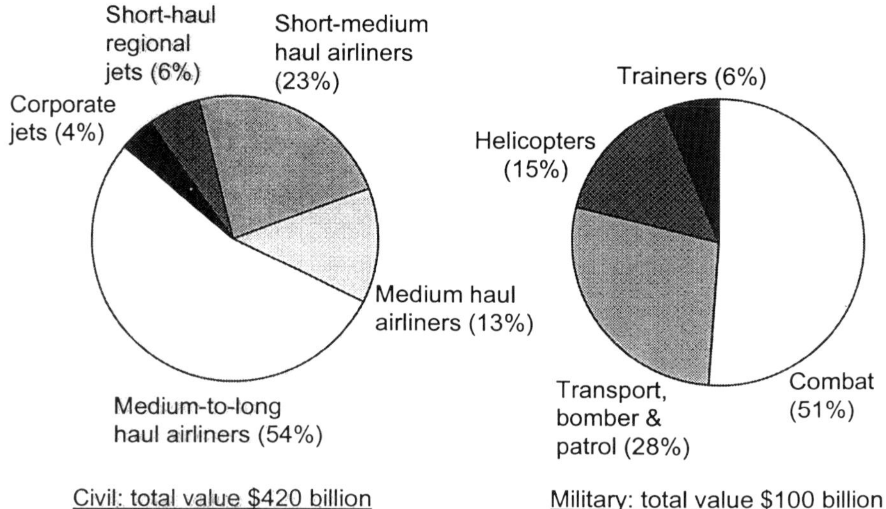

Figure 1 - Aero-engine markets 1998-2017

2.1 Civil

Over the last 25 years, global air traffic has grown by 6.6% per annum, helped significantly by deregulation, continually reducing aircraft operating costs and increased air travel for leisure purposes as disposable income has increased. Future growth is predicted to be 5% per annum over the next 20 years (figure 1), with fastest growth in the Asia-Pacific rim (6.3%), notwithstanding the recent disruption in this market sector. The North American market will remain the largest market in terms of value. Some 19000 aircraft, worth $1.1 trillion, will be needed during this period, requiring nearly 50,000 gas turbine engines, worth $420 billion.

Aero-engines producing over 50,000lbs thrust for medium-long haul high capacity airliners represent half of the future market by value. The other half requires smaller engines which, because of their lower cost, will make up the majority of engines delivered.

It is also worth noting that there is a structural change taking place in the regional jet market where turbofan powered aircraft are replacing turboprops. This sector is particularly cost sensitive and 'power by the hour' agreements are the norm causing the engine manufacturers to focus even more closely on life cycle costs.

In considering new designs, further enhancements to life cycle costs will be achieved by improvements in fuel burn, first cost and operating costs whilst satisfying increasing demands for lower environmental impact, in both noise and emissions. However, the balance between these design considerations is changing, with greater emphasis being placed on costs in relationship to performance.

2.2 Military

A new world order has emerged following the cold war which is characterised by local areas of instability with global economic and social consequences. This requires the major countries of the world to have a balanced and flexible defence capability to satisfy a whole range of potential threat scenarios, whilst smaller countries will continue to focus on regional concerns. This changed military scenario combined with reduced defence budgets mean weapons systems need to be procured faster and cheaper with the flexibility of cost effective updates to maintain capability.

Defence procurement budgets have generally stabilised at half the level prior to the cold war. Nevertheless, a substantial market remains, predicted to be worth $100 billion over the next 20 years, requiring engines for about 30,000 aircraft.

Combat aircraft represent half of the market, requiring engines worth $51 billion (27% in US) for a reducing number of airframes to replace many existing types, with increasing emphasis on multi-role capability. They must become more affordable, but first and foremost they will need to be sufficiently capable of performing their mission, demanding further increases in thrust : weight ratio and performance, whilst being more reliable, lasting longer and becoming more stealthy.

In the helicopter market, opportunities exist for medium transport helicopters, re-engining of existing aircraft and service life extensions. These will require a large number of engines, worth $15 billion, making this sector an ideal target for realising productivity improvements.

Sustained demand in the fixed wing transport and patrol sector requires engines worth $28 billion, where opportunities exist for using existing or planned commercial aircraft and engines to replace ageing products.

In the trainer market, worth $6 billion, a steady income will be enjoyed by those companies which have an existing product but a new product launch looks unlikely in the near future.

In summary, the need for flexibility in military capability will be met with a mixture of aircraft types and more flexible aircraft with affordability being a critical factor in determining the way ahead.

3 TECHNOLOGY CHALLENGES FOR CIVIL AERO-ENGINES

3.1 Reducing life cycle costs

The opportunity to substantially improve operating costs by improvements in fuel burn and weight, which for many years have made a large contribution, is diminishing unless major changes are made to aircraft and engine configurations, an issue discussed later (figure 2).

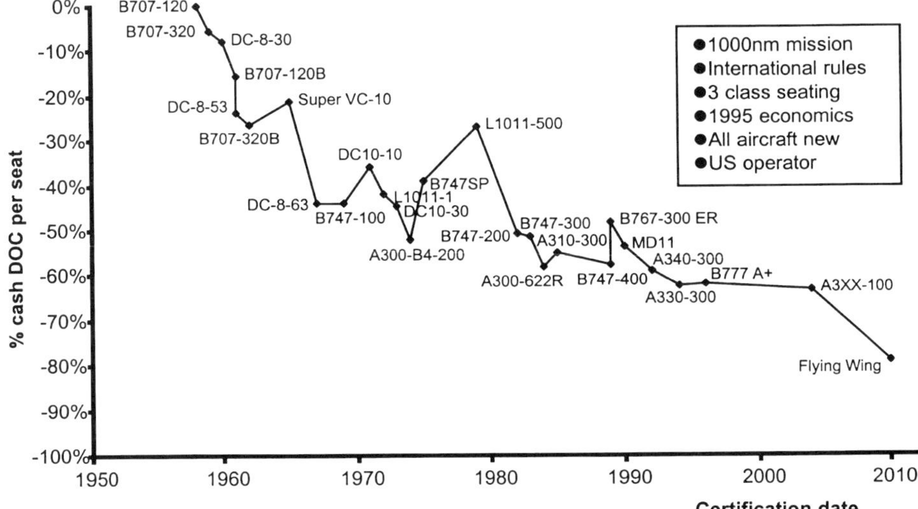

Figure 2 - Cash operating cost per seat mile

Significant improvements to engine life, reliability and maintainability have provided further reductions in life cycle cost, driven more recently by the development of twin engined long haul airliners with ETOPs capability; these aircraft required a step improvement in engine reliability from day one. To meet this challenge, manufacturers have optimised the processes used to design, develop and support the engine in service. The standard now achieved on the best engines will be expected, in the future, as the minimum acceptable standard for all new engines at entry into service.

As competition between manufacturers has intensified, companies have reduced product development costs and timescales. It is now possible to design and develop a new civil engine in about 48 months (to entry into service) through improved concurrent engineering processes supported by improved design tools and teamwork. The challenge is now to reduce these overall timescales even further by shortening design and manufacture timescales, and through better integration of the aircraft and engine development programmes.

The same competitive pressures are driving down unit costs and production lead times by 'lean manufacturing' approaches throughout the supply chain. Typically 70% by value of the engine is manufactured by suppliers who must participate fully throughout the product development and production phases for the full benefits to be realised.

However, in the future, a most rewarding area for further reducing costs is for engine manufacturers to provide a more complete aftermarket service. Here, technology can be used to reduce the cost of managing engines in service by using information from the engine health monitoring system, the known condition of the engine from its strip condition and knowledge of its design to define engine management plans which avoid unscheduled maintenance and meet the airlines demands for predictability (this is also known as 'Predictive Support Capability', i.e. *no surprises*). This information can then be used to plan maintenance schedules and control the production of spares to give a more stable and efficient use of resources by both the customers and suppliers.

With the technology becoming available it will be possible during revenue service for the engine to predict its own support needs which can then be used to order spares and control inventory, generate maintenance work scopes and schedule overhaul. Using neural networks these plans will be continually refined as more experience becomes available. These capabilities will become increasingly important as airlines move towards novel engine leasing arrangements and 'power-by-the-hour' type agreements with aero-engine manufacturers, where the lessor gets paid only while the engine is flying.

3.2 Meeting the environmental challenge

The aero-engine industry has always recognised its role in minimising the impact of air traffic growth on the environment. Legislation to reduce noise began as early as 1969 to be followed shortly afterwards, by legislation to reduce emissions of hydrocarbons, smoke, carbon monoxide and fuel leakage (figures 3 & 4).

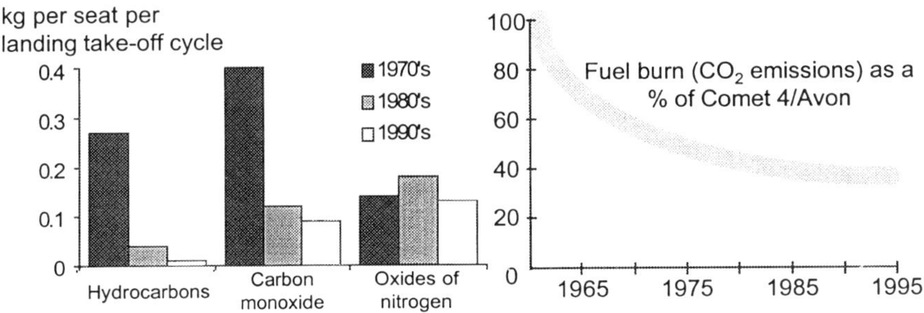

Figure 3 - Meeting the environmental challenge: Aero-engine combustion emissions

Since the mid-1980's the emphasis has been on substantially reducing emissions of nitrogen oxides around airports. Significant reductions have already been achieved despite the increases in combustion temperatures which have been necessary to improve performance. While NOx emissions are small, their presence at high altitude means that they will remain the focus of attention. Therefore, combustion technology must continue to be developed to negate the increase in NOx production accompanying future increases in combustion temperatures, in order to improve on today's standard.

With increasing concern over global warming, carbon dioxide emissions may become the next area of concern. Today, aero-engines contribute only 3% of all man-made CO_2 production and opportunities to further reduce CO_2 emissions depend upon improving fuel burn which, as already mentioned, is becoming more difficult to realise or by switching to alternative fuels such as hydrogen. The technical and economic difficulties associated with the latter suggest it is a long way off.

Although today's new aircraft are substantially quieter than the early jet aircraft, the increasing rate of take-offs and landings, and the evolution of bigger aircraft will increase the cumulative noise levels around airports unless further improvements are made in aero-engine noise reduction technology and operational practices. Aggressive local airport rules are now driving noise requirements with industry responding by designing to meet ever more stringent targets, particularly for large aircraft.

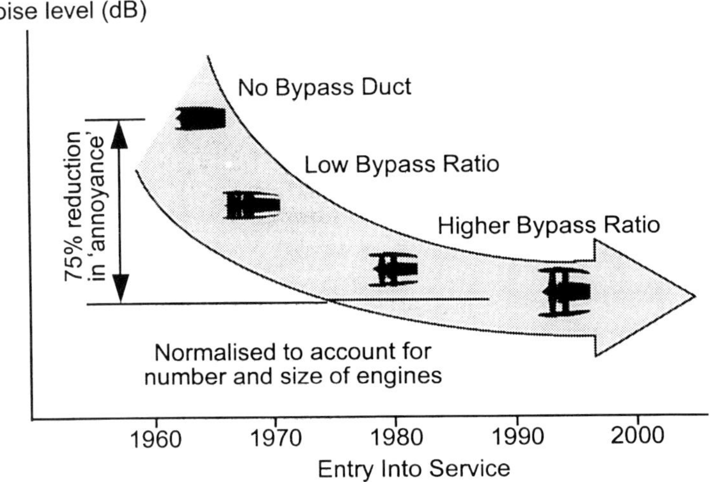

Figure 4 - Meeting the environmental challenge: Aero-engine noise 'annoyance' levels

The challenge for the engine manufacturer is to meet these targets without compromising either the payload range or fuel burn capabilities of the engine and hence deteriorating the position with respect to CO_2 production.

In summary, notwithstanding the significant progress made to date, the industry will need to continue to invest in technology to provide ever increasing value to the customer in all aspects of the engine design, manufacture and support whilst addressing these environmental challenges.

4 FUTURE CIVIL AERO-ENGINES

New technology will be applied progressively as existing products are developed and then in the longer term, through the development of radically new products.

Derivatives of the A340 and Boeing 777 aircraft will satisfy the market need, for long range operations up to 400 seats, while future derivatives of the Boeing 747 and A3XX will initially allow up to 550 passengers to fly a similar distance. Progressively, the technology applied to these aircraft, with further advancements, will be applied to other wide and narrow bodied aircraft.

In many ways the narrow body market offers a bigger challenge because the operating economics are more sensitive to cost and yet the cost per passenger mile of these aircraft is already higher than their larger counterparts. This makes it more difficult to launch new aircraft in this sector, particularly at the smaller sizes (e.g. 70-100 seats) despite the fact that they will exist in large quantities. Keeping these aircraft up to date will therefore be a bigger challenge.

One possible route for improving aircraft and engine economics is an 'all electric aircraft' where all the aircraft systems are electrically powered, driven by an all electric engine which makes wider use of electronic controls and integrates electro-magnetic technologies to offer major life cycle cost benefits at the air-frame level (figure 5). Predicted benefits equate to an improvement equivalent to 15% of fuel consumption.

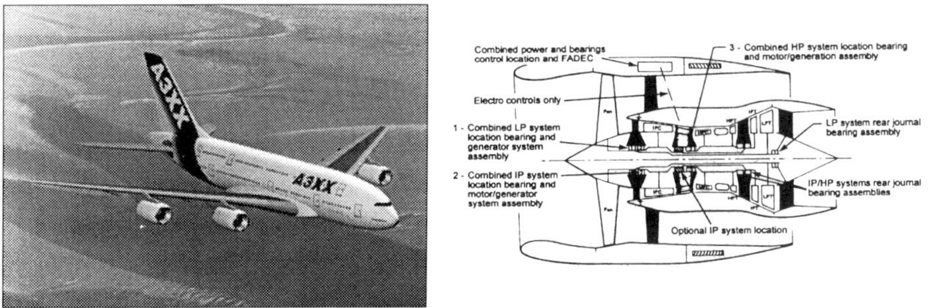

Figure 5 - Future civil airframe and engine configurations

The all electric engine concept progresses the evolution of propulsion system-airframe integration and meets the ever-rising electrical power demands of the aircraft. Embedding electrical generators (to supply aircraft power) inside the engine will be a first step towards complete integration of the aircraft and engine electrical systems, eliminating the need for heavy and unreliable accessories. Electromagnetic bearings will replace current technology, eliminating the engine's oil system. Engine operability will be improved as generated electrical power may be shared between shafts, enabling re-optimisation of the gas generator for surge control and enhancements to the relight envelope.

Eventually, however, more innovative aircraft designs will be needed to provide a step improvement in customer value.

A good example is the 'flying wing' concept, which may develop into the next big breakthrough in aircraft technology (figure 6). Preliminary design work has focused on medium to high capacity aircraft which integrates the wing and fuselage structures into a 'flying wing' in order to significantly reduce drag, thrust and fuel burn (benefiting the environment) and operating costs. Its viability depends upon the application of advanced flight control systems to overcome aerodynamic instabilities; such systems have already been developed for military aircraft.

Aircraft stability dictates that three or four large turbo-fan engines are installed at the rear of the wing, possibly with the inlets positioned above the wing to ingest a portion of the boundary layer and hence improve fuel burn, provided the profile drag-engine efficiency trade-off is favourable.

Greater benefits can arise by combining this aircraft concept with a new aero-engine concept which uses a contra-rotating aft fan driven by a two shaft gas generator. This engine design allows the gas generator to be re-optimised to improve specific fuel consumption and reduce engine weight. New light weight materials which offer greater stiffness would be required to support the rear engine structure. Engine noise will be shielded by the aircraft at take-off and landing making it possible to optimise the engine architecture for fuel burn and payload range requirements. Such an aircraft / engine combination could reduce fuel burn by as much as 30% compared with a modern conventional aircraft, or 40% if combined with the all electric engine technologies.

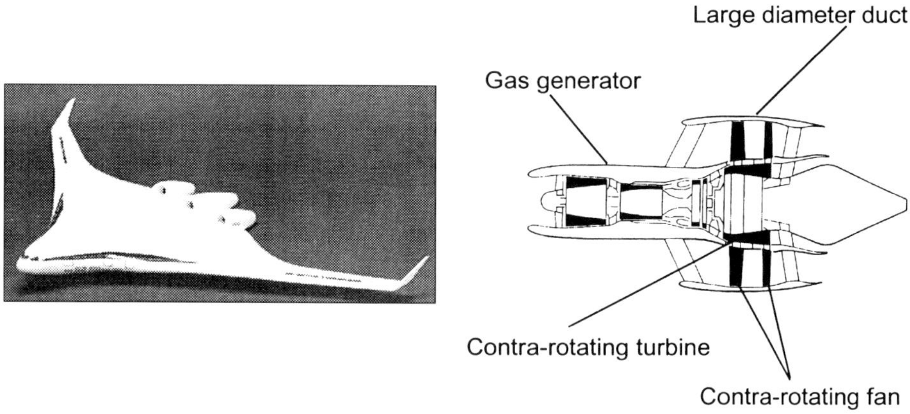

Figure 6 - 'Flying wing' and aft fan engine concepts

5 TECHNOLOGY CHALLENGES FOR MILITARY AERO-ENGINES

The predicted life-cycle cost associated with a new engine which meets the mission requirements will make it a key discriminator in future procurement programmes, challenging the engineering community to increase engine life and availability, reduce design complexity and improve ease of maintenance.

Flexibility for combat aircraft implies multi-role capability provided by CTOL/STOVL capabilities (e.g. JSF), supercruise and variable cycles. Capability for combat aircraft is primarily translated into the need for high thrust/power-to-weight ratios from the engines. With future engine cycles approaching thermodynamic limits in terms of temperature rise and hence limiting the thrust that can be achieved within a given envelope, the introduction of advanced materials and aerodynamics will be essential to realise smaller and lighter engine configurations. However, efforts to achieve a balance between affordability and capability may very well limit the rate of introduction of the more exotic advanced materials and hence weight reduction possibilities.

Helicopters and military transports significantly enhance flexibility by providing the ability to rapidly deploy forces overseas. Here, payload/range is an important consideration and so the emphasis will be on fuel burn and weight.

Reducing losses by maximising survivability of military aircraft requires low levels of 'observables', notably visible emissions (smoke and NOx), as well as infrared signatures and radar cross-sections, the latter two being military specific technologies. Cost effective solutions to these problems are priority.

There is an increasing demand from the military customer for extended 'out of area' operations, often from austere bases with only very limited maintenance capabilities. This is focusing a demand for levels of engine despatch reliability or availability akin to those achieved with civil engines. This concept of 'maintenance free operating periods' will shape the definition of future military propulsion systems, possibly leading to higher levels of system redundancy as has occurred in civil engines for long range twin engined aircraft.

Wherever possible these capabilities will be provided to existing aircraft types through mid-life updates. However, in the combat sector new aircraft will be required to fully satisfy future needs.

6 FUTURE MILITARY AERO-ENGINES

Within the European setting, existing combat aircraft will be replaced by Eurofighter 2000 which is now undergoing flight development leading up to operational service in 2002. The EJ200 engine is an advanced design with a thrust : weight ratio of 10 : 1. It is significantly simpler and has longer design lives than its predecessor, the RB199, and comparable in cyclic life usage with civil engines. New applications for the engine, such as the SAAB Gripen, will be found and the engine has significant growth potential.

Eurofighter will be complemented by the Joint Strike Fighter to replace the Harrier/AV8B which is the only operational aircraft with STOVL capability (figure 7).

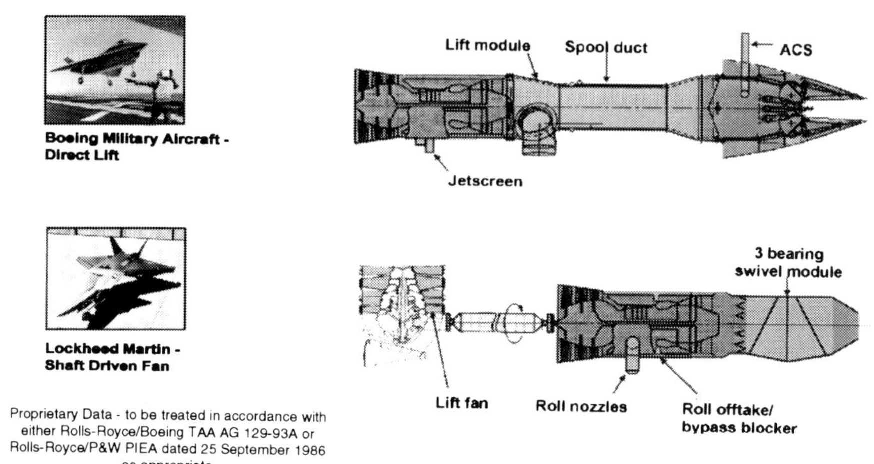

Figure 7 - Joint Strike Fighter propulsion systems

JSF is possibly the most important multi-role fighter project in the next 30 years. With a demand for 3000 aircraft, and an entry into service beginning in 2008, it is as challenging a timescale to that in the civil sector. JSF will be developed as a family of affordable air vehicle systems, using commonality of technology to achieve affordability. In this way, JSF will provide each of the military services with an affordable strike weapon system compatible with their unique needs, to replace several existing types of aircraft.

Most aircraft will be designed for CTOL, while a portion of the aircraft will have STOVL capability, which has proven its worth in several conflicts around the globe. Further, JSF will be capable of supersonic operation (unlike the Harrier) and increased weapons payload. The program is a good example of transatlantic collaboration, where the UK Government and industry is contributing towards the STOVL technology acquisition programme based on experience gained with the Harrier.

The Boeing concept is most similar to the Harrier design, using a vectored thrust, direct lift system with a jet screen and swivel nozzles as well as a series of attitude control nozzles.

The Lockheed Martin design uses a shaft driven lift fan and a 3-bearing swivel nozzle on the rear of the cruise engine to vector thrust downwards, and with additional nozzles to control roll. The lift fan is a two stage counter rotating unit, driven through a gearbox off the main engine LP shaft, and makes use of advanced fan design and manufacturing process technologies.

The GE F120 engine is being developed as an alternative to the P&W F119 as the main propulsion engine by a partnership between GE, Allison and Rolls-Royce. Rolls-Royce is developing the fan, while Allison are responsible for the combustor, LP turbine and accessory gear box.

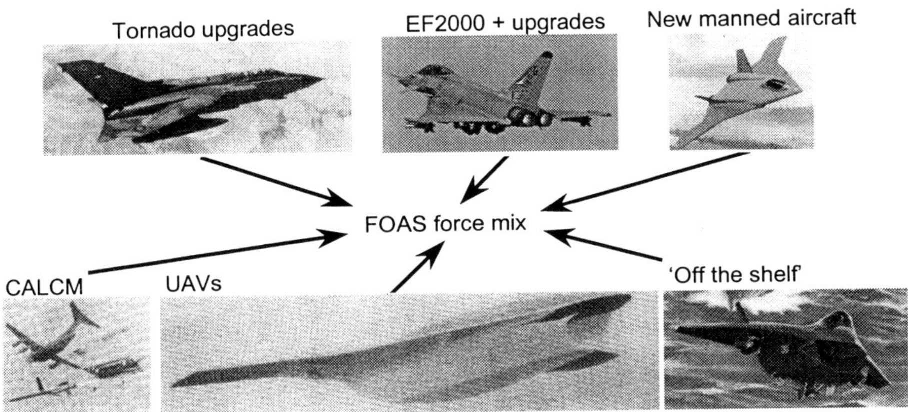

Figure 8 - Future Offensive Aircraft System

In the longer term the Future Offensive Aircraft System (FOAS) will be designed to meet the future RAF need to strike deeply behind enemy lines, to replace the RAF Tornado GR4, with entry into service in 2015 (figure 8). A manned aircraft would need to carry a weapons payload over a long range, and survive the hostile environment encountered over enemy territory. Manned aircraft are not the only option being studied, however, and the Future Offensive Aircraft *System* could be a mixture of unmanned UCAVs, manned strike bombers and transport sized aircraft with stand-off weapons such as cruise missiles. One advantage of UCAVs in this role is that by stripping out the pilot and related support systems, they can be designed smaller than manned aircraft and yet have higher 'g' capability. UCAVs may be feasible for strike missions by 2015, given the technological advances being made. The final decision will be driven as much by affordability as by the capability offered.

A consortium of companies are involved in designing the various elements of FOAS, making use of best practice and cost effective 'Smart Procurement' methods, which are essential to meet the benchmark for programme and price set by the JSF project. The customer has identified requirements for capability and affordability and placed the challenge on industry to determine how these can be achieved. At present, industry is assisting with the feasibility studies to determine how each concept meets the requirement, ahead of a down selection by the UK MoD.

7 EFFECTIVE TECHNOLOGY ACQUISITION

In order to deliver competitive civil and capable yet affordable military products, an effective technology acquisition programme must precede product development if short timescales and low product development costs are to be realised (figure 9).

In aero-engines, some technologies are application specific, but many technologies are common between civil and military applications. 'Dual Use' technology programmes offer the opportunity for more highly integrated research and technology activities supporting both military and civil products. For military projects, 'Dual Use' technology programmes can offer significant benefits to both governments and industry. For governments, they offer the opportunity for more cost effective application of new technologies as the demands from the civil market drive more rapid technology evolution and insertion into products, with service experience considerably in advance of military applications. For industry they offer the benefit of cost sharing.

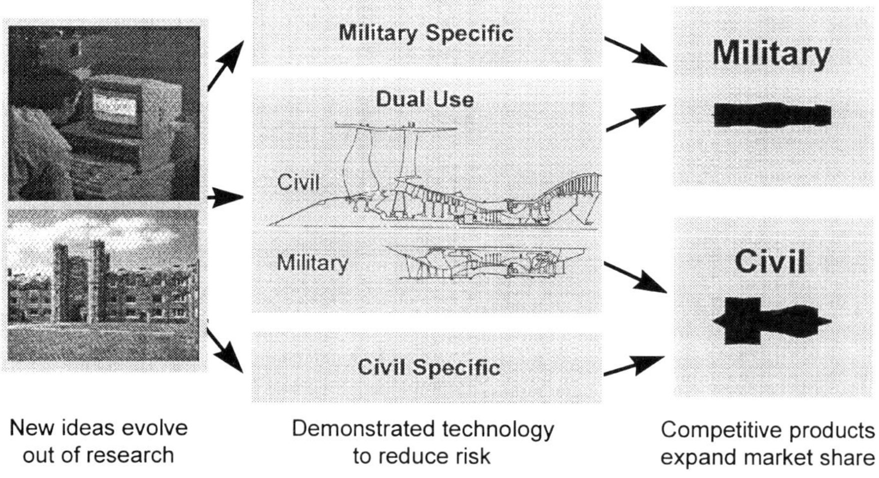

Figure 9 - Technology acquisition process

Demonstrating new technology must be performed in an environment which is representative of the engine. Technology Demonstrator Programmes using representative engines and rigs have proven their value on many occasions (e.g. XG40 for EJ200) and greater use of them will be made in the future to deliver proven and affordable technology 'just-in-time' for engine development

The cost of research and technology (R&T) is relatively small compared with the cost of product development, and this investment provides protection against the cost and risk of downstream engine development activities and product competitiveness. Growing competition between manufacturers drives the need to accelerate the technology acquisition process as the first part of reducing product development cycle times.

8 PRODUCT DEVELOPMENT

The product development process essentially comprises three stages; New Product Planning, Full Concept Definition and Product Realisation (figure 10). The deliverables for the process are a complete set of drawings, specifications, manufacturing instructions and technical manuals which have been fully approved by the appropriate authorities.

In the first stage, the market need is identified, design options are considered and the business plans are evaluated in order to identify the best concepts around which to acquire the necessary capability. FOAS is undergoing this process at present, involving all major suppliers, prior to a capability acquisition programme.

When the appropriate level of capability has been acquired, the concept is fully defined to include all system and sub-system specifications embracing all technical deliverables including maintainability and unit cost targets and the outline production schedules. All the suppliers would be fully involved at this stage.

A comprehensive risk assessment is carried out at the end of stage 2 and will form the basis of determining the subsequent product development programme which should aim to reduce the residual risks as rapidly as possible.

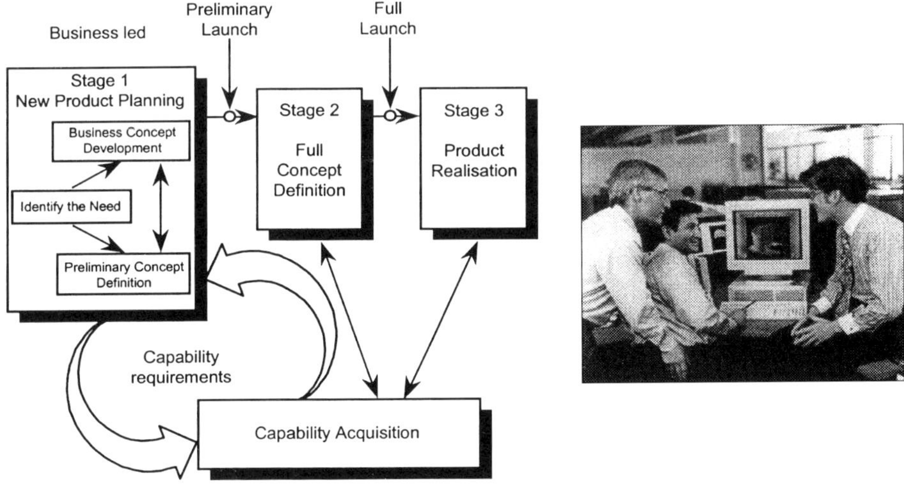

Figure 10 - Product development process and IPD teams

In the third stage the product is fully defined, manufactured and tested against the requirements set at the beginning of the programme. Some contingency should be allowed for re-design but this should be minimal if the previous stages have been carried out effectively.

Reduced costs and timescales are realised by ensuring that the technology is available 'just-in-time' and by the use of Integrated Product Development (IPD) teams who each manage different work packages within a product work breakdown structure, and with integration being co-ordinated by the project office. These teams are supported by modern IT systems, which speed up design analysis and data transfer so all members of the team are working on common data sets.

Formal reviews are carried out throughout the process by independent experienced engineers to ensure that best practice is being applied throughout the process. In this way, quality is continuously assured.

These processes have now been successfully applied to several civil engine product development programmes and clearly identified benefits have resulted. They are also being applied to military programmes with JSF and FOAS being the first opportunities to apply them from the start. The benefits are already being realised on the JSF programme.

9 SUMMARY AND CONCLUSIONS

Delivering value to the customer depends critically upon companies working closely with the customer to fully understand future requirements and identify solutions to customers' problems.

From this understanding, cost effective research and technology acquisition programmes can be pursued against clear timescales and objectives with innovation being stimulated by setting challenging yet realistic targets.

Provided technology is available 'just in time', product development programmes can proceed rapidly and smoothly within planned budgets and timescales often exceeding the customers' original expectations. However, these benefits depend upon having well defined processes, operated by Integrated Product Development teams supported by the latest Information Technology in order to speed up the design process and to link all people who are working on the project.

In this way, the aerospace industry will continue to deliver value to its customer thereby sustaining the growth into the 21st century that it has enjoyed this century.

Next Generation
Civil Aircraft

C545/045/98

Aerodynamic testing for the next generation of aircraft

M WRIGHT MA, MRAeS
European Transonic Wind Tunnel GmbH, Cologne, Germany

SYNOPSIS

New efficient aircraft configurations will require advanced experimental facilities. The greatest obstacle to the use of most wind tunnels is the need to extrapolate from the Reynolds number of the model in the tunnel to that of the full scale aircraft in flight. This can be overcome by the use of a cryogenic wind tunnel such as the European Transonic Wind Tunnel with the capability for testing at full flight Reynolds number. Validation and benchmark tests have shown excellent repeatability, agreement with existing data and high Reynolds number results which appear to be consistent with those obtained in flight.

INTRODUCTION

The next generation of aircraft will require significantly reduced life cycle costs. This will involve, in addition to reductions in development and manufacturing costs, major innovations in the fields affecting direct operating costs. It is sometimes said that aerodynamics is a classical discipline which is fully mature and hence that seeking further advances in that field is not a cost-effective way of improving performance. Aerodynamic efficiency is, however, far from being mature and fully developed. On the contrary it will have to be significantly improved in pursuit of ever more efficient air transport.

This, in turn, will imply the adoption of radically new aerodynamic design features and configurations. But the technical and commercial risks that would normally result from such an innovative approach are hardly tolerable in view of the trend towards fewer, ever more costly aerospace projects. Any significant modification of the aerodynamic configuration of a large aircraft, once the prototype has flown, is likely to prove impossibly expensive. Something must be done to allow the necessary innovation without incurring excessive risks even if this requires an increase in some part of the non-recurring cost of the initial design.

MANAGEMENT OF TECHNICAL RISK

The problem of justifying investments in the preliminary design phases, while having to control overall costs very tightly, is not new to designers and manufacturers of aircraft but it is acquiring increased importance as the world aircraft industry focuses on smaller numbers of bigger companies working on fewer large projects in an environment of intense competition. The key to success in these circumstances is an appropriate use of the most advanced theoretical methods and experimental facilities to support innovative designs hence ensuring a strictly controlled level of technical risk.

The aerodynamic design process has to be managed in such a way as to offer the possibility of a well prepared design in which the company can have full confidence even where it departs significantly from the concept of aircraft already flying. In the past piloted aerodynamic demonstrators have been used to confirm really new configurations but these are now usually considered to be too expensive for civil aircraft projects. One of the last aircraft to be used in this way on a European civil project was the modified Fairey Delta 2 as a demonstrator for the Concorde configuration in the late 1950's. Nowadays the aerodynamic design must rely mainly on theoretical methods including Computational Fluid Dynamics (CFD) validated by wind tunnel tests and backed up by specific tests of the developing configuration in the wind tunnel. Initial configuration studies can be carried out in relatively inexpensive low and high speed industrial wind tunnels but extrapolations to flight conditions must then be made with great care and may present some difficulty particularly for novel high performance configurations. The later design iterations must be tested and the final configuration validated in an advanced wind tunnel capable of accurate representation of the conditions in flight.

WIND TUNNEL CONTRIBUTION

Most modern wind tunnels have excellent instrumentation to measure accurately the forces, moments and pressures on the models. They can match the flight Mach number exactly and have more or less satisfactory methods for correcting for the effects of wind tunnel interference. There is however another important aerodynamic parameter which governs the relative contributions of inertia and viscous forces in a gas. It is called Reynolds number and it is not usually correctly represented in wind tunnels. Therefore the greatest obstacle to the use of wind tunnel testing for really innovative design of novel efficient aircraft configurations is the need to extrapolate from the Reynolds number of about 5 - 10 million of the model in the tunnel to that of 30 - 80 million of the full scale aircraft in flight. This obstacle can be partially overcome by testing large models in highly pressurized tunnels. By this means Reynolds numbers of about 10 million can be reached. The use of a cryogenic gas allows a further increase in density accompanied by a decrease in viscosity. Both of these effects contribute to an increase in Reynolds number and a cryogenic wind tunnel provides the only possibility for testing the configurations of large aircraft at full flight conditions. The European Transonic Wind Tunnel, owned by France, Germany, the Netherlands and the UK, and located in Cologne is a pressurized, cryogenic wind tunnel which, by using pressures up to 4.5

atmospheres and temperatures down to 100Kelvin, can test whole models at Reynolds numbers up to about 50 million and half models up to 80 million. The building of the tunnel between 1988 and 1992 required an investment of some 650 million Deutschemarks by the four Governments but this is easily justified by the contribution ETW can now make to the design of efficient aircraft by the aircraft industry world wide.

EFFECT OF REYNOLDS NUMBER ON AIRCRAFT CHARACTERISTICS

In general a low Reynolds number leads to an orderly or laminar flow pattern whereas a high Reynolds number flow tends to be turbulent. It was observation of this characteristic in the flow of water through pipes that led Osborne Reynolds to invent the non-dimensional number named after him in Manchester in 1865. As far as wind tunnel models are concerned the flow in the boundary layer close to the surface of the model undergoes transition from laminar to turbulent some distance behind the leading edge or nose of any part of the configuration. In flight boundary layer transition usually occurs at or near to the leading edge. This misrepresentation of the state of the boundary layer over much of the aircraft could lead to false estimates of aircraft performance, stall characteristics, buffet boundaries etc. If the transition position in flight can be reliably estimated the model boundary layer can be artificially tripped into a turbulent state by the application of carefully designed strips of roughness suitably positioned on the surface of the model. Figure 1 shows the variation of aircraft drag with Reynolds number as measured in ETW using transition fixing at the lower values of Reynolds number compared with a sophisticated prediction method assuming the two fixed transition positions at lower Reynolds number and transition at the leading edge for the highest flight value. As shown here, forced transition can give an approximate match to the main features of the flow in flight if carried out carefully but it can not fully represent the structure and thickness of the boundary layer on a full scale aircraft in flight.

DESCRIPTION OF ETW

ETW is a transonic tunnel with a slotted wall test section to give acceptably small errors due to the presence of the test section walls even in the transonic regime of Mach number. It is pressurized, cooled and maintained at temperatures down to 100K by the injection of liquid nitrogen into the circuit. The test gas is therefore itself pure nitrogen. Air would liquefy at the lower test temperatures. The tunnel can be pressurized up to 4.5bar the pressure shell being internally insulated so that it remains at or near the outside air temperature. The general arrangement of circuit is shown in Figure 2. Liquid nitrogen is sprayed into the tunnel through 230 separately controlled nozzles upstream of the 50MW compressor which drives the nitrogen round the circuit. This ensures adequate evaporation and diffusion of the nitrogen as it passes through the compressor on its way to the test section and allows the temperature uniformity in the test section to be precisely controlled. The gaseous nitrogen is exhausted to atmosphere through controlled blow-off valves which maintain the desired pressure in the tunnel.

Although the 2300 tonnes of steel in the pressure shell does not need to be cooled with the gas in the tunnel there are another 850 tonnes of steel within the insulation layer, mainly in the vicinity of the test section and this internal structure has to be cooled to a temperature near to that of the test gas. Starting from ambient temperature this cooling process is both time consuming and expensive so it is better to retain the tunnel temperature while making configuration changes and to change it as seldom as possible during a test campaign. For this reason special arrangements are made for transport of the model to Variable Temperature Check-out Rooms where it can be worked on while cold or itself warmed without having to warm up any of the tunnel internal structure. The arrangements are illustrated in Figure 3. The model is supported below a self-contained model cart which is transported between unconditioned Cart Rigging Bays, the Variable Temperature Check-out Rooms and the Test Section by a Model Cart Transporter travelling through a Transfer Hall built over the top of the preparation bays and test section. In the Dry Air Hall part of the transfer hall where cold models have to be handled the air is maintained at a very low dew point by the continuous injection of dry air.

CALIBRATION OF ETW - WIND TUNNEL CORRECTIONS

The tunnel has been calibrated to obtain precise empirical data concerning the corrections needed to allow for the presence of test section walls and model support structure.

As a modern wind tunnel offering the ability to test at real flight values of Reynolds number the calibration of ETW has to be more thorough and precise than that common in other wind tunnels which accept that they are fundamentally unable to offer such exact physical simulation of conditions in flight. Thus, for instance, the effects of the slotted test section walls have been established empirically using a highly instrumented model of an A-320 tested in ETW both with the usual slotted walls and with the slots blocked to create a solid wall test section. There is a large body of theoretical and experimental knowledge of solid wall test sections in various tunnels and this makes it an ideal datum against which to measure the effects of the walls. The wall interference corrections to Mach number, axial force and angle of attack are small but significant. For example the correction to Mach number shown in Figure 4 is a weak function of Reynolds number as well as Mach number itself and reaches maximum values of about 0.003 at high values of both these parameters.

Another example of the care that had to be taken in calibrating ETW is the determination of the far field effects of the model support structure including the sting. These corrections were initially obtained by testing axial probes which resemble in their downstream portion the straight stings which are used to support many models. Early results from ETW were corrected by this means which is accurate for straight stings used at low angle of attack. These corrections have now been extended to include a wider variety of stings including Z-stings and larger values of angle of attack by the use of a translating calibration rake mounted on the top wall of the test section. Typical results for a Z-sting are presented in Figure 5. The lower line shows that, at the point of model rotation, the Z-sting produces only small errors in flow direction which are a very weak function of angle of attack. Z-stings are sometimes assumed to be ideal for studying rear fuselage and tailplane characteristics but the upper line shows that this type of sting gives significant flow deflections over the rear part of the model and that they are now a relatively strong function of angle of attack. The necessary corrections, obtained from the translating rake tests must therefore be carefully taken into consideration when using a Z-sting mounted model for these kinds of investigation.

The near field effect of the sting on the rear fuselage of the model is strongly influenced by the geometry of the model. Such effects are sometimes calculated theoretically by aircraft designers who have ready access to computer simulations of the model configuration geometry. However this can also prove inaccurate in transonic slotted wall tunnels. Therefore ETW has acquired a twin sting rig shown in Figure 6. This rig can support the model from its wings while the direct effects of the presence or absence of the sting on the forces and moments on the rear fuselage are measured directly by comparing the results with no sting and with a dummy sting positioned near to the attachment point at the rear of the model. It is planned to perform the first tests with this rig during 1998.

VALIDATION OF ETW - BENCHMARK TESTS

A number of validation and benchmark tests have also been carried out and have shown excellent repeatability, agreement with existing wind tunnel data at modest Reynolds numbers and consistency with high Reynolds number results obtained in flight.

Figure 7 indicates the drag repeatability achieved in three test series in ETW. The parameters plotted are aircraft drag against lift and the repeatability is of the order of 2 drag counts in the three series which employed different internal balances requiring a complete breakdown of the model to exchange and even two different models of the same configuration. The repeatability for immediate re-runs of the same test is an order of magnitude better - less than one drag count. This means that the potential resolution of measurements in ETW is extremely fine so that the effects of relatively minor configuration optimisations can be measured accurately.

Figure 8 shows results from the same test campaign as Figure 1 and a preliminary version published earlier showed relatively good agreement of the lift measured compared with that measured at the same modest value of Reynolds number in the 8ft. wind tunnel at DERA Bedford. Figure 8 shows however the results after application of the recently measured wall interference corrections and the agreement is now excellent considering that two different models of the same configuration are being compared in two different wind tunnels using different internal balances to measure the forces on the model. The near field sting effects have been eliminated empirically from this comparison by subtracting the results obtained with fuselage alone. Thus ETW can match the accuracy of the benchmark wind tunnel at its own Reynolds number but ETW can also test the model of this aircraft at its full cruising flight Reynolds number of 30 million.

REYNOLDS NUMBER EFFECTS SEEN IN ETW

Reference has been made to the difficulty of representing accurately the Reynolds number effects on flow details by transition fixing in a low Reynolds number tunnel. Figure 9 shows the complex effect of shock wave position on the wing of a small military trainer configuration in ETW. The transition position was fixed in some of the lowest Reynolds number tests but this can be seen not to have given a good representation of the high Reynolds number shock position. Wing pressures were measured on the aircraft in flight at a similar Mach number although not at the precise flight conditions of Figure 9. The results at a similar flight condition are compared with those obtained in ETW at the same condition in Figure 10 and show very good agreement.

CONCLUSION

Future generations of aircraft will require the most advanced theoretical methods and experimental facilities if the technical risk inherent in novel, competitive designs is to be adequately controlled. ETW is an example of a modern high Reynolds number wind tunnel which has shown itself to be a unique and effective tool which can be used in the design of current conventional civil and military aircraft. It will gain even greater importance in the design and development of novel configurations in the future.

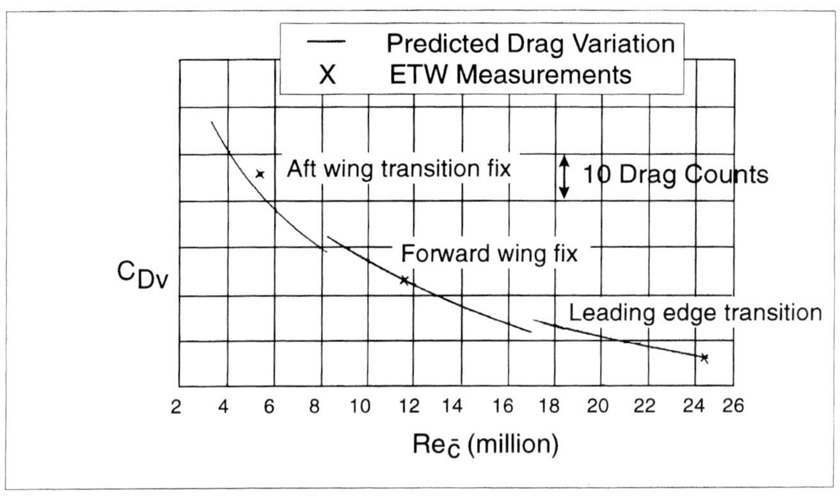

Fig. 1 Variation of Drag with Reynolds No at $C_L = 0.3$

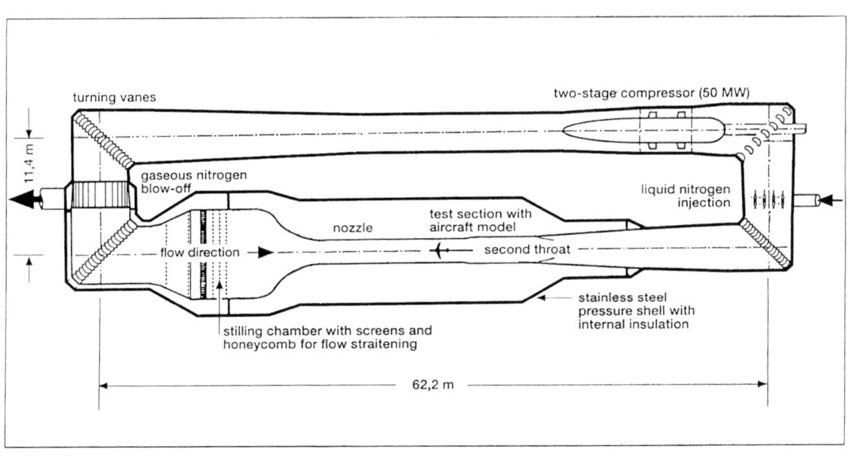

Fig. 2 ETW Aerodynamic Circuit

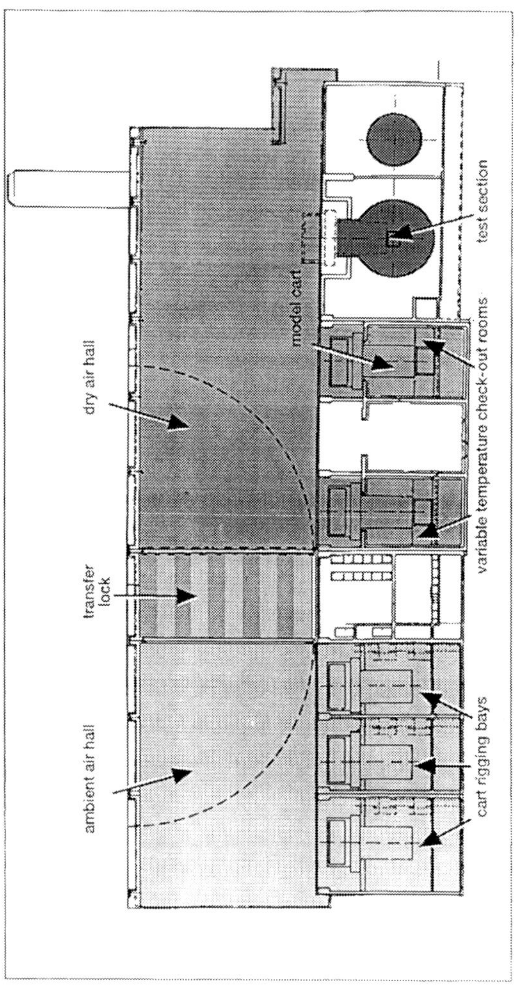

Fig. 3 Model Preparation and Transport

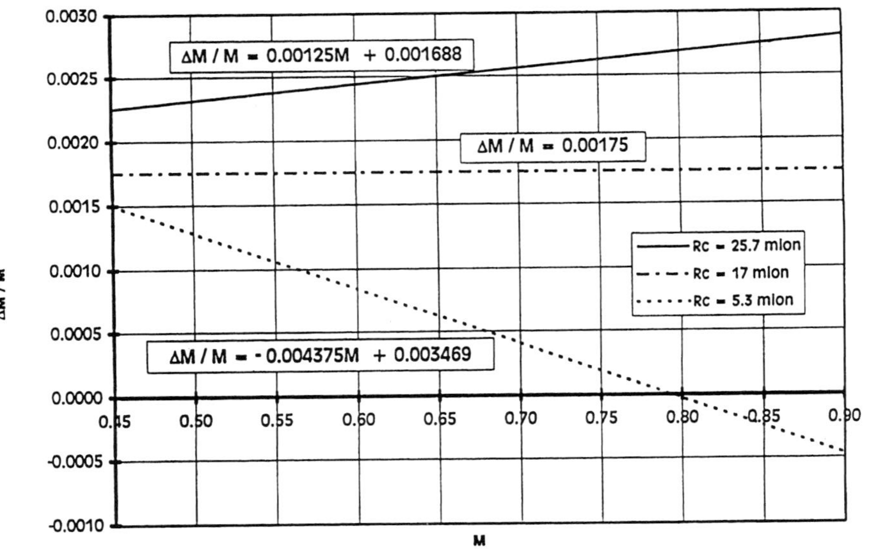

Fig. 4 - Model Tested in Slotted Wall ETW
Correction to Mach number based on solid/slotted comparison of wing pressures

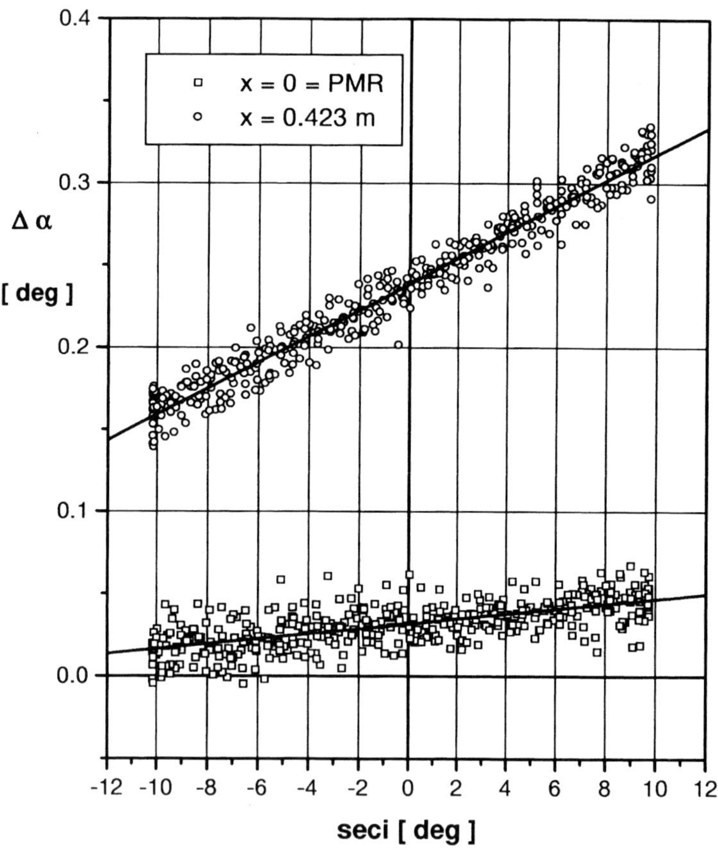

Fig. 5 - Translating Rake Results - Z-sting Generated Upwash

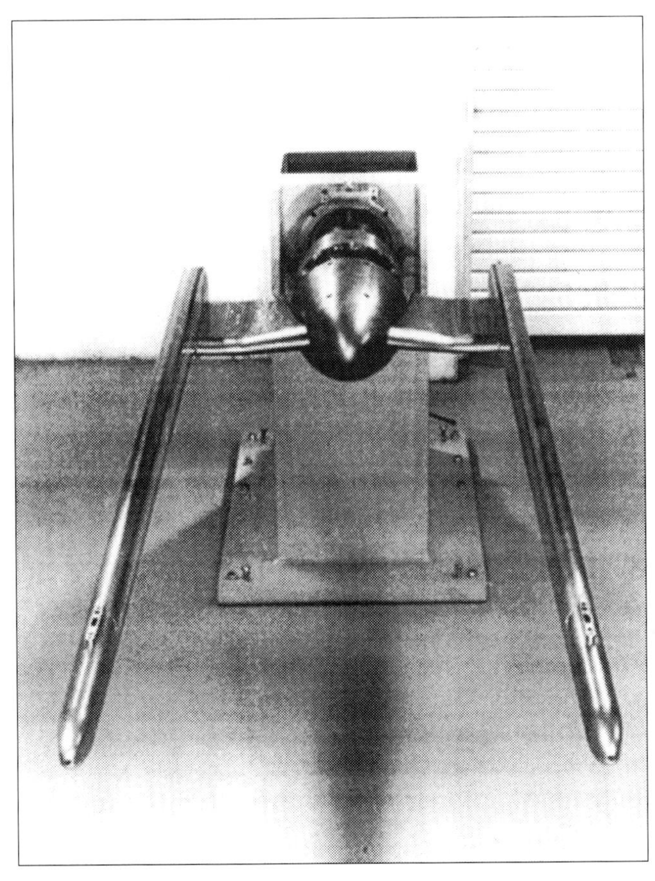

Fig. 6 - ETW Twin Sting Rig

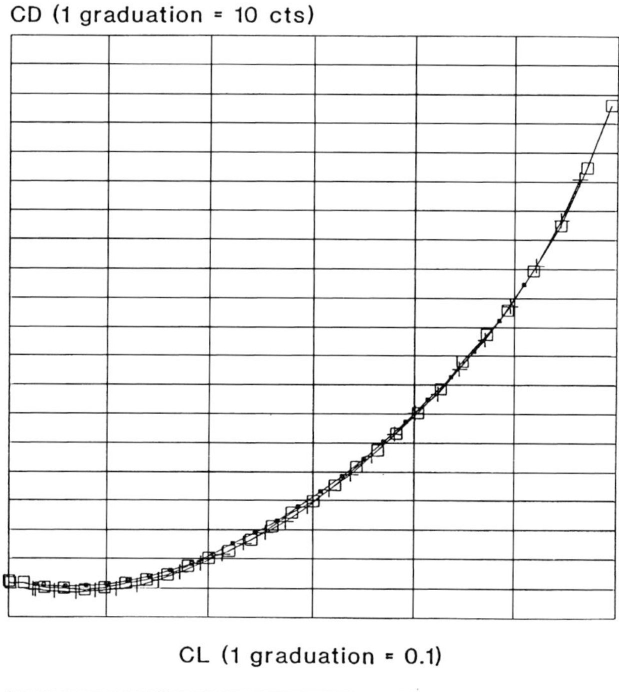

Fig. 7 - ETW Inter Test Series Repeatability

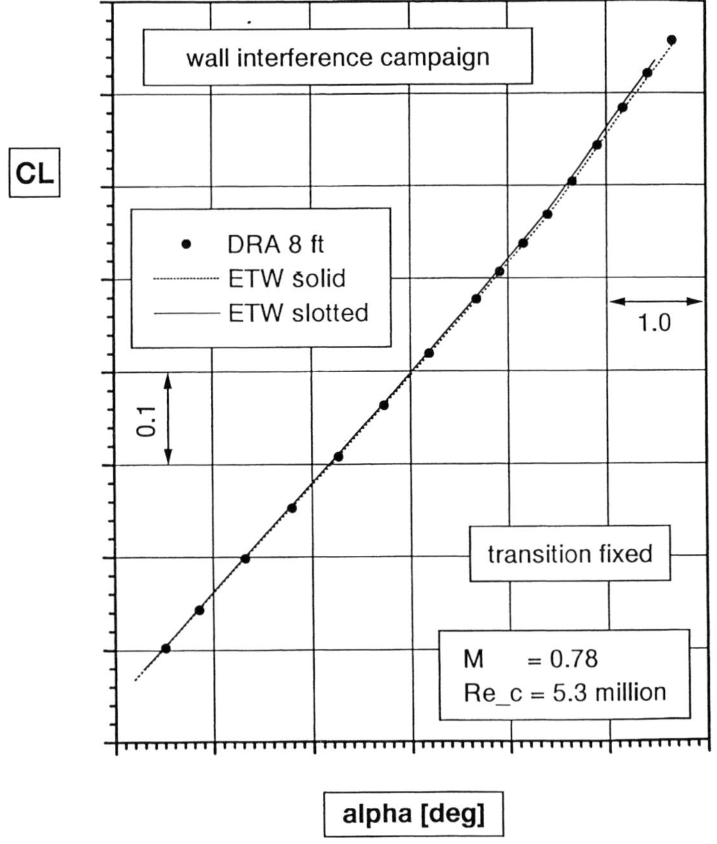

Fig. 8 Interfacility Comparison - Model on Straight Sting

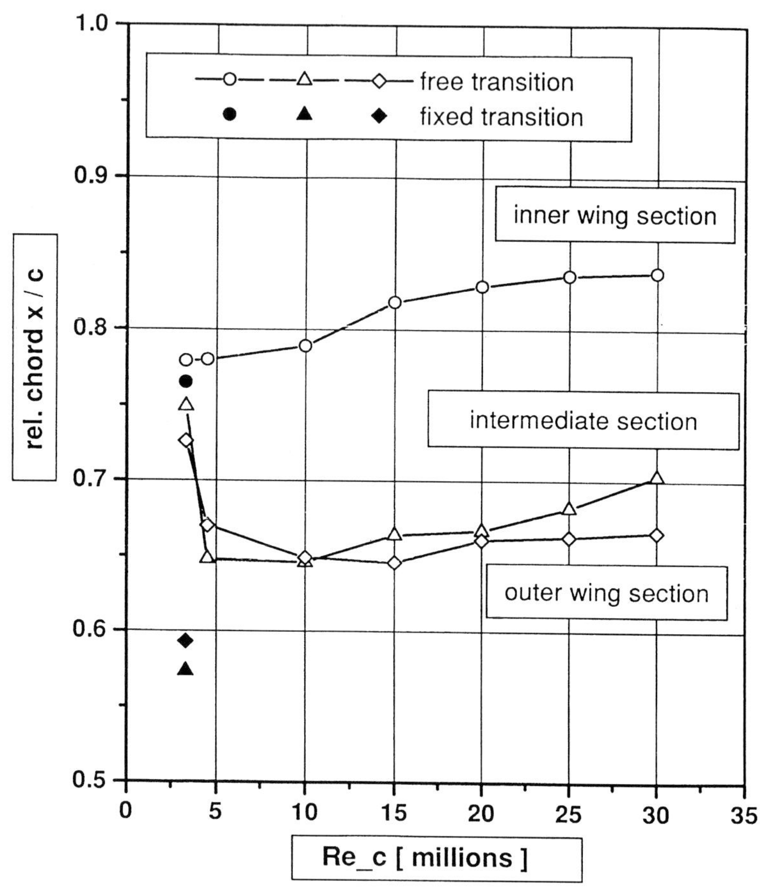

Fig. 9 - Shock position on upper Wing Surface

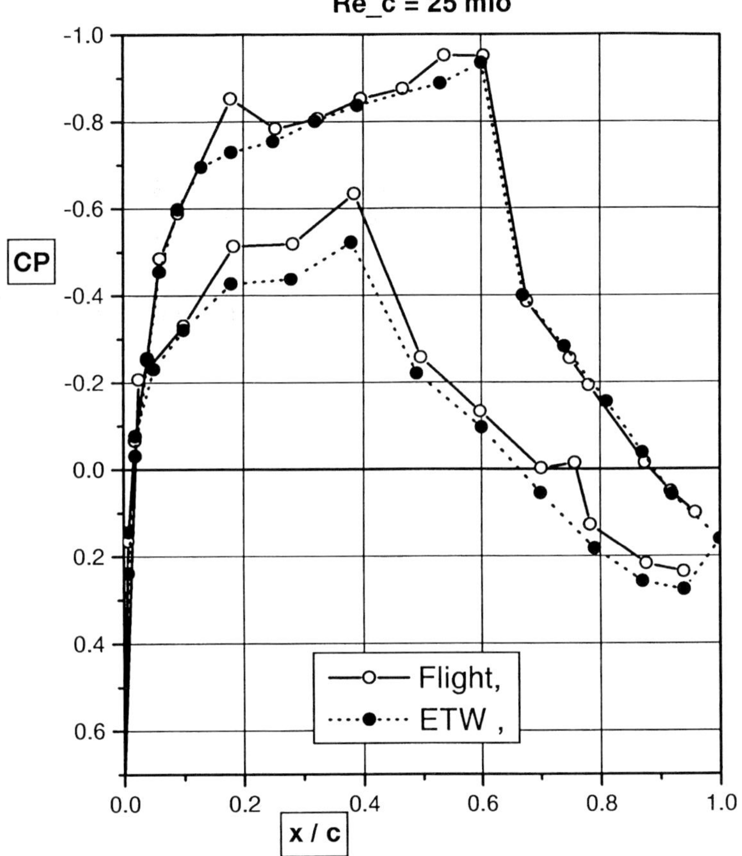

Fig. 10 - Wing Surface Pressure Distribution

C545/054/98

Innovations in airframe design and manufacture for the next generation of AIRBUS transport aircraft

J KOSHORST
Airbus Industrie, Blagnac, France
M BLACK
British Aerospace, Airbus Limited, Bristol, UK

SYNOPSIS

The next generation of AIRBUS Transport Airplanes will consist of airplanes larger than any airplane before.

They will be more efficient through economy of scale and by way of technological advances. The economic challenge to meet the airlines' expectations is for a 15 to 20 % reduction of operation cost per seat-mile, relative to its only existing competitor the Boeing 747-400.
The technological challenge for meeting these requirements will demand new solutions in the field of airframe design and manufacturing.

To improve structural efficiency new materials such as Al-Li alloys, GLARE and advanced composite materials complement or compete with developments of current aluminium alloys. This will be combined with new manufacturing techniques such as automated assembly of metallic structures and automated lay-up of composites. Novel use of, for example, welding techniques, large extruded panels, adaptive creep forming and high quality casting will be used to improve efficiency and drive down costs.

These advanced design concepts while producing highly reliable and more efficient structures at reduced manufacturing costs will also generate more user friendly maintenance requirements and contribute significantly to an improved operational efficiency.

The paper will outline the main efforts currently being undertaken in the field of structural technology to produce an attractive next generation of transport airplanes.

1. INTRODUCTION

The technological challenge to develop the future Airbus transport airplane has to be very carefully evaluated. The A3XX, which is planned to enter service in 2004, will initially be some 40% larger than today's largest airliner, the B747, and will grow to be some 60% bigger.
The development history of large aircraft shows that major steps in size were seldom successful.(Fig. 1)
The reasons were mostly technical problems, sometimes operational or economical aspects,sometimes missing market opportunities. As an example the BOEING 747 project had great difficulties in the early days: based on then existing technologies it encountered weight, flight control and airworthiness problems from the beginning. Consequently, engine thrust had to be raised more quickly than initially planned and fuel consumption was higher than expected. By the end of 1973, the B747 eventually had become an excellent aircraft and has now been exercising a monopoly position for more than 20 years.

To manage this risk, Airbus Industrie has identified the major design and manufacturing problems which need to be resolved before the program becomes a reality.
The structure will have unusual mass and stiffness distribution characteristics with relatively high flexibility, so there will be a high sensitivity to weight changes.
Handling and stability will be largely affected by the control of high mass moment of inertia.
Runway requirements will increase the complexity of the landing gear.
FAR noise regulations will present a particular problem as today's limits seem to be tailored for the B747 category aircraft.
Emergency evacuation and **ground operation** requirements complete the picture.
Clearly, all these problems increase on a non-linear basis with increasing aircraft size and Airbus Industrie needs to develop technological solutions which are not fully available at present.
Especially in the area of

material technology
structure design and manufacture
structure analysis
major efforts are necessary to address the weight problem which heavily affects **direct operating costs (DOC)**.

2. THE AIRBUS TECHNOLOGY PROGRAM

The economic challenge is to meet the expectations of the airlines for a 15 to 20 % improved operating cost per seat relative to the 747-400. A 15%+ target will be applicable to the basic version of the A3XX and a 20%+ target to the capacity increased version. Our analysis shows that, using the latest technology incorporated in the A340 together with the economies of scale, would result in about 10 % improvement relative to the 747-400. The remaining "shortfall" will have to be made up by additional direct operating cost savings derived from new technologies and resulting in reduced weight, fuel consumption, maintenance cost and cost of ownership.

The part of materials and structures in the Airbus Technology Program is essential to obtain the desired improvements.

3. THE MATERIALS CASE

When the Airbus A3XX makes its first flight at the end of the year 2003, in order to enter into service in 2004, it is expected to have an initial max. take-off weight of approximately 530 tons.
From the above, about 140 tons will consist of Aluminium alloy structure, about 35 tons will be of carbon fibre reinforced plastics, titanium structure and high strength steel parts will account for roughly 15 tons and 13 tons respectivly.
This excludes the 4 engines of 80 000 lbs of thrust each which are not accounted for in this picture.
Assessing the material distribution as a percentage of the total structure weight and comparing today's Airbus A340 with the future A3XX (Fig.3) one can immediately draw the following conclusion:

- Aluminium will still be the most widely used material in airframe design with a 68 % share;
- there is a growing interest in Titanium alloys resulting in a reduced utilization of high strength steel;
- the use of CFRP for the manufacture of large transport aircraft has apparently stabilized;
- the balance is a small fraction of miscellaneous materials.

Advanced Al-alloys of the 2xxx and 7xxx series with improved properties in strength and fracture toughness could generate a weight reduction of approx. 5 to 8%.
Al-alloys of the 6xxx series, such as 6013 from ALCOA and the 6057 from Pechiney are under investigation for their effectiveness for welded applications.

Al-Lithium alloys still take an important part in the Airbus R&D program for its outstanding weight saving potential mainly due to its lower density.
Beside the remaining traditional western supplier ALCAN a further chance is seen in the cooperation with Russian material suppliers. Al-Li alloys have been widely used in Russia for applications on fighter airplanes. Conversion of material standards to transport aircraft requirements is in hand.

GLARE, a newly developed Fibre Metal Laminate is an other strong candidate to be used in the design of fuselage structures.
Between 15 and 28% of weight reduction over todays design practice could be achieved in typical areas of a fuselage structure.

Titanium will progressively replace high strength steel mainly in engine pylon and landing gear structures.

Advanced alloys with improved strength and damage tolerance properties show a weight reduction potential up to 15% in such areas.

Composite materials have been used in all Airbus aircraft with the aim to reduce weight and to improve structural efficiency. Stabilizers, the outer wing, floor beams and the rear pressure bulkhead are all potential candidates for composite material application with achievable mass reductions in the order of thousands of kilograms (Fig.4)

For A3XX the stabilizers will be the largest composite element built so far for the current Airbus fleet. For comparison the horizontal stabilizer is similar in size to the wing of the Airbus A310 twin aisle airliner.

To minimize risk comprehensive development programs have been undertaken to make sure that all necessary technological development issues are resolved at the appropriate time.

Advanced composite structures with exceptionally large sizes require the development of new manufacturing concepts to reduce costs, improve reliability and enable the expected production rates.
Current automated composite manufacturing processes have already been adapted in that new Automatic Tape Laying machines with higher lay up rates and with the capacity to process larger width tape material have been developed and are in production trials. Further improvements in deposition rate are being sought using resin infiltration of stitched non-crimp fabrics. An initial demonstration for wing structures was demonstrated recently and is widely known as the AMCAPS box (fig 5)
Upgrading of existing facilities includes larger autoclaves and ultrasonic inspection facilities to handle large size components.

On the engineering side complex software packages have been developed to improve the electronic link between design and manufacturing groups to reduce design to build cycles. Structural optimization technics which permit to consider all design constraints simultaneously help to produce lighter structures and improve the level of safety.

4. THE DESIGN CASE

There is no doubt that the cost of producing the "next generation" of transport aircraft has to be the lowest possible, so that the aircraft will sell at a price which enables profitable airline operation. Advanced design concepts need to be developed to achieve the goal of reduced manufacturing costs.
Current design practice for Aluminium structures is based on the use of semi-finished products and mainly four categories are purchased in the following product form:

- forgings 6 %
- extrusions 13 %
- thin sheets 17 %
- thick plates 64 %

Clearly the use of thick plates keeps a major share within the total amount of Aluminium products needed. However very large quantities of the raw material being machined away in most current applications, ending up in an extremely high **buy-to-fly** ratio.

For example, as previously mentioned a roughly 140 tons of Aluminium alloy as part of the overall flying structure weight of the A3XX. Using current design principles about 980 tons of semi-finished products in form of plate, sheet, extrusions and forgings would have to be bought for this purpose.

This results in an overall buy-to-fly ratio of 980/140 equal to R ~7. Thus, 840 tons per aircraft of high cost Aluminium alloy end up as scrap material representing an initial value of approximately 13 million of USD. For the next generation of advanced Aluminium alloys about 30 % price increase could be added to that figure.
The overall aircraft buy-to-fly ratio may be 7, but in some cases the ratio can go much higher; for a machined center wing spar where more than 95 % of the material is milled away.
However, some improvements have already been made since the design of the first Airbus as shown on the example of the floorbeam (Fig.6).

Initially, the floor beam has been machined out of a thick plate of 128 kg. The end product had a weight of 8,9 kg which represents a buy-to-fly ratio of R ~14,4.
Subsequently, it was machined out of an extruded profile. This reduced the buy-to-fly ratio to R ~4,4.

Cost effective design of Aluminium components would tend to reduce the utilization of thick Aluminium plates and to increase use of extrusions and forgings or to enter into entirely novel manufacturing technologies in order to keep the buy-to-fly ratio as low as possible.
Extensive research and technology programs have been initiated some years ago throughout the Airbus system to acquire and validate the necessary advances in the most promising manufacturing techniques as
- aluminium castings
- integrally extruded wing and fuselage panels
- welded structures.

Just to mention some predominant examples:

Casting is the most consistent "near-net-shape" process. The cost saving potential of the investment casting process is considerable for Aluminium alloys. Due to the good surface quality, practically no finish processing is required .
The successful example of an A320 bulk cargo door, which has been cast as a pilot study by a French and a German foundry, has now resulted in an industrial application study for the A340-600 Airbus passenger door.

Extruded panels for fuselage structures with integral stringers or alternatively welded-on stringers compare very positively with today's riveting process in terms of weight and cost reduction. On riveted structures there is a substantial additional volume of stringer material and additional skin thickness to compensate for the rivet holes.

Currently, the Airbus partners traditionally producing fuselage structures have made investments in large CO2 laser beam welding machines capable to produce integral panels, about 4 m wide and about 10 m long. Initial test panels are being manufactured to optimize the process and to collect statistical data. High welding speeds up to 15m/min and a high degree of automatization allow to reduce manufacturing costs by 20% compared to the automatic riveting process.

For wing structures the technology for **heavy and very long extruded panels** is under investigation. This principle has been used for many years by the Russian aviation industry, mainly on large transport airplanes designed by Antonov.
The large An 124 wing structure, for example, is built up from 44 extruded panels, up to 28 m long (Fig.7). Indeed, machining is still necessary to a certain amount but first investigations show a double profit which could be obtained from reduced buy-to-fly values as well as increased structural efficiency over a conventional riveted structure.

Airbus wing design provides outstanding aerodynamic performance. However achieving this requires use of complex double curvature wing panels, particularly near the root end where thickness is greatest. This presents an exacting manufacturing challenge, whether considering our current skins with fastened stiffeners or integrally stiffened panels. A process called **Adaptive Creep Forming** (ACF) is being developed by AI partners in consortium with various research organisations, which provides for rapid panel forming to a precise contour with software control of adaptive tooling. The ACF process also supports our automated wing box assembly project. This system, currently at the demonstrator stage, is to provide a jig-less, tool-less high throughput assembly, using machine visual feature recognition and laser position scanning. This manufacturing process promises maximised use of factory facilities. It offers increased production rates and flexibility to introduce any developed or new products that are within the machine geometrical capacity, in principle by software changes alone.

Welding of thin fuselage structures, using the laser beam technology, has been mentioned above; for welding of thick wing panels the friction stir welding technology is currently investigated with very promising results.

Friction stir welding is a non-fusion solid phase welding process based on research in a consortium with the TWI. It is a continuous hot shear process. A non-consumable rotating tool made from a material harder than the work piece is passed along a joint between two closely butted sheets (Fig.8).
The friction heat creates a plasticized region around the immersed welding pin which consolidates behind forming a solid phase bond.
The novel process causes a more limited heat affected zone than conventional welding. This allows retention of a high proportion of parent metal properties for high strength heat treated Aluminium alloys.
It is also suitable for long joints and high tracking speeds, using machines similar to those for milling. It offers the possibility of low cost, high integrity assembly of integral wing structures.

5. CONCLUSION

These few examples of innovation in airframe design and manufacturing give an indication how Airbus can improve overall aircraft efficiency.
Advanced metal and composite materials combined with new manufacturing technologies are the possible means to face the economic challenge for meeting the expectations of the airlines for a 15% to 20% improved operating costs over current large transport aircraft.

One should not deny, however, that significant efforts will be required in innovative technical and industrial approaches, in a world in which progress has become an increasingly self generating process.

Acknowledgements

Jean Roeder Senior Vice President, Airbus Industrie (former)
Jurgen Thomas Senior Vice President, Large Aircraft Division, Airbus Industrie (former)

Karl-Heinz Rendigs Daimler-Benz Aerospace Airbus

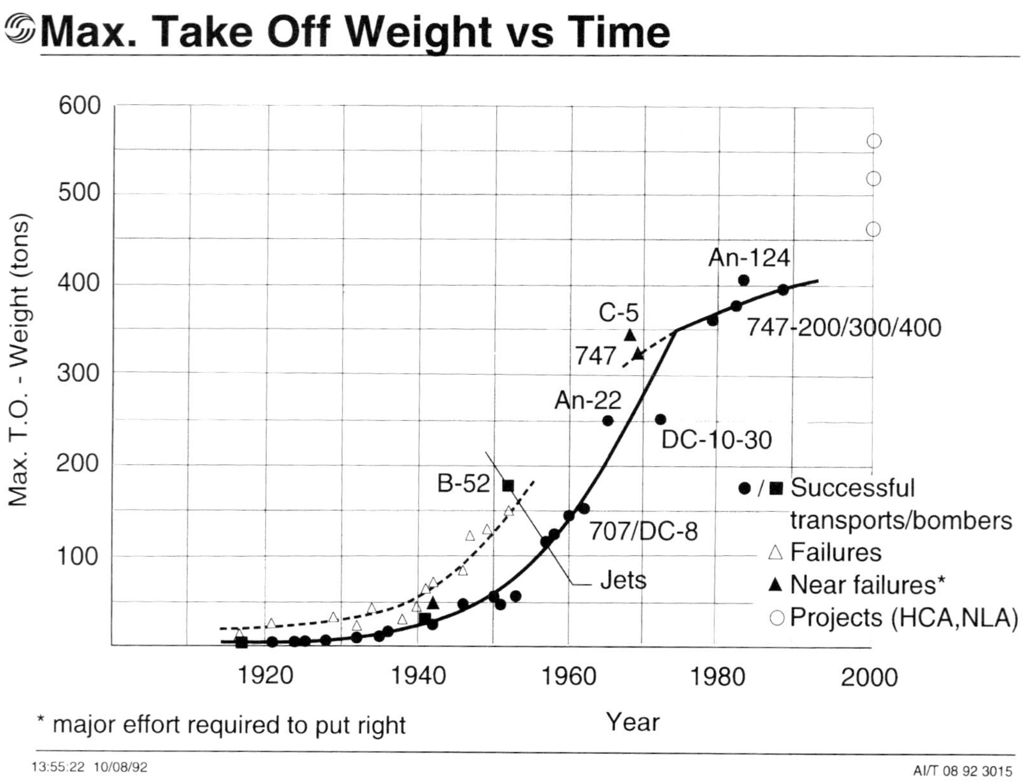
Max. Take Off Weight vs Time
Max. T.O. - Weight (tons)
600
500
400
300
200
100
1920
1940
1960
1980
2000
Year
An-124
C-5
747
747-200/300/400
An-22
DC-10-30
B-52
707/DC-8
Jets
● / ■ Successful transports/bombers
△ Failures
▲ Near failures*
○ Projects (HCA,NLA)
* major effort required to put right
13:55:22 10/08/92
AI/T 08 92 3015

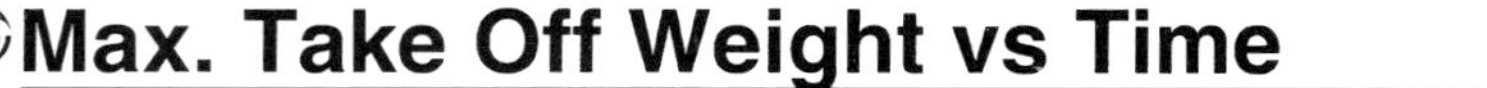

Figure 1

AIRBUS
A3XX OPERATING COST TARGET
Airline target
747-400
A3XX-100
≈ -7%
Size effect and today's technology
Improvements
A3XX-200
≈ -12%
Size effect and today's technology
Improvements
Relative seat mile cost (COC)
Better
-5
-10
-15
-20
-25%
Size effect, configuration optimisation and introduction of cost effective improvements will allow to achieve airline target.

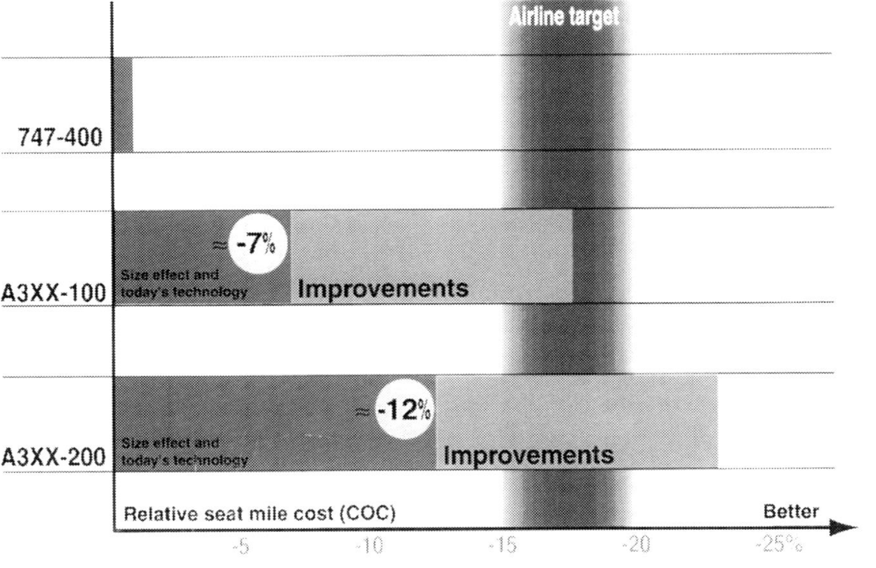

Figure 2

AIRBUS

STRUCTURAL MATERIAL DISTRIBUTION IN AIRBUS TRANSPORT AIRCRAFT

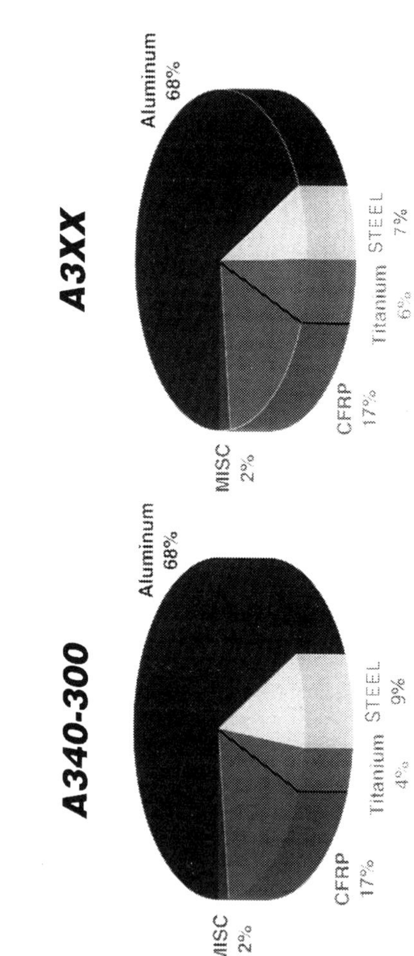

Figure 3

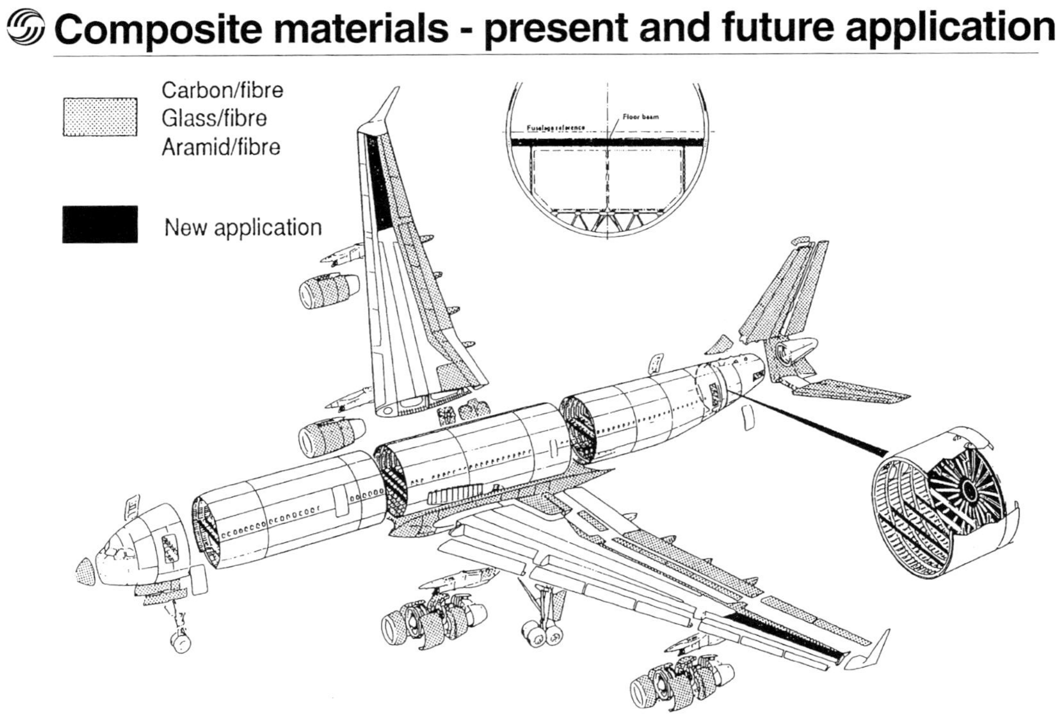
Composite materials - present and future application
Carbon/fibre
Glass/fibre
Aramid/fibre
New application
Floor beam
Fuselage reference

Figure 4

Figure 5

AIRBUS
MANUFACTURE OF FUSELAGE FLOOR BEAM
BEFORE MODIFICATION - A300
SEMI-FINISHED PART-PLATE: 128 kg
FINISHED PART: 8,9 kg
BUY/FLY RATIO: R=14,4
AFTER MODIFICATION - A340
SEMI-FINISHED PART-EXTRUSION: 39 kg
FINISHED PART: 8,9 kg
BUY/FLY RATIO: R=4,4

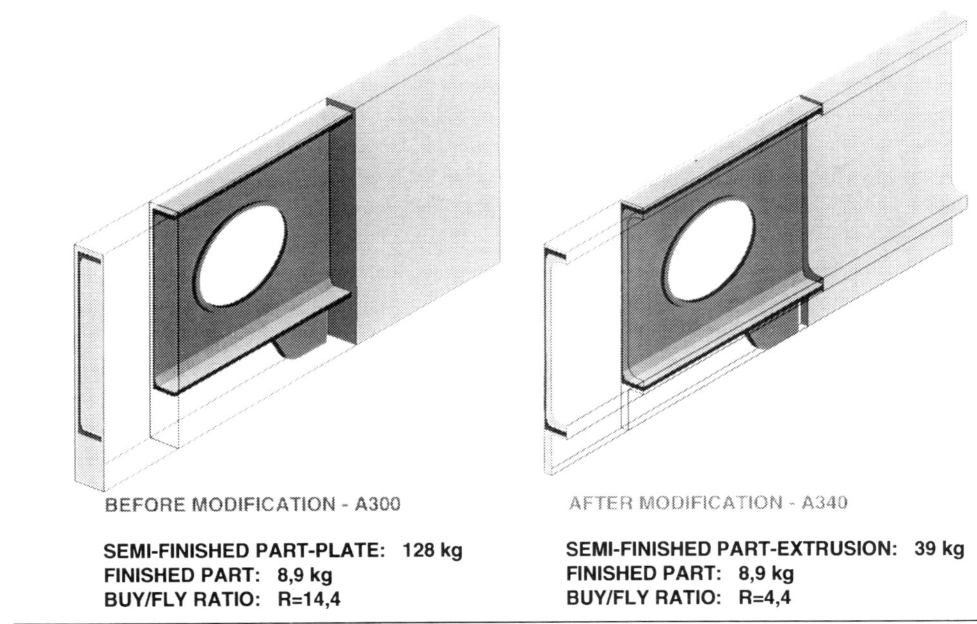

Figure 6

AIRBUS

HEAVY AL-ALLOY EXTRUSION FOR WING PANELS

Figure 7

AIRBUS

FRICTION STIR WELDING CONCEPT

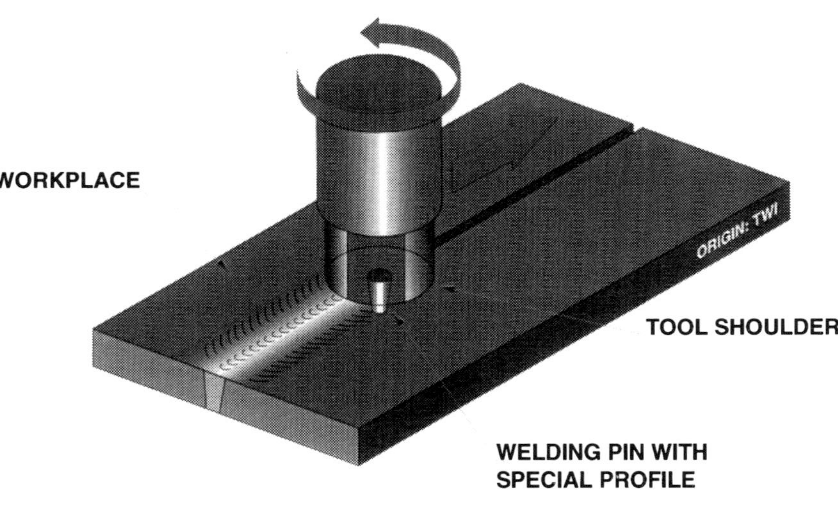

Figure 8

C545/041/98

Composites in airline service

R J HEATH BSc, MRaeS and **M HALLAM**
British Aerospace Airbus Limited, Bristol, UK

1) SYNOPSIS

Large civil aircraft have to be operated intensively by airlines to maximise profits and remain competitive in today's business environment. Similarly the major manufacturers are competing to supply the operators with products which meet their demands of minimum purchase price and operating costs. Composites play an important role in this market and the following script discusses current perceptions based on in-service experience of airlines, on going work within the industry to resolve customer concerns, and potential future applications of the material and the approach which one company, British Aerospace (BAe), is taking.

2) INTRODUCTION

Composites have contributed to the evolution in the efficiency of airframe structures over the last 15 years. There are very few truly revolutionary technologies, like the jet engine, and continuing improvement is the sum of many small steps. On today's generation of commercial aircraft, the composites provides a weight saving of the order of 1% of the Operating Weight Empty. The competitive environment in commercial aviation drives us to seek further ways of reducing Direct Operating Costs (DOC). A typical breakdown of DOC is shown in fig 1. The largest contribution is first cost followed by fuel. It would be convenient if we could concentrate our research and technology in the largest sector of the 'pie'. It is unlikely that this approach would result in a large enough improvement in DOC so we have to try to improve in all sectors. The implication for structures is that we have to reduce weight while reducing the overall manufacturing cost of the aircraft.

One option is to increase the use of composites. Current generation (A320 and A340) typically have around 15% by weight of composite in the structure. If the main wingbox was to be composite, this proportion would increase to 40%. However, to realise the potential weight saving presents a number of formidable technical and economic challenges. The solutions to these challenges are beyond the current state of the art.

At BAe Airbus, the Composite Wing Programme is carrying out research into the most effective way of making a wingbox from carbonfibre composite. The importance of reliability and maintainability of the structure in service is recognised and a team has been specifically tasked to investigate technical issues relating to the in service environment. The two parts to this investigation are understanding the customer's (airlines) needs and investigating in service damage threats; each complements the other.

3) LIAISON WITH POTENTIAL CUSTOMERS:

As the development costs of large technically advanced aircraft continue to escalate, the need to have a product which meets the requirements of the customer and hence will sell successfully in the market becomes ever more critical. The substantial investment which is required to introduce a composite wingbox onto the next generation of large civil airliner demands that the product must achieve customer approval and if possible exceed expectations as soon as it enters into service. It was therefore decided that contact would be established with civil aircraft operators, potential customers and repair bases, the primary objective being for us, the Original Equipment Manufacturer (OEM), to understand the in-service environment with respect to any issues relating specifically to the product. A range of operators worldwide were selected to include both those with large fleets and those with smaller numbers of aircraft. It was anticipated that most of the composite experience would be based on thin laminates typically on removable structure as this is where the majority of the non metallic parts exist at present. All feedback from these discussions is recorded and influences the work content of the research programme where relevant. Continued liaison is envisaged to take place as the design of the structure evolves and customer participation is encouraged to reduce the risk of new technologies to the OEM.

The industry introduced advanced composites cautiously to ensure the capabilities of these new materials hence the majority of carbon components used on current civil aircraft is of removable structure (ref fig 2). Moving aerodynamic surfaces, radomes, fairings, and engine cowl doors are all typical examples of composite components and in service experience is showing that there are advantages and disadvantages for both materials. For example, composites eliminate the corrosion problems which metallics suffer (good design and manufacturing practices must be exercised at carbon to aluminium interfaces), but are more prone to erosion on exposed edges while honeycomb structure, of either material, is generally more complex to repair than monolithics.

For structure specific to wings, trailing and leading edges contain many lightweight composite parts which, due to their location, are vulnerable to impact from service vehicles and other hazards while on the ground. (This is equally applicable to metallics but less so for a wingbox which is protected by these 'bumpers'). If damaged, it is generally accepted that

such components are more costly to repair than their equivalent metallic parts. Damage assessment, preparation, cleaning and curing of bonded patches are time consuming operations and it is of paramount importance to an operator that this type of work can be completed within the scheduled maintenance downtime. In addition, availability, standardisation and interchangeability of repair materials needs to be improved and this again risks leading to longer downtimes; storage of the materials if required is expensive and often have to be purchased in large quantities. These are some of the concerns which have been raised by the aircraft operators to the OEMs and as a positive response, an organisation called the Commercial Aircraft Composite Repair Committee (CACRC) consisting of OEMs, airlines, composite repair facilities, material manufacturers, and regulatory agencies was established to address the problems.
It's task groups cover:

Repair Materials
Repair Techniques
Inspection
Design
Facilities
Training
Airline Inspection and Repair Conditions

This work is progressing.
It should be remembered that the CACRC is focused on the current in service problems and so is primarily concerned with thin components, often of sandwich construction (ref for example SAE AE-27, the 'Guide for the design of durable, repairable and maintainable aircraft composites' whose design case studies are mostly sandwich parts). As in service confidence in composites grows, more primary structure is introduced so that today for example, we find the Airbus vertical fins and horizontal tailplanes of composite as well as the outer wingboxes of Aerospatiale's ATR72s. However this construction is monolithic and generally of a greater thickness than the secondary structures. Experience has shown that these structures are extremely tolerant of in service hazards with the result that any increased unscheduled maintenance costs due to more complex repair processes (compared to metals) are insignificant considering the limited number of occurrences. Nevertheless, efforts must be continued to be made to improve the current repair techniques. To quote the aforementioned publication, 'The vertical stabiliser box has demonstrated that composite component design can successfully meet commercial aircraft structure performance objectives including durability, maintainability and repairability.'
As may be expected, it has been found that there is a different emphasis placed on various aspects of composites between airlines. For example, the more widespread use of composite primary structure is very much encouraged by some, others more cautiously. Some customers will carry out all their composite maintenance in-house, others will outsource it to a third party. Stripping of the paint concerns some, others not and similarly bonded or bolted repair techniques have their supporters.

4.1) Investigation of In service Threats:

As with most industries, commercial operators are subjected to a more and more competitive business environment, one consequence of which is the intensive utilisation of the aircraft themselves to maximise revenue. Aircraft turnaround times of only 30 minutes ensures that risks of impact from catering trucks, stairs, service trucks, towing vehicles, passenger buses etc are ever present. Reliability of the product is therefore of paramount importance as any unexpected downtime reflects directly in the profits of the airlines. This has a greater impact than in years gone by, because today, a passenger is easily able to take a flight on a competitor's aircraft. However, maintenance of the aircraft, be it planned or unplanned, is a cost which the operators have to incur in order to keep them in an airworthy condition. It is the OEM's responsibility to understand the working environment of the aircraft so that both the Airworthiness Requirements are met and the maintenance costs of it's products are kept to a minimum. **The operator requires a design which is reliable and tolerant of in service threats.**

There are many and varied threats to the aircraft in service as they are used in a hostile environment. Investigations are continuing within BAe to recognise, understand and find engineering solutions to these threats which are acceptable to the Airworthiness Authorities (to meet regulations), ourselves (to minimise design and manufacturing costs) and the customers (to minimise initial purchase price and life cycle costs (LCCs)). Liaison with the operators is again used to verify that all the threats have been recognised and that they are realistic. Some examples of the investigations carried out by BAe are discussed below.

4.2) Threats under consideration

The approach taken by BAe was to establish a list of all the potential threats which risk damaging a wingbox in service. The threats were then prioritised so that those which potentially have the greatest influence on the wingbox design were investigated first. The programme is continuing. The work includes an evaluation of the tolerance of the wingbox to:

a) Impact from tyre debris

b) Impact from engine debris (both large and small)

c) Engine Fire

d) Lightning Strike

e) High Intensity Radio Frequencies

f) Local Heating of Structure from Fuel Pump Dry Running

g) Impact from Birdstrike

h) Hot air impingement from duct burst

i) General impact (hail, dropped tools, 'hangar rash', refuelling nozzles, runway debris etc)

j) General Environmental Effects

Conclusions can already be drawn on the results of some of the above investigations. For example:

(a) Impact from Tyre Debris.
When a representative section of tyre debris (0.82kg at 95 m/s) strikes a lower stiffened skin panel of the dimensions which would likely be used for an A320 size of aircraft, no damage occurs. The only evidence of the strike taking place is a rubber witness mark on the painted surface. (ref fig 3)

(c) Engine Fire.
Resistance to an engine fire by a composite skin, again dimensioned for an A320 type aircraft, is at least as good as that of the equivalent metal structure which is currently manufactured. 'Resistance' has been investigated in 3 ways:
Penetration of the skin by flame
Residual strength of the structure during and after the fire
Transfer of heat to the fuel vapour within the tank by metallic fasteners
(ref fig 4)

(i) General Impact.
An attempt has been made to zone the wingbox structure in order to understand the occurrence and magnitude of impacts from 'normal' or everyday sources. This has been done by investigating the historical database of all reported damages on metallic wingboxes for the entire fleet of Airbus aircraft and assessing the damage sizes and their associated energies. Having established a realistic assessment of the threats, an equivalent composite wingbox may then be designed such that it has suitable resistance to such impacts.

5) OTHER AREAS OF CONCERN

Other areas of major concern from a maintenance point of view are:

5.1) Repair of damaged structure.

Despite evidence showing that thick monolithic laminates are extremely damage tolerant, there will always be the potential requirement for embodying a major repair. Unusual examples of historical events resulting in major damage include:
- Collision with a hangar wall
- Penetration of a leading edge by scaffolding
- Jacking an aircraft into a hangar roof
- Collision with scaffolding after jumping chocks

It must be remembered that unlike the majority of the current composite parts, removal or replacement of a wingbox is not an option. Hence it must be recognised that a structure of this type will require major repair at some time and the challenge exists to find techniques which restore adequate and durable strength to the structure within a timescale that causes minimum downtime for the operator.

There is very little experience at the airlines of major repairs being carried out on thick composite monolithic structure where, for example, thicknesses may be greater than 1 inch in some areas of an Airbus type wingbox. Whereas thin panels are normally repaired by bonding, this technique will meet with additional difficulties as the structural thickness grows. Scarfing at currently accepted angles will mean that the original damaged area may grow significantly in size and run into adjacent structural features which will complicate the repair process still further. Consistent heat application and consolidation will be difficult unless carried out in multiple lay ups, however, the downtime for completing successive cures may become unacceptably long. Finally the assurance of bondline strength is still a question which is often raised and will have to be answered if applied to wingbox structure which has the additional complication of being exposed to longterm fuel contact. It may be that bolted patches (either composite or otherwise) using metallic type processes are the better method if major repairs are required whereas minor damage can be restored using bonding techniques.

5.2) Corrosion at interfaces with metallic components.

Corrosion of metallic structures causes the aircraft industry vast expense in maintenance due to inspection, repair or attempts at it's prevention. Undoubtedly, one of the main advantages of composites is the enormous scope for reduction of this burden, however, there will always be metallic components on the aircraft and the potential for corrosion at the interfaces with carbon must be taken very seriously. There are already many airframes existing which have completed high flight cycles and many years in service without suffering such damage. Adequate protection schemes exist to prevent these problems and careful design and maintenance should be sufficient to realise a significant saving in costs. Particular care must be taken at major interfaces such as an outer to inner wingbox joint and the design should allow reasonable access for inspection.

5.3) Stripping for repainting.

Stripping and repainting is a regular maintenance operation for the airlines either to renew the cosmetic appearance, to change the company logo or due to change of owner. An aircraft is typically repainted every 3 - 5 years although it is quite common for the wings to be done at every other overhaul. Concerns have been raised that this will be more expensive on a composite wingbox as abrasive methods are required rather than chemical stripping. The sophisticated equipment which is required for such abrasive methods (dry ice, wheat starch, water jet, lasers etc) requires a large capital investment which the operators are reluctant to spend.

Chemical stripping risks damaging the composite and is not a practical option at present however research is continuing to develop paints such that the top coat is easily removed by a mild stripper but the primer remains in place. Alternative solutions may be resin with colour pigments or adhesive films which can be renewed without paint removal.

6) CONCLUSION

(Ref fig 5)
Opinions about carbon composites within the aircraft industry are both positive and negative and are based mainly on components of thin laminates and/or sandwich construction. In order to gain in service experience and hence increase confidence in their performance, they have often been used to replace metallic secondary structures in locations which are vulnerable to impact. Consequently damage occurs on a regular basis (as it would with metallics) and repairs are required. Moisture ingress has caused problems with very thin laminates (typically of 2 plies) however good design, by specifying a minimum of 3 or 4 plies, should provide the solution. It is recognised that these types of composite parts are more time consuming to repair than their metallic equivalents and so incur higher unplanned maintenance costs. This situation is made worse due to allowable damage limits of composites being conservatively small. These drawbacks have been recognised and the aircraft industry is working together to improve the situation.

When thicker composite monolithic laminates are employed however, in service experience is showing that their tolerance to the in service environment is excellent and if used to manufacture wingboxes, will be protected from the majority of impact sources. Nonetheless, there will always be the infrequent occasion when action is required to repair major damage and the resulting unplanned maintenance cost may indeed be higher than it's metallic equivalent. However, when considering the overall expected reduction in LCCs to be gained by the operators due to advantages in fatigue, corrosion, reduced scheduled inspections and fuel burn (and these advantages are already appreciated by the airlines), we should continue to make positive efforts to realise the potential benefits of composites in large scale applications on commercial aircraft. The importance of LCCs is becoming more appreciated by the industry generally but we are still in the early stages of having the tools available for assessing and monitoring them. Consequently, it is difficult to predict exactly the magnitude of the financial benefits to be gained due to the replacement of a metallic component by a well engineered composite one.

The manufacturers must recognise that their customers will only accept large scale application of composite technology if economical benefits can be demonstrated both on initial purchase price and life cycle costs. This is one of the challenges facing the OEMs at present and it is their task to fulfil the requirement by intelligent design and the correct application of the material.

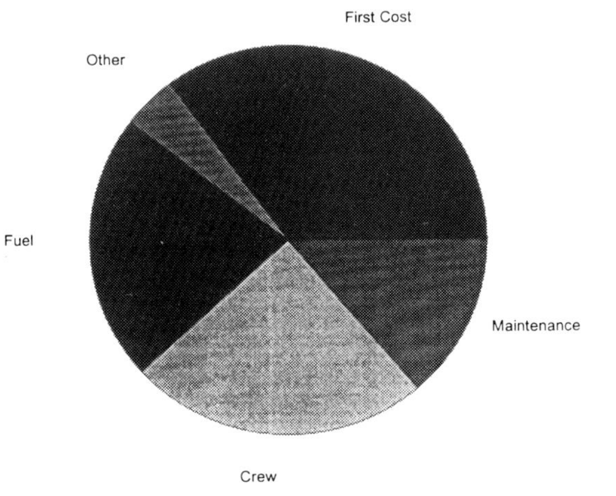

Figure 1 Direct Operating Costs

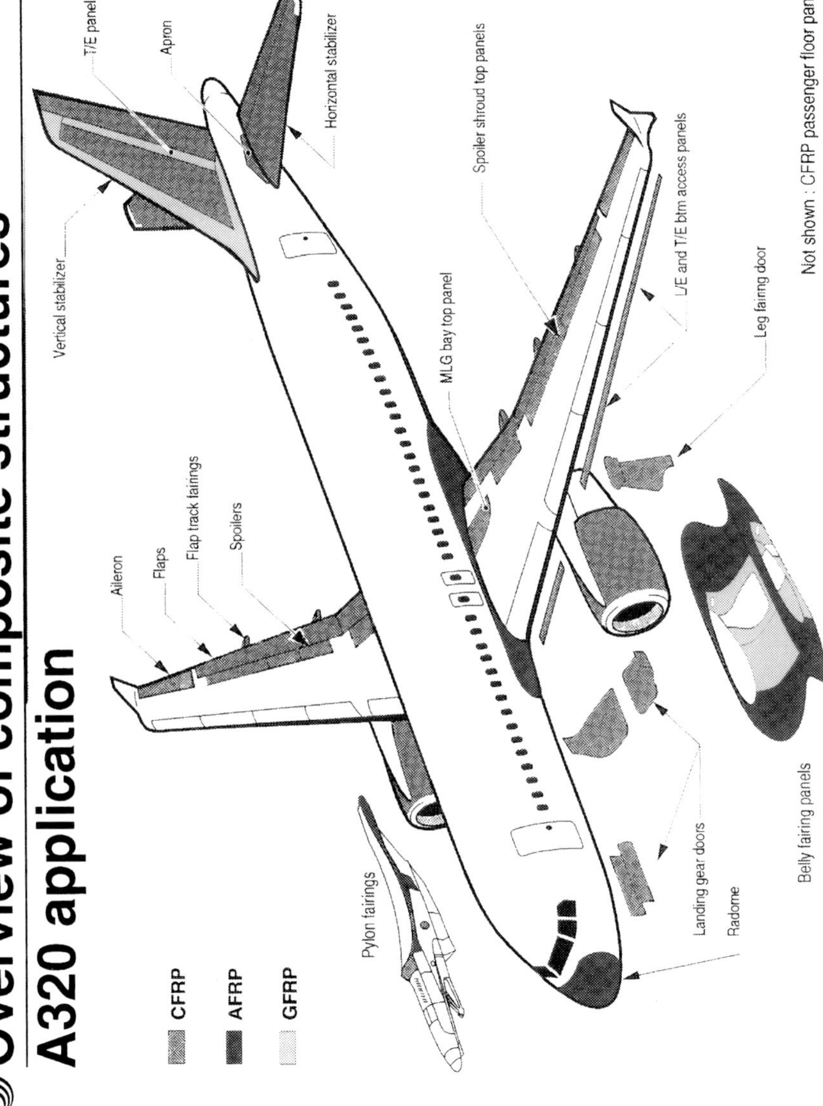
Overview of composite structures
A320 application
Fig 2
CFRP
AFRP
GFRP
Pylon fairings
Radome
Landing gear doors
Belly fairing panels
Aileron
Flaps
Flap track fairings
Spoilers
Vertical stabilizer
T/E panels
Apron
Horizontal stabilizer
Spoiler shroud top panels
MLG bay top panel
L/E and T/E btm access panels
Leg fairing door
Not shown : CFRP passenger floor panels

BRITISH AEROSPACE AIRBUS LTD

Fig 3a

TEST ARRANGEMENT FOR TYRE DEBRIS IMPACT

BRITISH AEROSPACE AIRBUS LTD

Fig 3b

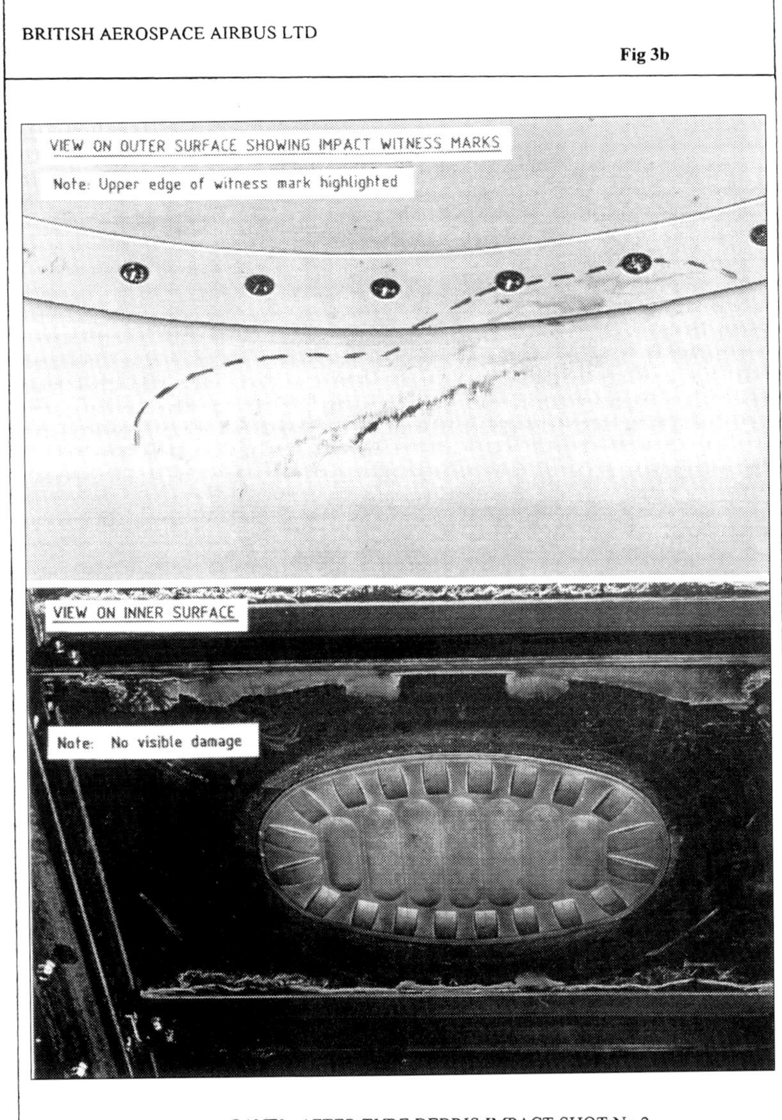

PANEL AFTER TYRE DEBRIS IMPACT SHOT No 2

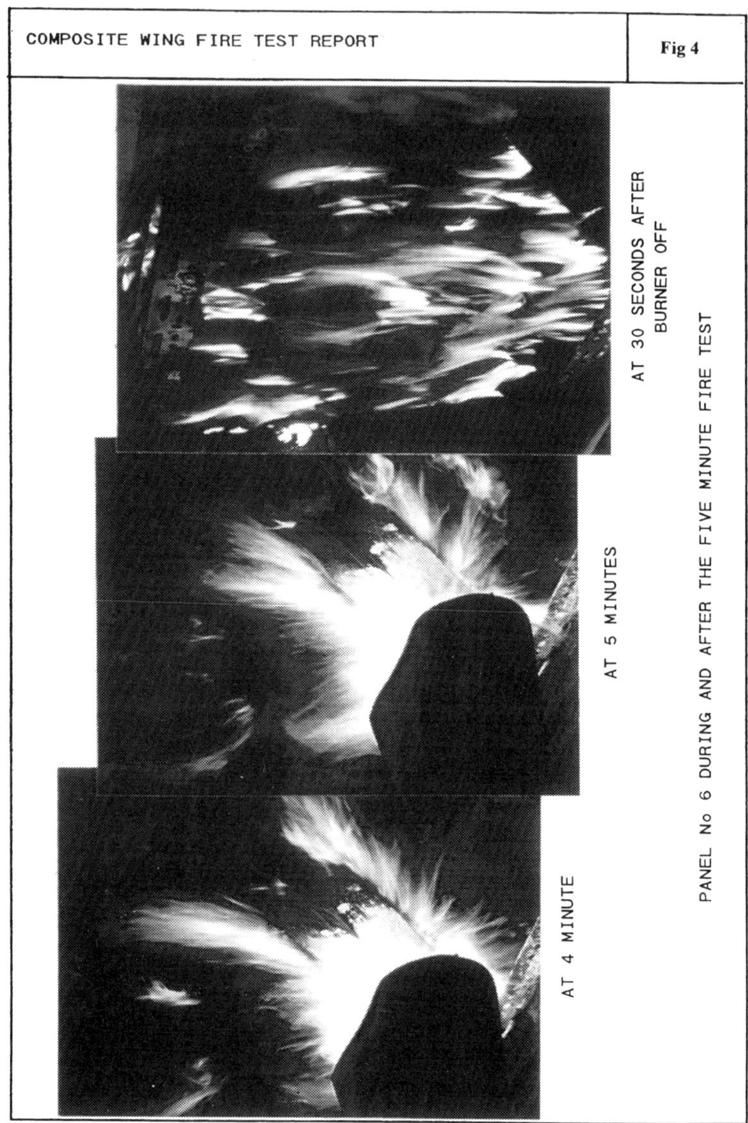
COMPOSITE WING FIRE TEST REPORT
Fig 4
AT 30 SECONDS AFTER
BURNER OFF
AT 5 MINUTES
AT 4 MINUTE
PANEL No 6 DURING AND AFTER THE FIVE MINUTE FIRE TEST
COMPW112

Current Perceptions

Figure 5

- Negative
  - Vulnerable locations
  - Low damage tolerance (ADLs)
  - Water ingress
  - Free edge & hole delamination
  - Surface degradation
  - Galvanic corrosion
  - Removable, thereby prone to damage
  - Repairs complicated, expensive, time-consuming
  - Operators not willing to invest in new facilities
  - Lack of standardisation of repair materials
  - Fear of the unknown (Black art to outsiders)
  - Increased insurance premium (technology risk)
- Positive
  - Recognised performance advantages
  - Current primary structure appears to be very damage tolerant in service, e.g. Vertical fin, HTP, ATR72
  - Fatigue & corrosion advantages
  - Those working within the composites industry are confident
  - Reduction in inspections

Life Cycle Costs: LCC is the main influence in the purchase of an aircraft.

C545/055/98

Driving down product introduction costs – the contribution from people, processes, and CAE

J A JUPP MA, FRaeS, FEng and **A SCOTT** BSc
British Aerospace Airbus Limited, Bristol, UK

1. SYNOPSIS

All partners of Airbus Industrie have reacted to the ever increasing business challenges in various ways, from committing to form a 'Single Company' through to major internal cost reduction programmes. This paper describes one contribution towards ensuring Airbus Industrie proceeds to a commercially successful future, that of the Engineering Team of British Aerospace Airbus.

The importance is stressed of improving cultural issues and working processes as well as the increasing use of Computer Aided Engineering 'tools' to ensure step change improvement in the costs and timeliness of introducing new and developed products such as the A3XX.
Whilst much has been achieved, there is still much left to do!

2. INTRODUCTION

2.1 British Aerospace Airbus

As well as our main manufacturing role of supplying the wing structure for each Airbus aircraft since the first A300, which flew nearly 26 years ago now, BAe has also retained responsibility for overall design and integration of the wing in all aspects. The expertise and experience of the teams responsible for all the British jet transports, the Comet, Trident, VC10, BAe 1-11, Concorde and BAe 146 were utilised in support of Airbus Industrie in designing the wings that have played a significant role in the Airbus success to date, and finally being consolidated in the Engineering Team at Filton.

From a single product focus (Concorde) through the 1960's and 70's, by the beginning of this decade the Filton team were managing several Airbus product streams (as well as Concorde and BAe 1-11!), all with their different variants and different aspects of development and support work, both for manufacturing and in-service problems. During that period the Filton Engineering Organisation had not evolved significantly at all. It rightly had a proud technical tradition ensconced in major Design and Technical Functions who dominated a matrix organisation of small project core teams. With the completion of the main design effort for the A340 and A330 (which first flew in 1991 and 1992 respectively), the wind of change was about to blow!

2.2 The Business Imperative for Change

Of the Airbus Partners, British Aerospace was the first to feel the full pressures of Commercial realism after it was privatised in 1982. As well as designing and building the aircraft, we were actually expected to do it profitably! As the business volume represented by Airbus started to become more and more significant the spotlight within British Aerospace trained onto us in the late 1980's with increasing demands by the shareholder to reduce the cost base and move into profitability.

While the Design Team at Filton "had it's head down" on the major development programmes for A320/A321 and A330/A340, it was left to get on with the job and at the same time was becoming very introverted, with little recognition of how businesses were developing in the rest of the world. The team saw themselves as reasonably successful, with due cause, given their contribution to the Airbus products, and the business was moving towards an operating profit, so why change? Certainly the usual imperative of change or die did not apply! However, things were already moving in the other parts of British Aerospace that hadn't the strength of the Airbus order book behind them and the scope for improvement was all too clear to the top management of the Company.

The initial driver for change was thus the first one in figure 1. Although "British Aerospace Airbus" as we now were, was becoming profitable, it was certainly not giving the return required of a successful company. This was exacerbated by Driver No 2 with Government support for Research continuing to reduce in real terms and the GATT trade agreements limiting future assistance loans to 33% of launch costs, ever increasing the pressure on the company's ability to sustain its long term future. Thus it was that "EROS" was launched in Airbus Engineering in 1992, the "Engineering Review of Operations and Strategy". This was one of several initiatives throughout the Company as a result of a conscious decision to keep ownership of the improvement initiatives with the line management. It was agreed that facilitation by outside consultants would be essential and following a 'Beauty Parade' as we called it, Coopers and Lybrand were enlisted to help us in mid 1993.

Coincidentally it was at about this time that the final Drivers for Change (see figure 1) were becoming all too apparent. The Airlines were descending into their worst ever recession with losses about to become higher than their accumulated profits since they had been in business! Further orders were drying up and some commitments were being cancelled. Airlines were thus becoming ever more demanding – driving their aircraft harder (utilisation climbing from 7 to 12 hours per day), delays through unreliability becoming totally unacceptable and price levels starting to be driven downwards. Boeing were responding to this – they had started their own cost reduction programmes and had announced that product costs would be reduced by 25%. They began to take contracts at prices which assumed cost improvements yet to be achieved (creating their own internal imperative for change). Thus a clear message could be put in front of everyone, including Engineering. Not only were improvements of order 15-20% in the cost base required to make an acceptable business return, but DOUBLE this improvement was going to be necessary to compensate for the reduction in income levels or we would all be out of business!

3. TOWARDS OUR BUSINESS GOALS – THE ENGINEERING CONTRIBUTION

For the remainder of this paper I am going to refer to the three major domains for improvement, which in our view must be pushed along together to achieve the vision of substantial improvement in the performance of any business:-

- People
- Process
- & Technology (Tools)

This is captured in Figure 2; they all interlink and cannot be dealt with in isolation.

Technology in this sense is not product technologies but the (mainly Information Technology) tools and infrastructure that allow our people and processes to work even more effectively. In our view, for example, the cardinal sin is to try and imbed new IT Systems without having addressed the people and process issues; it is doomed to failure. The attempts to use "Knowledge Based Engineering" software at Filton over many years is a case in point – we will return to that.

3.1 "EROS"

In fact, in the EROS Programme, we reacted so much against the "tail wagging the dog" that sorting out our IT systems, although by no means forgotten, was put very much in third place whilst we concentrated on "People" and "Process".

Describing our facilitated improvement programme in detail would be the subject of a paper in its own right. However, it was a crucial step on the road and a summary is appropriate here. EROS had four main themes:-

1. Vision – Where did we want to be, what would it look like?
2. Process – What were our core process, what should they be (step change)
3. Local Improvements – "Quick Hits" within our current organisation that would engage many people quickly (incremental change)
4. People – Underlay everything we did, changing the hearts and minds and the whole culture of the way we behaved.

In the "People" domain, the two main platforms were "Visioning Events" and the "Playing To Our Strengths" (PTOS) initiative. The former was a process of rolling out through the whole of the engineering population the business environment and imperatives we were facing, the themes we felt would be part of the future vision and the beginnings of the new organisational framework we saw as being necessary. This was highly successful, but proved the benefit of very careful preparation, ensuring that 90% of the proposed way forward was agreed and worked out by the senior management team beforehand and working through a cascade process to win the support of the population.

PTOS was a process of self-assessment and structured interview for all the senior members of Engineering, that is all those involved in supervising tasks or a group of engineers, around 20% of the population. This went with a re-definition of roles into task management, resource management and technologists, whilst recognising that any one person might have competencies in more than one area. The Engineering functions were totally re-organised and de-layered to recognise these role requirements and at the same time the budget ownership changes. Budgets at this point in time were still to be owned by the Chief Engineers, but a work-breakdown cascade was introduced below them, that is, project focused, with task managers appointed with budget responsibility and full accountability for delivering the tasks to meet specification, time and cost, using the necessary resources from any Engineering Function. This was a fundamental changes with the Function Heads no longer having accountability for task completion – a bitter pill to swallow for some! However, it was a necessary change to achieve one of the main vision elements, multi-functional team working giving the necessary focus and communication to achieve one of the step changes we were seeking.

The re-organisation reviewed above paved the way for the most significant change of all, in the "Process" stream.

3.2 Step Change in the Development Process

As part of "EROS" a cross-business process sponsor team was set up, chaired by the Director Engineering, but including the Operations Director and 'voices' of other parts of the organisation. Right from the start this area was seen as needing to evolve to become totally cross-business; Engineering could not continue in isolation.

Several core processes were identified:

- Develop Derivative Product (e.g. Design Weight increase for range/pay load improvements)
- Continued Product Development (e.g. cost reduction modifications)
- Support Product (e.g. Service bulletins)
- Acquire New Technology (Research)
- Define New Product (Future Projects)

(At this stage our primary work was developing the product lines now in the market place rather than all-new aircraft projects).

A very important aspect that our consultants pressed us on at this stage was the concept of "On-Line Pilots". Our immediate thoughts had been to put a team together to come up with a new "process-based" way of working off-line, perhaps as far as "Best Practice Guide", and then embed it into the Project teams. At a famous "Bowling Alley Debate" (intended as an informal team-building get together) our consultants convinced us to "go for it" on-line, that is just to get on with it on a real live project. Of course they were able to help from their past experience which significantly mitigated the risk of what we did.

So we entered our first "Design Derivative Product" (DDP) on-line development project, the 217 ton Maximum Take-off Weight version of the A330 (then at 208T MTOW) which involved significant drawing re-issue for structural strengthening, systems and undercarriage changes and re-certification. By that time the project was already underway, with some of the 'Quick Hit' local improvements within Engineering already embodied. However, the project team was totally reorganised "on the run" into a true cross-business multi-disciplined team and the cross-functional task breakdown cascade that had been built up in Engineering was transferred totally business-wide. A new project team was appointed, with representatives from the central Airbus Projects Office, Procurement and Manufacturing brought in.

The Project team was effectively a team of team leaders; each member (except for the Project Leader) leading a "Deliverable focused" team – such as responsible for the wing box or leading or trailing edges; or the fuel system or undercarriage (figure 3). The concept was "from cradle to grave" i.e. the team was not just responsible for handing over data for manufacture, but also responsible for the modified wing through assembly at Chester, on to equipping at Bremen and into the final assembly at Toulouse. Their job was only to be completed when the modified deliverables were "mature" and were achieving the same delivery standards as earlier versions (in practice the team was de-mobilised after delivery of the second wing). In this regard we took a major step beyond anything Boeing had achieved on the 777 (where, for example, they still had Engineering and Manufacturing team co-leaders). Another crucial point that was recognised was the importance of an "Integration Team" working across the "Deliverable" teams to "own" the wing specification and to ensure the deliverable teams did not sub-optimise.

This was the next 'Culture Shock' for Engineering. Not only did engineers sometimes find themselves working for a team-leader from, say, Procurement, but many were pulled out of their functional homes into a co-located product team for a significant period of time. The Integration Team Leader was effectively taking the role of "Chief Engineer' for this project, but he only had responsibility for the specification or definition of the project, not for accomplishing the engineering aspects of the tasks, which were of course embedded in the deliverable teams.

The team was fully supported by the Process Sponsor Team and proper team building and mobilisation was carried out. The results were outstanding and fully justified the slight risk of going on-line, with the immediate 'ownership' by the team who actually had to deliver a business result! Drawing re-issues for all reasons (errors, load changes, late manufacturing/tooling inputs etc.) were substantially down, re-tooling costs were reduced, concessions were down, and above all, for the first time ever the Design standard for initial build (Stage 0 Modification closure) was closed off before the wing went into the final assembly jigs! Flushed with this initial success we continued the on-line pilot process with our next development project, this time a Design Weight increase on the A321 Single Aisle aircraft (the "89 Ton" project). The Sponsor Team also set themselves the target of progressing through a total of three "pilots" and then confirming that that was how all work would be organised in future. One initial link that was made was that workshops were held to transfer the learning (both positive and negative!) from the "217 Ton" team to the "89 Ton" team, as well as to the Process Sponsor team who were overviewing the improvement being achieved. It was clear that most work was going to be necessary on the cultural aspects, in getting the understanding of what was being achieved and the acceptance of the new ways of working throughout the whole business.

Without going into details, the "89 Ton" team were equally successful, as was the third and final on-line project, the next version of the A330, a shorter, much increased range variant, the A330-200, with Maximum Take-off Weight again increased significantly to 230 Tons.

In each case, the total 'Non-Recurring Cost" budget for these developments was set at what we called "Stretch Goal" standards relative to a project estimate based on historic achievements up to and including the year 1993. These "Stretch Goals" were bought into by the teams at the start, even if they weren't quite sure how they would be achieved! Each project NRC's were multi-million pound sterling budgets and in each case the Stretch Goal was set at around 60% of the 1993 standard. With some small variations, in each case the Stretch Goal was substantially achieved if not beaten! The savings to the company were, to say the least, significant!

What was the reason? Clearly the often quoted ones were indeed borne out, captured in four words:

• Focus

• Motivation
• Communication
• Empowerment

The teams were properly mobilised, they were given the responsibility to get on with the job and given the necessary support when required. (They were also formally de-mobilised with due reward and recognition for what they had achieved). Co-location clearly helped communication and motivation. Were there any downsides? Yes – nothing is perfect! Although contact with the main engineering functions was by no means lost, communications between peers and other project teams suffered. Thus in some instances 're-inventing the wheel' was seen, and/or solutions started to diverge from standard practice. The teams did feel responsible for getting on with the job and were perhaps reluctant to call in more experienced people from outside the team. Is this a misinterpretation of the word 'Empowerment'? When asked, we have always endeavoured to make it understood that teams are "Empowered" to do the most efficient job for the business. This will inevitably sometimes if not always mean calling on experience and skills from outside the team.

4. THE NEXT STAGE OF THE JOURNEY

Two major further changes were afoot whilst the three "DDP" pilots were going on (over the period 1994 – 1996). Firstly, the Functional Improvement Programmes were being brought together in the framework of a Company "Performance Improvement Programme", the top level process model for which is shown in Figure 4. The EROS Process Sponsor team evolved into the Company "PIT 3", the Process Improvement Team "pushing" improvement in the "Develop New Technology and Product" core process stream, with "DDP" as the platform for further work. The "People" improvement aspects of EROS were absorbed into "PIT 2" driving one of the main supporting processes "Develop Employee Contribution". Since the "DDP" initiative had already taken on a total cross-enterprise perspective, it continued to roll forward under the Company programme. The second change was organisational. This simply captured at a Company level what had already happened in the DDP pilot projects. The operational functions of the Company, Projects, Engineering, Manufacturing and Procurement, with some support functions notably IT followed later by Human Resources, were brought together under a new post, Director of Product Operations. Other key posts were the "Product Executives", who were to be responsible to the Director of Product Operations for their product business, such as the BAe share of the Airbus "Single Aisle" family (A320, A321, A319). A typical organisation chart for the Product Executive management team is shown in Figure 5. The concept was a 'Project-led Matrix', with Engineering, Manufacturing and Procurement being on the 'Resource' side of the matrix. Several members of the Product Executive's Management team would also retain a responsibility to a functional director. For example, the Chief Engineer would still retain a reporting link (albeit a "dotted line") through the "Chief Engineer Airbus" to the Engineering Director for the technical integrity of the product and for his technical development. The role of the "Chief Engineer Airbus" was seen as key. This was to ensure a forum existed for exchanging technical experience between the projects and agreeing technical policies and standards which would apply to all Airbus projects. It also gave a focus within Engineering for developing that resource within the Company which would become the future Chief Engineers and Product Integrators.

The Programme Manager within the PE's team was to be responsible for delivering a portfolio of projects (such as "89 ton") run on DDP lines, which was now to be the norm with a "Best Practice" Guide (one of several) issued as a live document to be continually updated. The Project Chief Engineer, as for DDP, was now just responsible for the technical definition, as well as of course contributing to the PE's management team. He therefore had an important task to also ensure technical integration across all the 'DDP' projects running in the product family, working through the leaders of the Integration Teams reviewed earlier.

5. SO WHERE ARE WE TODAY?

"Product Operations" has now been running for two years, and although the benefits are unquestionable, neither are matters anything like perfect. There is still much to do in getting people to understand roles and responsibilities on the two axes of the matrix. Also in understanding "Natural Team Working" and when it is appropriate to co-locate resources in the Product "DDP" style teams or when working in a peer group is more appropriate.

However, there can be no doubt that in embarking on our journey of change, we have progressively instilled in our people a culture of continuous improvement. Thus we can claim to have largely overcome the biggest barrier to performance improvement - that of an in-built reluctance to change!

We are becoming increasingly aware that there is much more to be gained from focusing on cultural issues. The nature of the drivers for change discussed earlier and the realisation that processes are vitally important to reducing costs and lead times meant that our people development in the past had tended to concentrate on multi-functional team working and the creation of the Product Operations organisation. We are now undertaking a programme which concentrates on the 'softer' side of people development as we recognise that attitudes and behaviours have a major effect on the motivation and performance of people and teams. Elements of this programme includes all aspects of leadership, customer satisfaction (internal and external) and team working. The programme is already proving beneficial to both our people and the business and is creating enough interest to be the subject of a paper in itself....watch this space!

Returning to the "Development Process"; as already implied, "DDP" has become "The way we do things here" and, so is the "On Line Pilot" process.

Figure 6 indicates how our initial three DDP projects have rolled forward both "downwards" into the CPD (modifications) process and "upwards" or upstream into new projects as the focus of our business is changing. The Process Sponsor Team has also evolved, through NPI (figure 6) "New Product Introduction", to now become "PIP", the Product Introduction Process Sponsor Team. In fact, we now believe that every major project will become an "on-line pilot" with dual accountability for product delivery and improving our processes. This was captured for the first time in the top level "Purpose Statement" for our Nimrod 2000 Wing Team (figure 7). This team, as well as helping the BAe Military Aircraft Business Unit by supporting the Nimrod 2000 project, was also responsible for pushing ahead with our DDP learning and the next step change in our objectives for reducing the time and cost of the development process, new "Computer Aided Engineering" Tools.

6. COMPLETING THE PICTURE, THE IT AND CAE CONTRIBUTION

6.1 "ACE"

After the completion of a comprehensive benchmarking process, the end of 1995 saw an historic contract signed by the Airbus Partners to procure at least 1,500 seats of common CAD/CAM/CAE and enterprise data management software from a single supplier – Computervision.

To support the new tools being introduced into the business, it was recognised that a cross-partner framework was needed for hardware, software, networks, processes, tools, methodologies, organisation, training & support. This was the goal of Airbus Concurrent Engineering (ACE).

During the initial ACE start-up period BAe Military Aircraft won the Replacement Maritime Patrol Aircraft (RMPA) contract to supply the Royal Air Force with the Nimrod 2000. Due to its experience in large aircraft, BAe Airbus at Filton was given responsibility for design of the wing, centre torsion box and some aircraft-level systems.

This was the first opportunity to use the new CAD/CAM/CAE tools on a complete wing programme in BAe Airbus. The programme has and is contributing enormously to developing the people/processes/tools needed for ACE, for BAe Airbus itself, for the current A340-600 programme, and for future programmes such as A3XX.

6.2 Overview of Electronic Product Definition Tools (EPD)

EPD enables the creation of a complete digital product definition, and allows this to be shared across an organisation throughout the product life. This allows an enterprise to concurrently create, manage, share and reuse electronic product information in a collaborative environment.

The basic Computervision™ EPD™ design tool in BAe Airbus consists of CADDS 5™, a parametric solid modeller and OPTEGRA™, the data management tool. Parametric modelling means that in model construction, parameters appear on the screen to show the desired leading dimensions of a 3-D geometric entity. It is possible to change these parameters when the design needs to be modified, and share the information to all relevant specialists, simultaneously, at each design iteration, without the time and cost impact of "drawing" revision and re-issue. The re-sizing of the component, and any consequential changes through the assembly is carried out automatically by the CADDS 5™ software.

OPTEGRA™ consists of two sub-modules. A visualisation tool that allows large assemblies to be viewed without complexity of the full geometric model, and a product data management tool, that allows parts to be attached from CADDS 5™ into a product structure tree. Figure 8 shows an example of this in action. Attached onto each node in the tree are project defined attributes. These are used to filter relevant information from the tree, and thus can highlight components and assemblies in the attached visualisation packages.

6.3 Nimrod Wing Team

Following the DDP principles, the Integrated Design Build Teams for Nimrod were set up in a co-located, secure area (this being a Military Programme). Each IDBT has a Team Leader commercially responsible for each work package from initial concept through to first-build. Across these streams are the technical integrators ensuring that design is to the up-to-date specification, that functional specialists can be brought to bear in areas of technical difficulty, and that quality and consistence of standards across the design and manufacture cycles are to an acceptable level. This can be seen in Figure 9.

The main structure of the wing was split into 'Zones' (e.g. centre box, inner wing, outer wing, etc.). Systems are allowed to route across structure in the concept phase, and then in the engineering phase (when concepts are frozen), they are broken into work packages within the zones.

Under the team leader it became apparent due to the nature of the EDP tool itself that a central point of contact within the work package area was needed to co-ordinate the creation of the product structure tree, provide a foundation for communication, and to deal with issues such as data exchange and manufacturing liaison. Thus the key roles of "Zone Controllers" and "Zone Integrators" were born again reflecting how the tools and technology can affect processes.

As the definition evolves through the "gates" of the "Concept" and "Engineering" phases, controlled by a design review process, the product structure tree evolves to reflect the design-as-build structure, instead of the traditional functional-based breakdown. This means that it is vital for the Zone Integrators to liaise with their manufacturing personnel to orient the tree to reflect the latest build philosophy. This is helping further to overcome the traditional cultural problems in getting manufacturing and design to communicate closely, as any changes can be seen immediately and acted on. The advantage is that the product structure reflects the jig build and therefore provides a common frame of reference for manufacturing and design.

True to the "On-Line" concept we have now developed, the "Nimrod Experience" is already being transferred into the A340-600 team working on the heavily adapted A340 wing for this new 400 seater project launched in December 1997.

This leads to the next major CAE development which was once again researched off-line but brought on-stream through its own 'pilot process' with the A340-600 team.

6.4 Knowledge Based Engineering

Knowledge Based Engineering (KBE) has been a much used buzz word throughout a number of industries. In common with other businesses, KBE had it's technical enthusiasts within our organisation and they had tried to implement it on a number of occasions. It was only when a KBE team was set up, focused on looking at the potential of KBE within the Product Improvement Domains, for a real hard Project requirement, that "Buy In" by the business was achieved.

To BAe Airbus, the core of KBE activity is the creation of Generative Total Product Models which enable "Analysis Driven Design" with, for example, detail design, stressing and costings "Rules", or Knowledge, input to the initial EPD solid model definition rather than following in subsequent iterations

BAe have been monitoring KBE software tools for a number of years before choosing ICAD 18 months ago, as the most mature product in the marketplace. With several years of ICAD experience in research projects, a KBE team was set up to define a strategy that would deliver a range of business benefits. The strategy had to cope with the different requirements of Initial or Conceptual work and the very immediate demands of project teams, like the A340-600.

During the aircraft conceptual design stage, program timescales and costs are small and of secondary importance. Here the KBE strategy focuses on enabling a far higher level of product definition with an increase in design iterations. The Business benefits by being able to understand more design solutions to a higher level of definition than before. This yields higher confidence and a consequent reduction in risk associated with proposed design solutions.

For project teams such as a A340-600, cost and timescales are of critical importance. Thus here the strategy has been matched with these criteria so that the product is delivered to a higher level of quality in a shorter timescale whilst maximising the available resource's productivity.

The KBE strategy enables the project teams to develop the applications required on a given project. The KBE integration team then also ensures that the development and integration of KBE applications meets other project teams' requirements, so that A300-600 applications also meet A3XX needs for example. The management support, documentation, KBE best practice and training are dealt with by the KBE team, ensuring overall business benefit.

6.5 Application Example, A340-600 Rib Foot

At this point it is worth considering an example of a typical KBE application, to illustrate some of the points raised so far and the benefits of the approach. The rib foot is a structural component through which the wing skins, wing stringers and wing ribs are bolted together, as illustrated in Figure 10. Each rib foot is subtly different and there are about 1,200 on the A340-600 wing. To design all of the rib feet using classical methods costed about 12,000 man hours per design iteration.

An ICAD rib foot application was developed at a cost of 800 man hours. This tool does 99% of all rib feet and takes about 10 man hours to finish the A300-600 rib feet, generating the EPD Solid Models in Parametric CADDS5 form. The tool integrates stressing programs, Finite Element representation, cost analysis and manufacturing requirements into the model and so all the resulting rib feet meet structural and manufacturing requirements and recognise costing repercussions.

6.6 KBE Impact on the Design Process

Traditional philosophies of project planning will need to be radically re-aligned to suit the process possibilities due to the impact of KBE technologies. Project plans will need to focus on the process of achievement alone. For example, the A340-600 wing box rib feet; whereas traditionally milestones concentrated on designs being visibly progressed during each phase of a project, now the result is taken as a given and the process of achievement is analysed and confirmed guaranteeing the outcome. Time can be given to the evaluation of multiple iteration developments before settling on the final design solution with the data for manufacture compiled in minutes rather weeks or months. This reinforces the concentration of effort on concept and away from detail, thus freeing up highly qualified designers to where they can add value, in optimising the design and continuously updating the "rules" in the KBE Software.

Once the KBE rib foot tool had been developed, its utilisation truly enabled the Concurrent Engineering process. Stress manufacturing and design experts could simultaneously see how each of their requirements affected the product and more importantly, see the effect of their requirements on others.

Often in the past, the workload on projects would only allow designers to implement the design processes and designs that they were used to. This makes generation and incorporation of innovative processes difficult to achieve. With the reduced timescales achieved using automated KBE Software, not only are manpower requirements reduced but designers are freed up both to generate innovations for incorporation into an improved tool and to work on more interactive and intuitive work. Continuous Process Improvement is thus supported inherently by KBE.

7. IN CONCLUSION – THE FUTURE VISION

Finally, I return to the three "Product Improvement Domains", People, Process and Technology – Figure 2, I hope I have shown in this paper the absolute imperative for progressing in all three domains. In 1995 British Aerospace Airbus set itself some overall Business Improvement Goals, relative to "1993" standards, which included:-
50% reduction in cost base
50% reduction in lead times
- to be achieved by 1998.

Whilst these were mainly aimed at the "Develop, Procure Parts and Manufacture" core recurring process (see figure 16), the "Product Introduction Process" owned the following consistent improvement Goals:-
50% reduction in Product Launch Costs (for a given standard of task)
30% reduction in Time to Market
30% reduction in Recurring Costs (due to PIP)

To date, I must be honest and say that we will not fully achieve these goals this year! However, we have seen sufficient real improvements to indicate we can and will achieve these goals in the not too distant future. Further, we will share our learning with the other Airbus Partners, and no doubt learn from them, to ensure that as Airbus Industrie moves towards a Single Company it will be able to call on World Leading Design and Development Processes. This has already started through the "Airbus Concurrent Engineering" (ACE) programme referred to earlier, now well under way.

In summary, we believe our "Product Operations" model of Product Teams working within an agreed strategic framework, supported by Resource Teams responsible for developing resources, processes and knowledge has much to commend it. Further, that every new development project should always be an "on-line pilot", with dual responsibility for delivering the project and developing working practices. That is not to say that "off-line" improvement does not have its place – in fact it being essential to develop IT tools to a certain stage in a "Laboratory" environment, before capturing the imagination of the Product Teams and going "on-line".

The CAE tools we have been embedding through the Nimrod 2000 and A340-600 teams are now demonstrating what huge improvements are possible and what fundamental changes are going to happen in the design process – we have only scratched the surface so far. Not only do they bring specific cost reduction and time compression, but they truly enable concurrent working and allow the benefits of team working to be realised many times over!

The future is exciting, even if the end game is not yet clear! We believe we are on the right track and have put most of the building blocks in place to develop a radically different development process in the next year or two, delivering real additional value to the Airbus business. Most important of all, we are certain that we have the best people, who are highly motivated, skilled and experienced and who are more than capable of meeting the challenges ahead.

8. ACKNOWLEDGEMENTS

Our sincere thanks are due to the many colleagues who have contributed to the improvements that have made this paper possible! Also to those who have contributed to the paper itself: Bob Landeg (for the concept of fig. 2), Mike Grealy, Bill McLundie and Andy Powell, and finally to British Aerospace PLC for permission to publish.

- Shareholder expectation
- Reduced Government funding and GATT
- More demanding customers
- Increased competition
  - Boeing manufacturing improvements
  - Aggressive pricing

FIGURE 1: DRIVERS FOR CHANGE

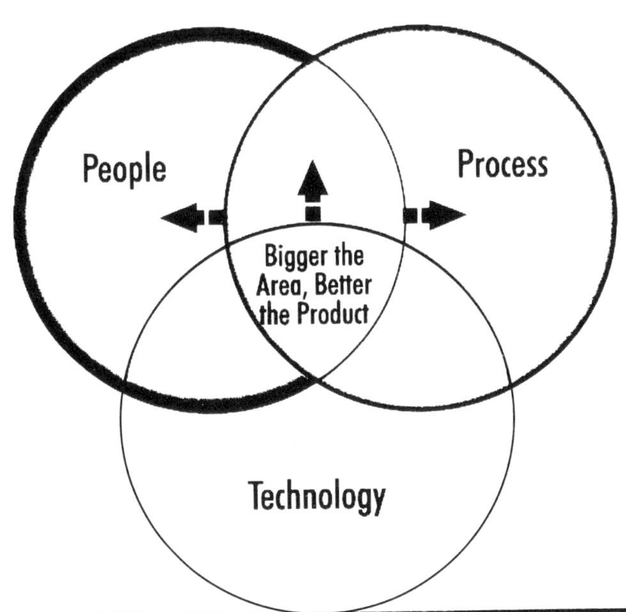

FIGURE 2 : PRODUCT IMPROVEMENT DOMAINS

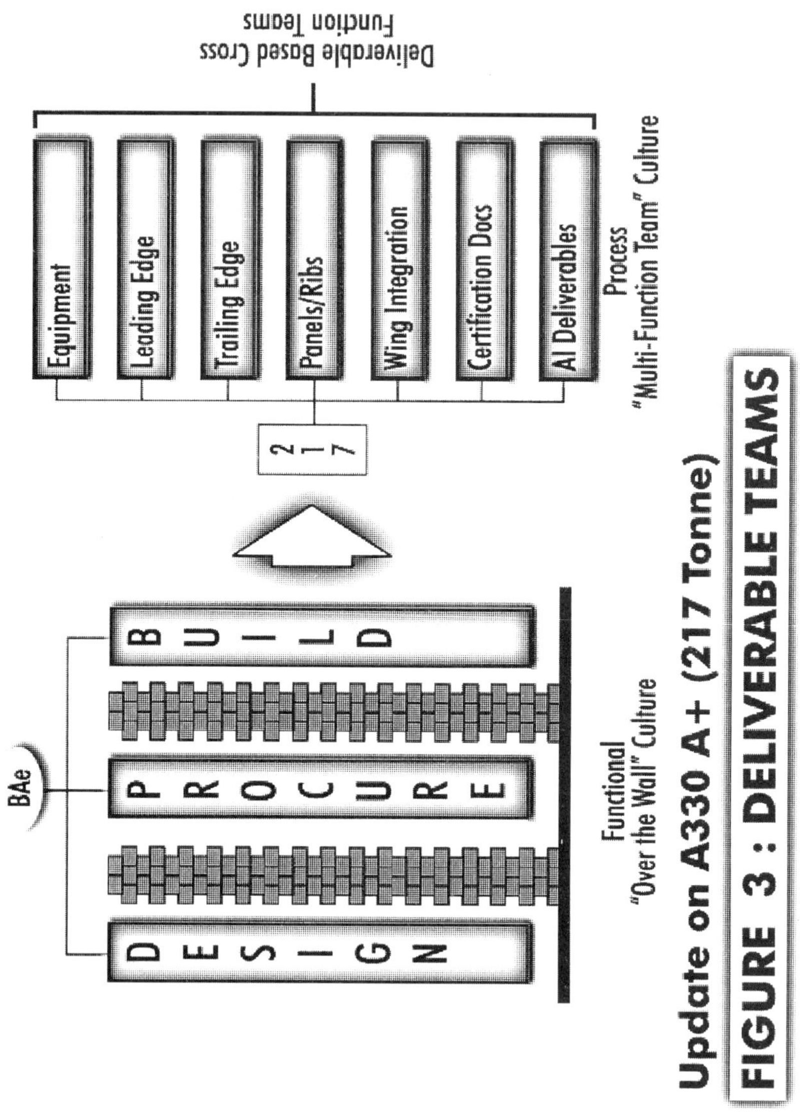

Update on A330 A+ (217 Tonne)

FIGURE 3 : DELIVERABLE TEAMS

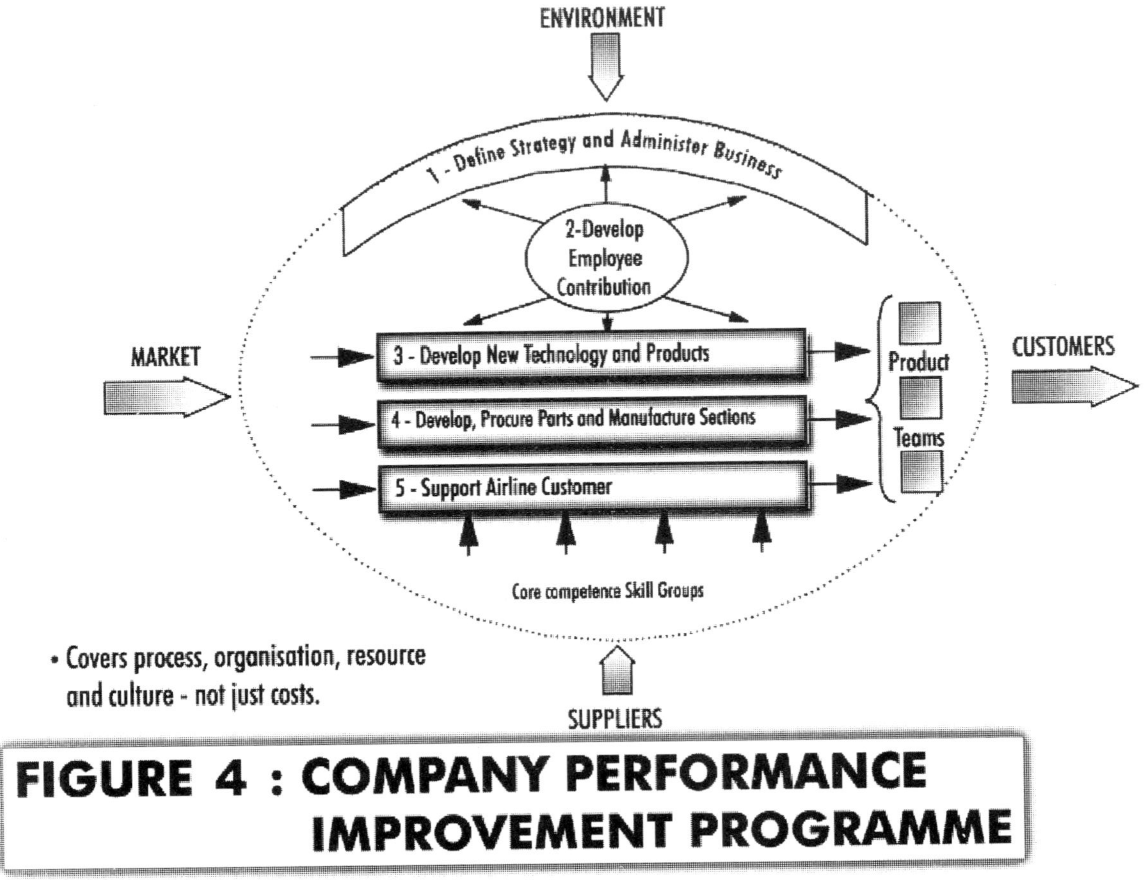

FIGURE 4 : COMPANY PERFORMANCE IMPROVEMENT PROGRAMME

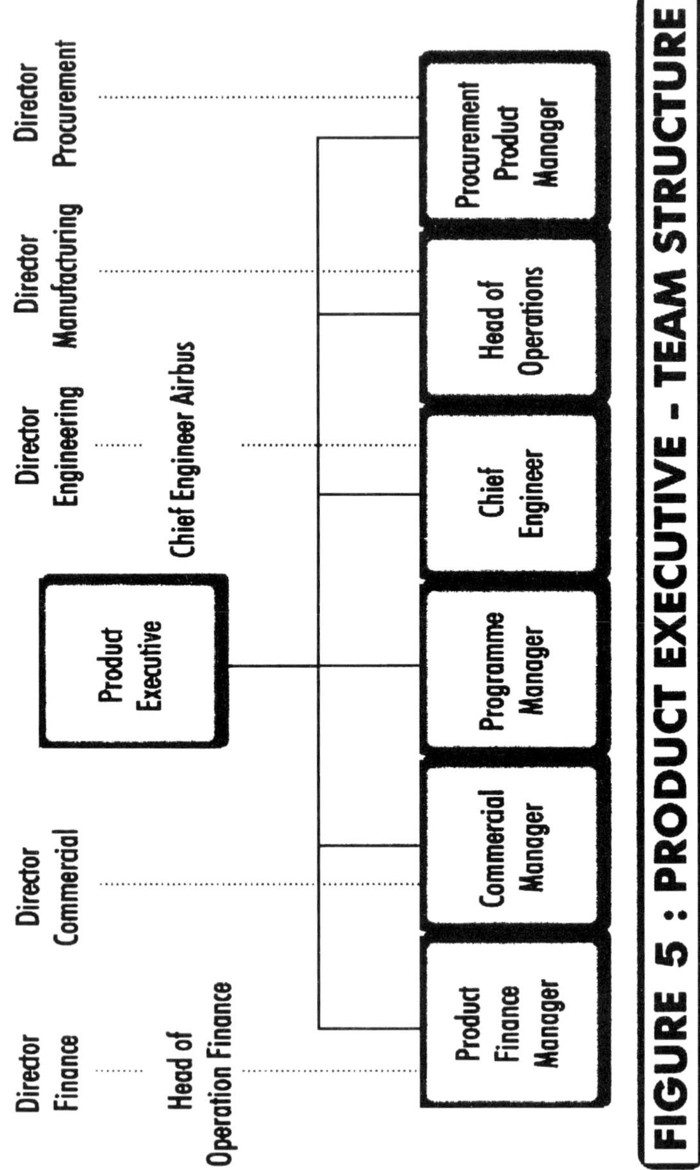

FIGURE 5 : PRODUCT EXECUTIVE - TEAM STRUCTURE

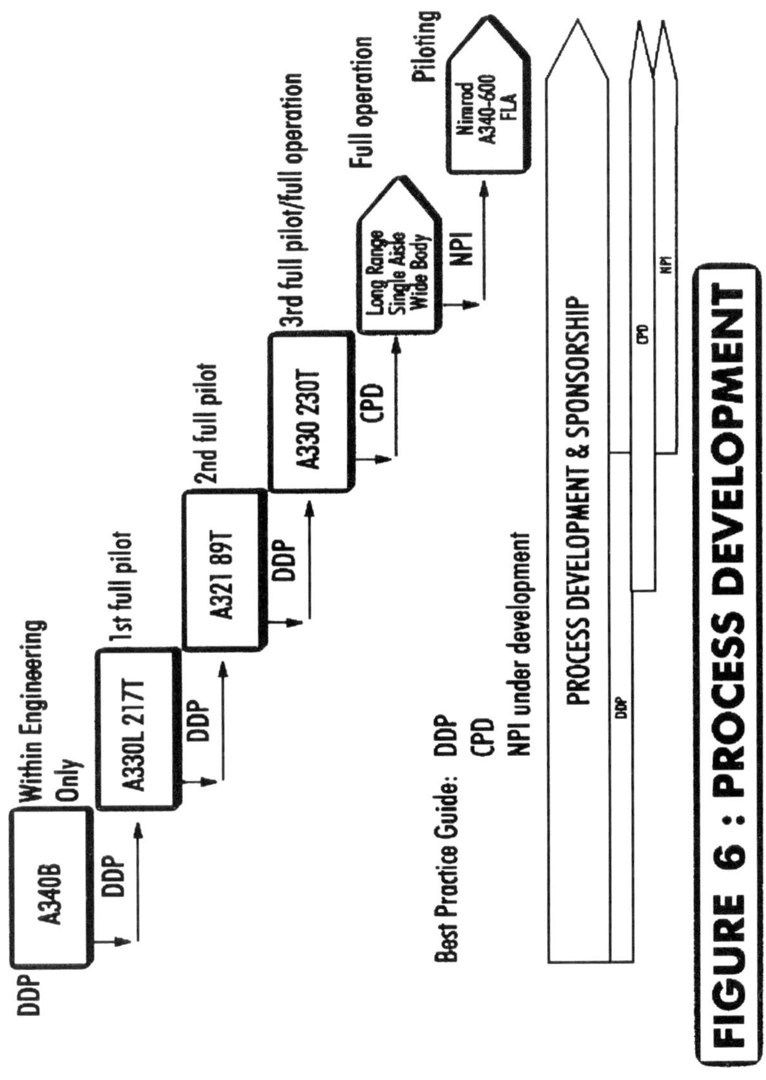

FIGURE 6 : PROCESS DEVELOPMENT

To deliver the Nimrod Project consistent with Our Business Objectives whilst maximising development opportunities

FIGURE 7 : P.O.R.T. – NIMROD TEAM PURPOSE STATEMENT

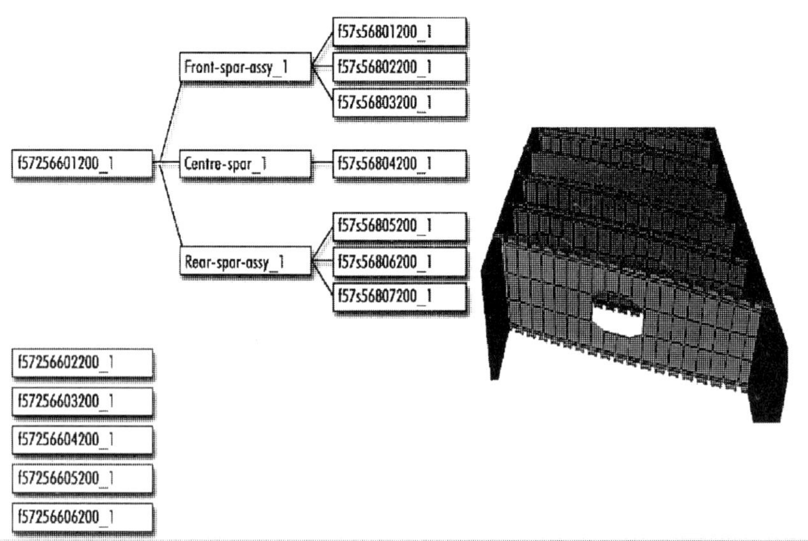

FIGURE 8 : OPTEGRA PRODUCT TREE & VISUALISER

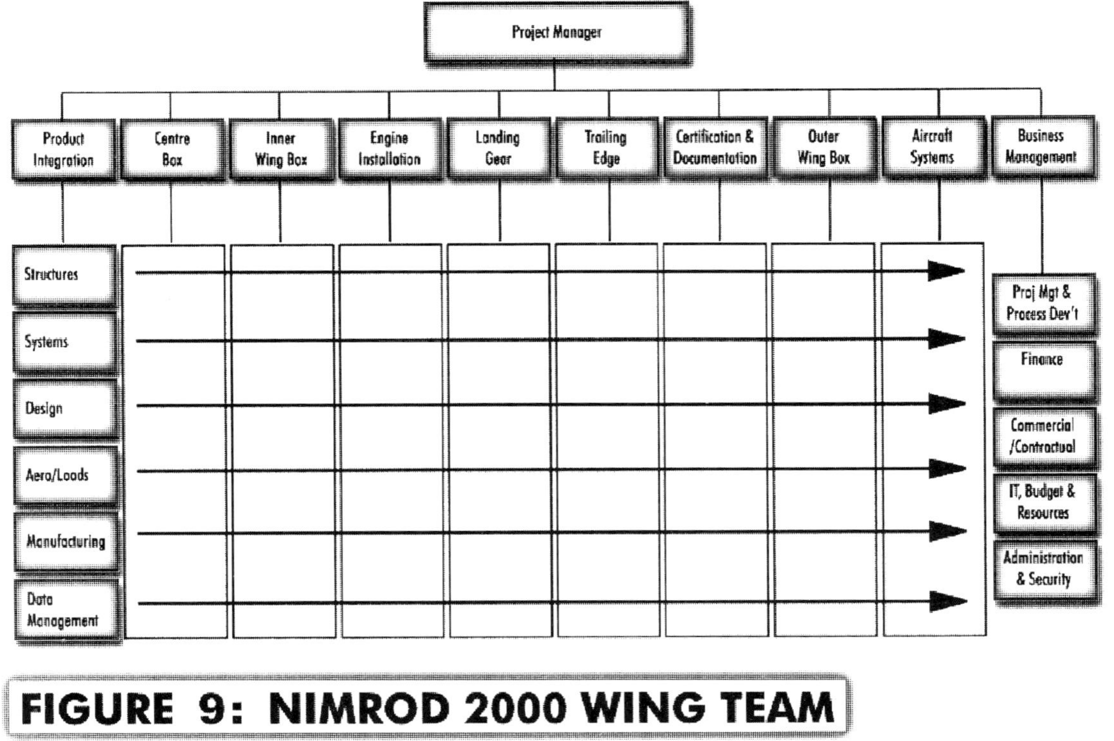

FIGURE 9: NIMROD 2000 WING TEAM

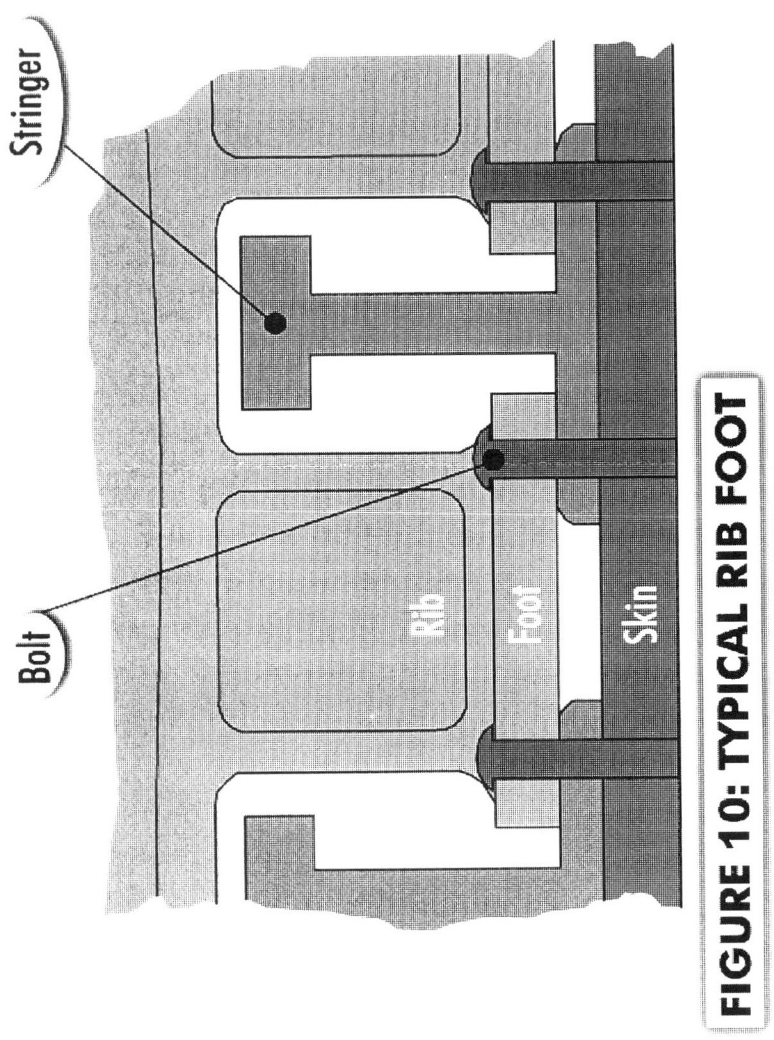

FIGURE 10: TYPICAL RIB FOOT

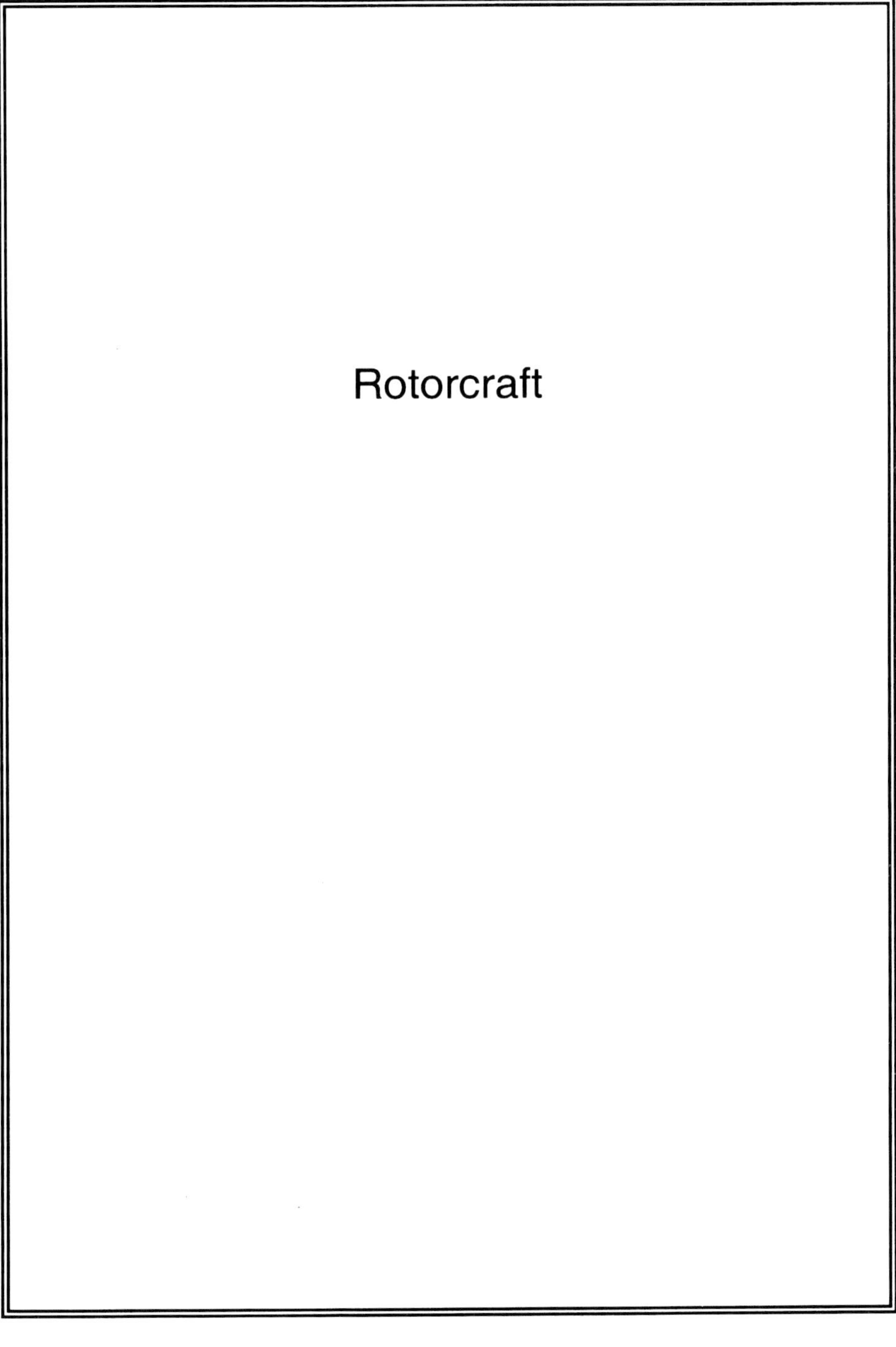

Rotorcraft

C545/067/98

A perspective of future rotorcraft technology

P N GODDARD MSc, FEng, FRAeS, MIMechE
GKN-Westland Helicopters Limited, Yeovil, UK

SUMMARY

This paper is introduced by considering the changes in the market that will impact future rotorcraft technology development. The main implications for organisations engaged in rotorcraft research, design and development are identified. A perspective of future engineering process technologies is provided together with a review of the prominence and importance of an integrated systems engineering approach to design. A perspective of future engineering product technologies is undertaken by reviewing the priority research and technology programmes at GKN Westland. This review will focus on three particularly important programmes: the next generation rotor blade, avionics and systems developments, and the advanced compound helicopter. Finally, by way of example, the successful application of new product technology to the Anglo-Italian EH101 helicopter is summarised.

LIST OF ABBREVIATIONS

ACH	Advanced Compound Helicopter
ACSR	Active Control of Structural Response
AFCS	Automatic Flight Control System
BERP	British Experimental Rotor Programme
DERA	Defence Evaluation & Research Agency
EDM	Enterprise Data Management
IPT	Integrated Product Teams
KBE	Knowledge Based Engineering
PDM	Product Data Management
MoD	Ministry of Defence

INTRODUCTION - CHANGES IN THE MARKET-PLACE

A number of trends are clearly discernible in the market-place for rotorcraft. These trends will have a profound impact on the technologies that must be developed and implemented by the rotorcraft community if the product is to increase its share of the world's aerospace market.

Technology developments over the last two or three decades have been supported by relatively large and stable defence budgets associated with relative economic and geo-political stability. The future is much less certain and could well be less stable; defence budgets have been consistently decreasing and we cannot yet be confident that they are now levelling off. Future military operations are characterised by change and unpredictability as witnessed by political events and the many conflicts across the world over recent years.

Military planners consistently emphasise the need for increased operational flexibility and adaptability from the equipment that is to be employed in the future. The inherent utility nature of the rotorcraft, as witnessed by the plethora of operational roles and associated equipment fits that have been employed, augur well for its attractiveness to the military user in the future.

Projections for the civil market appear to be very variable between one set of forecasts and another. The view that we hold is that in broad terms future demand in this market is likely to exhibit modest growth in both volume and revenue terms. Clearly for the civil rotorcraft market to further develop, the reduction of operating costs and increased safety will be the primary requirements followed by improved performance, comfort, external noise reduction and high reliability/despatch readiness.

One fundamental trend developing in all markets is thelongevity of aircraft lives. For the future, we can clearly see new designs being delivered into the market-place being in service for up to 40 or perhaps as much as 50 years.

In broad terms we can clarify four clear needs in the future rotorcraft market-place:-

i) The need to harness technology that emphasises cost-effectiveness rather than simply capability to meet customer requirements.

ii) Whilst it will continue to be a requirement to meet customer operational needs, these are likely to be broader and less specialised, emphasising flexibility and adaptability with multi-role capability to maximise cost-effectiveness.

iii) Economic pressures will be unrelenting and an increasing emphasis will be placed on dual use military and civil technology.

iv) Increasing longevity will require technologies that are readily applied to existing products at acceptable levels of cost.

v) Finally, there will be a strong and consistent need for both military and civil operations to develop and apply technologies to improve safety.

ENGINEERING PROCESS TECHNOLOGIES

Market and economic pressures are such that successful organisations will be continuously striving to:-

i) Significantly reduce the time to bring products to the market (both new products and product updates);

ii) Sustain excellence in the design of the product;

iii) Reduce product non-recurring costs.

Traditionally, any review of rotorcraft technology would tend to focus on the "product aspects", e.g. new hardware or software developments, improved analytical techniques, new means of testing, etc. Whilst these product aspects are very important, and will continue to be so, "process aspects" of Engineering Research Design and Development will become increasingly important demanding reduced "time to market", improved "product quality" and reduced "non-recurring costs". These process aspects are considered as new technologies in their own right and will be important attributes for future competitive success.

At GKN Westland we are embarked on an Engineering Process Improvement Programme to deliver the changes necessary. There are six main elements to this programme as described in the following sections.

Organisation and Working Practices

Centres of Excellence, Integrated Product Teams and Helicopter Systems Engineering are the basis upon which our organisation will develop in the future. The common theme pervading all our future activities will be Concurrent Engineering.

Centres of Excellence for Rotor Systems, Mechanical Systems, Integrated Avionics, Structures and Materials, and Product Qualification/Certification will be progressively developed and implemented. Close integration with Manufacturing and the supply organisations will be a prime consideration. In particular, longer term partnership arrangements with key suppliers is to be aggressively pursued.

Integrated Product Teams (IPT's) comprising co-located personnel from Engineering, Industrial Engineering, Suppliers and the Build Shops are the key to obtaining concurrency and further reducing timescales. Furthermore, IPT's provide the basis for ensuring that the Design Engineer is provided with best possible data and advice at the point at which the design is frozen, thus ensuring that the design is manufacturable in the most economic fashion.

The very comprehensive nature of customer demands and product specifications will increasingly require that a Systems Engineering approach to Product Design be undertaken in the future. This aspect is more fully covered in section 3.

Design Tools

3D solid modelling will become the norm throughout our Engineering organisation. Electronic mock-ups and electronic assembly modules will significantly benefit the way in which product kinematics and product space management will be analysed. These features will directly contribute to the improved quality of our engineering output, i.e. Right First Time. There is every indication that an overall 15% to 20% compression of time to first product can be achieved by large-scale application of solid modelling.

The full integration of design with the manufacturing and assembly computer tools, including the generation of component process and assembly instruction manuals, is the aim.

Knowledge Based Engineering (KBE) systems linked to 3D parametric design packages show promise where repetitive types of design work are common. The technique may also be applicable to situations where time between designs is large (say 10 years plus). KBE may provide a form of corporate memory where techniques and lessons learnt on one design could be used to assist the next designs, even though personal expertise may be lost in the intervening period. The increasing time periods between designs, and the limitations of resources enabling personnel to continue practising these skills, will increasingly require that this form of learning from one programme be captured for application to future programmes.

Product Data Management

Information (data) in a wide variety of forms represents a vital but hidden and often unrecognized asset to any organization. The obvious embodiment of this information to the engineering organization is the finalized design definition. But this is merely the culmination of the engineers' efforts in accessing, combining and appending to diverse data streams such as product requirements, knowledge of physical processes, and experience of previous designs. The underlying business process provides the framework by which these data sources are combined.

Design information on new products is now almost universally captured in digital form and increasingly, the associated data relating to methods and processes. But like a library without an effective index, the value of that data is significantly compromised unless the data is captured together with its associated structure and interrelationships. Since most engineers today spend most of their time accessing and manipulating data, effective access is paramount. Much time can be lost searching for data and more still lost if work commences using outdated information.

Product/enterprise data management (PDM/EDM) systems offer the means to capture this data in a structured way. The systems can, however, only be as effective as the underlying information they contain. Business processes must be reviewed and optimized if the promised benefits of improved design quality and reduced timescales are to be achieved. Improved quality results from the capture and consistent implementation of the optimized processes, ensuring access to the correct version of the correct information. Reduced timescales result from 'right first time', minimizing the need for remedial work and virtually instantaneous transmission of data from one individual to another.

Importantly, for most large projects in today's aerospace industry, PDM/EDM systems offer the underlying framework for delivering these benefits where international collaborative programmes demand multi-site concurrent working. For prime contractors concentrating upon system integration and programme management, an effective encapsulated business process can help to minimise risk and provide competitive advantage.

Management Systems and Controls

Systems to ensure control of the engineering process need to be more comprehensive (i.e. covering the whole of the process) whilst forming an integral part of the company's overall business system.

A four-stage process with clearly defined events separating the end of one stage and the commencement of the next is being implemented in our organisation. The four stages cover project planning, full concept definition, system realisation (i.e. development and production) and in-service support.

A major challenge for the industry is the development of measures of performance for each stage of such a process broken down to the detailed level at which specific tasks are being undertaken. Only by developing and introducing such a system can changes and trends in productivity be accurately assessed as a basis for management action.

The Culture of Engineering

Significant reductions in time to market, sustaining excellence in product design and reducing product non-recurring costs can only be achieved in organisations where continuous improvement, innovation, entrepreneurship and technical excellence are integral elements of the culture.

It is very clear from the extensive benchmarking of best practice that we have been undertaking that developing such a culture is a pre-requisite of ensuring deep and lasting changes and improvements to the performance of the organisation and the personnel within it. Suffice to say that this is a very substantial issue that presents a major challenge to the whole industry.

Integrated Computer-Based Design Tools

Computer-aided design and analysis methods have long promised to reduce times for product design and analysis, but expectations have not always been realised.

The advent of high functionality 3-D design tools, seamlessly integrated to analysis tools, offers the potential to deliver the promised benefits. When 3-D modelling and finite element analysis are closely integrated, a truly iterative approach to design becomes practicable. Other analysis tools can be incorporated in this same loop; at GKN Westland, for example, Company-written dedicated software is being introduced to provide an integrated solution to blade design.

However, the full benefits are only achieved if the 3-D product definition is used not only throughout the engineering function, but elsewhere - for manufacture and support - both within the Company and with suppliers. But these benefits can only be achieved if the new technology is accompanied by changes in the Company business processes to allow effective concurrent engineering.

PRODUCT DESIGN - AN INTEGRATED SYSTEMS ENGINEERING APPROACH

Ever demanding requirements will need to be met by future rotorcraft (both military and civil). The competition is increasing across the world and new suppliers will be entering the market (particularly from the Far East); customers will expect clear visibility of specification compliance (and from the very earliest stages of new design), and the risks (essentially commercial) of non-compliance are increasing.

The design of a new rotorcraft, or a major product update and development, is a very complex and comprehensive task; there are very many functions and attributes that have to be fully addressed and innumerable trade-off studies that have to be undertaken as the design progresses.

It is widely appreciated that the out-turn costs and associated programme risks are substantially locked in over the very early phases of the design of the product, i.e. the feasibility and project definition stages.

In the future, an increasing emphasis will need to be placed on "total systems integration" as the key to successful design.

At GKN Westland we will be ensuring that our future product designs are led and managed by integrated systems engineering teams, supported as necessary by computer-based integrated system design tools and process controls. These teams will need to be led and staffed by the very ablest of engineers who will be able to exercise the necessary judgement, trade-offs and innovations required in developing new designs. These teams, ultimately leading and integrating comprehensive IPT structures will be at the heart of Product Design and will live and breathe the product as it successfully progresses from feasibility into service.

A major challenge for rotorcraft design organisations is the selection, development and retention of a sufficient number of these engineers to resource future product design needs. This requirement is exacerbated by the increasing time periods between one new design and another where practising engineering skills over the intervening period is very difficult unless programmes of technology development can be introduced to continuously refresh engineering capabilities.

ENGINEERING PRODUCT TECHNOLOGIES

Priority Programmes At GKN Westland

We are now embarked on a number of research and technology programmes at GKN Westland that will refresh our technology base and support our product market plans into the early part of the next century and beyond. There are nine areas of key technology activities:-

i) Rotor systems, particularly the next generation rotor blade.
ii) Integrated avionics systems developments (aircraft and mission).
iii) Advanced rotorcraft configurations.
iv) Advanced transmissions.
v) Flight control, essentially active control technology.
vi) Noise and vibration reduction.
vii) New materials and structures.
viii) Survivability.
ix) Reliability and maintainability.

Given this range of key technologies, three fundamental points of policy apply to our activity:-

- Firstly, all the developments are strongly focused at product development and future new project application.

- Secondly, increasingly the technology will be developed and acquired through collaborative and partnership arrangements. These arrangements encompass research organisations, helicopter manufacturers and the wider sub-prime suppliers (e.g. engine manufacturers, avionics suppliers, software houses, etc.).

- And thirdly, the harmonisation of military and civil technologies, commonly referred to as dual use.

Three research and technology programmes are at the heart of activities and these are now considered in turn.

The Next Generation Rotor Blade

Enormous advances have been made with rotor blade technology over the last two decades. In the U.K. the state of the art is the rotor that incorporates composite materials and advanced aerodynamics applied on the EH101 as shown on Figure 1. This technology has also of course been retrofitted on the Lynx.

The advanced aerodynamic blade widely referred to as BERP (British Experimental Rotor Programme) first flew on a Lynx twelve years ago and demonstrated a 37% increase in usable blade loading in cruise conditions. The World Speed Record was taken in 1986 on a Lynx with the advanced rotor; at 400 kph (249 mph) straight and level flight, the record is still held today.

Main rotor blade technology development in the U.K. will now build on research programmes conducted over the last ten years in GKN Westland and DERA (Defence Evaluation & Research Agency) and the first phase of a six-year Next Generation Rotor Technology Demonstration programme commenced in October 1997 under joint GKN Westland and U.K. MoD support. This activity aims to reduce the risk of applying generic new technologies that offer improved main rotor cost-effectiveness through a programme of technology development and cost/operational assessment. It will prove both the technologies and the design rules for application to helicopter projects in the timeframe up to 2020+.

GKN Westland now employs mature composite main rotor blade manufacturing processes, but a wide-ranging review is being conducted to determine how state-of-art composite manufacturing processes or optimisation of the current processes can provide cost improvements, for which very ambitious targets have been set. In addition, customers' experience of operating composite main rotor blades is being collated to ensure that their future needs for improved reliability, maintainability and repairability are met.

Improvements in lift, power and vibration performance are being sought from new families of advanced aerofoils and planforms and aeroelastic tailoring. Smart rotor studies are also underway. Survivability is an increasingly important requirement and appropriate measures are also being studied. New generation composite materials are being investigated, focusing on increased stress/strain capacity, toughness and decreased environmental degradation factors. Finally, alternative de-icing, lightning strike protection and erosion protection solutions are also under consideration, with the emphasis on lower cost and increased repairability.

Integrated Avionic and System Developments

The rapid advance of avionic system technology has underpinned the development of rotorcraft operations, and the avionic systems installed in future helicopters will play a vital role in meeting increasingly demanding operational needs. Areas where new developments are likely to have a major impact on future rotorcraft operations and capability include:-

- The core avionics system with the introduction of integrated modular avionics technology;

- The cockpit with increased emphasis on integrated human centred design;

- Day/night all-weather systems fusing information from databases and sensors to improve the pilot's situation awareness; and

- The flight control system, with widespread adoption of fly-by-wire, and the introduction of enhanced control augmentation.

For the next generation of new and updated products we must reduce the cost and lead time of the avionics systems whilst further increasing capability and providing more flexibility to allow in-service modifications and updates.

It is widely accepted throughout the avionics industry that the key to reducing costs and lead times for military systems is the introduction of open architectures and re-usable, modular hardware and software. The introduction of open standards and modularity will help to reduce development cost and timescales (by re-use of modules), reduce production costs (by standardising across a wider range of aircraft types) and reduce in-service support costs (by reducing spares holdings and minimising second line maintenance).

At the heart of this new generation of systems will be a series of data networks, linking modules to sensors, displays and actuators as well as to each other. These networks will operate at far higher speeds than those in current systems and fibre-optic implementation may therefore be inevitable.

Architectures supporting open, modular systems will allow greater exchange of data between system elements, paving the way for higher levels of data fusion to achieve maximum performance from a given suite of sensors.

As mission requirements have become more severe and operations have been extended into increasingly difficult environments, the demands on the crew have risen dramatically. The new man-machine interface technologies provide an enormous opportunity for facilitating the interaction between the crew and the rest of the system. Further advances in technology, such as larger displays and virtual cockpits, will also provide opportunities for improving crew interaction with the vehicle and its systems, paving the way for further increases in operational capability.

A major consideration which must be addressed by prime contractors when proposing to introduce many of these new technolgies into rotorcraft systems is that of establishing the merits of each technology; if benefits cannot be demonstrated in quantifiable terms, then there is little prospect for the required investment being made available. In some cases, quantification of the benefits is relatively straightforward, for example in prediction of weight savings, maintainability improvements, etc. However, where the benefits arise mainly in the areas of rotorcraft/mission system capability, it can be much more difficult to show to what extent an improvement in capability results in an operational benefit.

For example, the use of synthetic environments (Figure 2) to allow simulation of representative battlefield scenarios provides a potential solution to this problem. Simulation technology now allows high fidelity representation of an individual aircraft's characteristics and environment, and this is being combined with the latest Information Technology developments to allow simulation facilities to be linked together to form the multi-player scenarios of the real battlefield. This approach to simulation not only allows a more representative evaluation of the capability of a new technology at an early stage in the development process, but allows the method of using a new technology or capability to be optimised before the quantitative evaluation of the merits is performed. This avoids the weakness of conventional operational analysis, which tends to be constrained by the use of tactics based on current technology capability.

Synthetic environments and distributed interactive simulation can therefore be seen as taking increasing importance in the processes of selecting and developing avionics and systems technologies, and defining required system performance for future rotorcraft.

Advanced Rotorcraft Configurations - The Advanced Compound Helicopter

The concept of compounding seeks to redress limitations encountered in current helicopters by enhancing basic, fundamental, vehicle parameters. Prominent amongst these are high speed values of Lift to Drag ratio, propulsive efficiency, agility and vibration. Improvements in these quantities lead directly to the prospect of higher productivity, superior operational effectiveness and reduced maintenance. These in turn offer improved competitiveness and the opportunity to expand the range of helicopter applications.

The fundamental nature of the concept also appears to provide a very robust solution to the problem of improving current helicopter capabilities. This is evident in the versatility displayed by compound design solutions when applied to a wide range of roles, including applications for which helicopters were not previously considered eligible. This theme also leads naturally to the idea of a family of vehicles, maybe modular, whereby differing variants of a basic design are matched across a demanding set of requirements. Whilst exploiting significant commonality of design, development and manufacture this also minimises the inevitable compromise traditionally exposed when multi-role vehicle programmes are undertaken. For example, high altitude, low speed missions favour lift compounding, whilst in contrast, hover dominated missions, are not best served by lift compounded vehicles, due to the penalty of wing download. Those roles that combine low speed/hover and high speed dash requirements, such as SAR, etc. are well suited to thrust only compounding.

For cruise dominated applications such as off-shore oil support, then the improved lift to drag ratio of a wing and superior high speed propulsive efficiency of a dedicated propulsor can combine to mean that a lift and thrust compounded solution offers very worthwhile overall benefits. This configuration may also be well matched to applications such as escort where, it may be speculated that at some time in the future, high speed agility and sufficient speed to regain the convey after an engagement will become crucial.

As presently conducted, reconnaissance involves a great deal of hover/low speed flight and is well suited to a conventional helicopter though requirements for increased speed of response or higher productivity could modify this finding.

The idea of compounding is not new and indeed this should not be surprising, since it simply seeks to exploit the basic laws of physics. What is believed to be true is that, in common with so many good ideas in the past, various of the critical enabling technologies have now reached a level of maturity at which the exploitation of the concept is ripe to proceed with an acceptable degree of certainty. Despite undoubted challenges in getting this exploitation right the fundamental risks are felt to be low and the potential payback is high.

Much effort is currently being channelled into studies to examine the need for the construction of a Technology Demonstrator, based on a Lynx Airframe - Figure 3. The Lynx was chosen as being an eminently suitable candidate aircraft on several counts, including the fact that it had already flown at speeds approaching 220 knots. As a result of this flight test experience it was known that the behaviour of the aircraft under these conditions was very satisfactory and the risks associated with further expansion of the flight envelope were minimised.

To improve the speed of a Lynx by something of the order of 50% over and above service release clearly implies a substantial increase in installed power. Rolls-Royce Plc have been partners in this ACH study from the very early stages and the RTM 322 engine, which they manufacture in conjunction with Turbomeca of France, offers the required level of installed power. The installation of these much more powerful engines is a relatively simple and straightforward matter.

For the purpose of demonstration a simple, light, low cost and low risk solution was sought to the provision of propulsive thrust. To meet this requirement, Rolls Royce has designed a variable area nozzle to fit onto the back of the RTM322 engine.

Although exact figures are open to debate a number of target requirements are in view for the performance of a future advanced rotorcraft and the ACH Technology Demonstrator will help to underpin these.

Examples include:-

- Speeds of the order of 250 knots
- Lift to drag improvements of 25-50%
- Propulsive efficiency increases of up to 50%
- Productivity and payload/range bettered by 20% plus
- Life cycle cost reductions of 10-20%
- Operating altitudes of 20,000 ft plus

All of these represent substantial improvements over current rotorcraft capabilities and augur well for the expanded exploitation of helicopters in civil and military roles.

APPLICATION OF NEW TECHNOLOGY TO THE ANGLO-ITALIAN EH101

My preceding remarks relate to a perspective of the future. However, our confidence in the future and the new opportunities and challenges that we may look forward to are very much based on the benefits that the application of new technology has brought to today's products.

The EH101, designed by the partnership of Agusta and GKN Westland, and for which aircraft number 14 is now on the build line, employs new technology extensively. The principal new product technologies employed are:-

- Advanced aerodynamic composite blades;
- A composite elastomeric hub;
- An active vibration control system (ACSR);
- Composite and aluminium lithium structural components;
- Health and Usage Monitoring;
- An all digital AFCS;
- An Electronic Instrument System; and
- An Integrated Avionic System employing Dual Data Buses.

To highlight the benefits of just one of these items, as an example, let us consider the main rotor. If we had retained the aerodynamic technology of the previous generation of helicopters, the aircraft would have needed to employ two extra blades to match the performance of the new rotor. These extra blades would have had a significant consequential impact on increased installed power requirements, and increased empty weight and overall weight for the missions to be performed. This improvement is a direct consequence of the 37% improvement in cruise blade loading available from the advanced aerodynamics employed.

CONCLUSIONS - CHALLENGES FOR THE FUTURE

A number of conclusions and challenges may be drawn from this review of technologies for rotorcraft in the future:-

First, global market and economic changes are leading to changes in operational priorities and needs that emphasise cost-effectiveness and flexibility, rather than simply performance.

Second, economic pressures require that technology developments are clearly planned and focused on both new products and product updates. With an emphasis on longevity in service, product updates are growing in importance.

Third, collaboration and partnership arrangements between industries and research establishments must develop in response to ever-increasing cost pressures.

Fourth, whilst the development of product technologies must be sustained, an increasing emphasis will have to be placed on process technologies. The focus on process technologies will be driven by market and economic pressures to reduce "time to market", "improve product design" and "reduce non-recurring costs".

Finally, of course, the prime motivator for us all is the advancement of technology to increase the size and penetration of the market for rotorcraft.

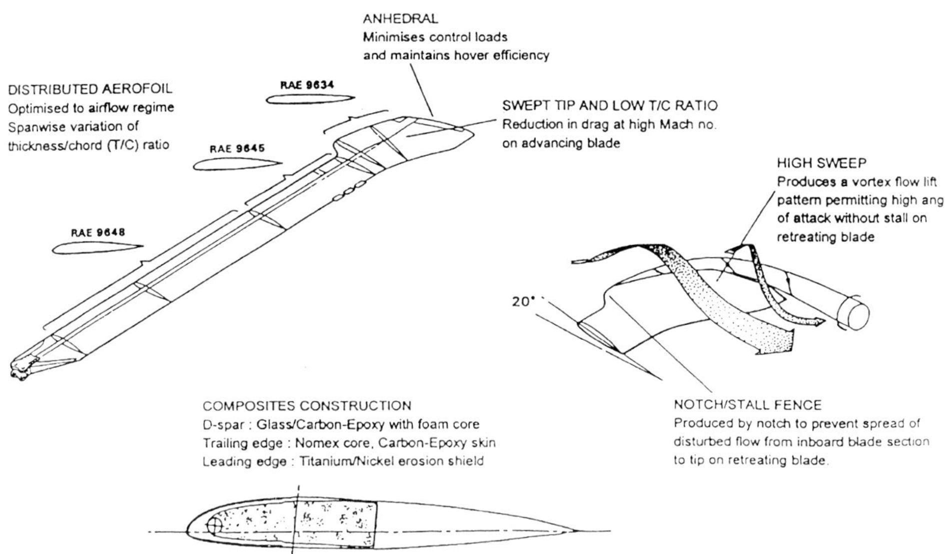

Figure 1. BERP III Rotor Technologies

Figure 2. Simulation of Battlefield Scenario utilising Synthetic Environment

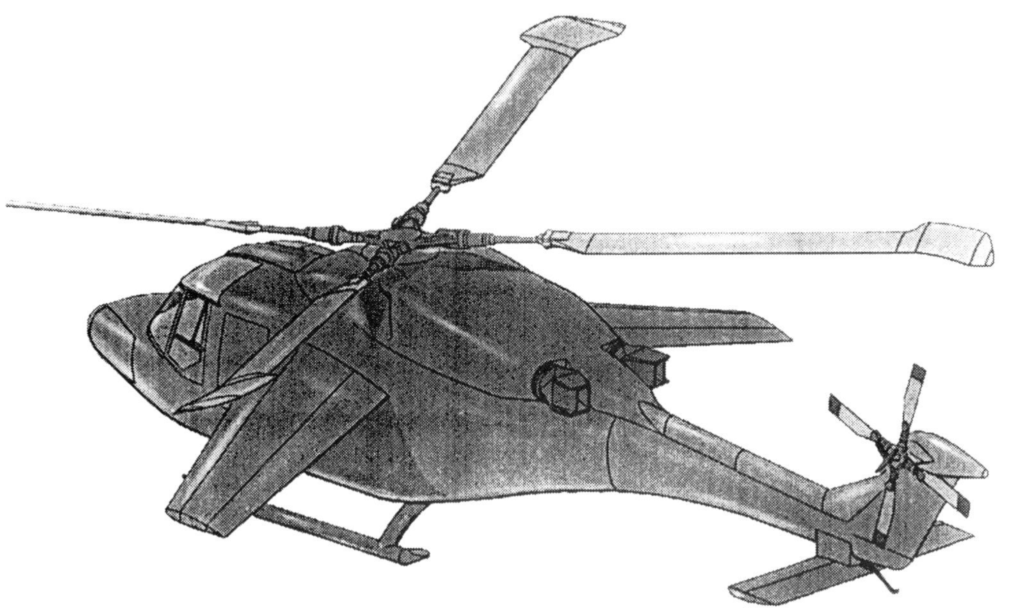

Figure 3. Lynx Based Compound Demonstrator

C545/027/98

Validation of a new generation of active vibration control systems via rig and flight tests

T KRYSINSKI
Technical Directorate, Eurocopter, France

1. INTRODUCTION

The reduction of helicopter vibrations has traditionally been a difficult task to achieve. The oscillatory motion of the fuselage has been a concern for several reasons :

- crew's and passengers' fatigue,
- high-cycle fatigue of different components, low reliability and high maintenance costs,
- low performance of different weapon systems (Difficult to use sights, difficult to point missiles,....)

The major new development programs still present high risks as far as helicopter dynamics are concerned. The main industrial motivations for the improvement in helicopter vibrations are :

- helicopter acceptance in the future (comfort, weapon system platform stability) will impose low vibration levels (0.10 g $\rightarrow$ 0.05 g $\rightarrow$ 0.03 g),
- extended flight envelope (speed, load factor), wide range of payloads and fuel loads, requiring high performance antivibration devices for easy flying,
- dynamics problems during development can lead to costly development delays and impose fundamental modifications in the aircraft design.

The objective of every helicopter manufacturer is to design the new rotorcraft so as to allow « flying right from the drawing board » with minimum development time. In this paper, the following issues will be discussed :

- the origin of helicopter vibrations,
- methodology overview,
- description of new generation of semi-active isolation systems,
- validation methodology and perspectives for production application.

2. ORIGIN OF HELICOPTER VIBRATIONS

There are many causes for helicopter vibrations such as rotors, shaft gears, engines. These vibrations have an almost constant frequency due to the constant speed of rotating parts. The frequency range for comfort is from a few Hz to a few hundred Hz. There are also random vibrations from the air flow exciting the tail surfaces called « Tail shake ».
The different sources of vibrations are pointed out on Figure 1.In this paper, we will consider main rotor vibrations only. Vibratory response of the blade at its passing frequency is a natural ehaviour of any rotor. In hover, the aerodynamic loads acting on the blades are constant as a function of azimuth and no vibratory loads are generated on the hub.

In forward flight, the airload on the blades varies during rotation due to the relative wind and incidence imposed by pitch. The loads on each individual blade are periodic at the frequency which is a multiple of one-per-rev.

The dynamic response of the blade is dependent on the fundamental blade characteristics like blade natural frequencies, damping and mode shapes. The dynamic loads can be amplified or attenuated by the blades dynamics and transmitted to the rotor hub. The rotor is a filter with some cancelling and some reinforcing components.
The basic mechanism of helicopter vibrations is shown in Figure 2. Another mechanism of vibrations appears when blades are not identical. Blade dissimilarities can be induced by manufacturing errors or various damages. Many inspections are made during the manufacturing process (weight, mechanical properties, mold temperature, holographic inspection ...). Each blade is still slightly different from the other and cannot be fitted to the aircraft before being balanced. Blade defects may be of different types : weight, spanwise and chordwise c.g. position, spanwise weight distribution, airfoil shape. All can have consequences on the dynamic behaviour and on cabin vibrations. These non-isotropic rotor vibrations can become important for modern helicopters when basic N-per-rev (N - Number of blades) is overreduced. In these conditions, the resulting non-isotropic levels through the airframe can become similar to N-per-rev. vibrations levels and their association produces a beating phenomenon at low frequency which can be disturbing for the crew.

While discussing the helicopter vibrations problems, one must mention the tail shake phenomenon. The tail shakes are the non-harmonic vibrations at frequency below 10 Hz in most cases. These vibrations are due to the response of first fuselage bending modes excited by the airflow from the main rotor, aircraft engine cowlings or front fuselage. Most development programs faced the problem of tail shake and the solution was found afterwards by external aerodynamic optimization. In this overview of helicopter dynamics, we should mention the problems of transition. Transition is the high vibration problem at low speed which is due to the excitation of the rotor either by its own wake or by that from the pylon. The induced velocity flow-field going through the rotor during transition is very complex, which makes the analytical analysis very difficult. The transition problem in most cases is solved by dynamic tuning of the rotor, introduction of hub absorbers or isolation system on the interface with the fuselage.

3. OVERVIEW OF HELICOPTER VIBRATION CONTROL METHODOLOGY

Vibration control methodology is based on the following three issues :

- rotor dynamics
- fuselage dynamics
- choice of appropriate antivibration device.

ROTOR DYNAMICS

In this chapter, we will discuss the minimization of helicopter vibration from the main rotor since it is the primary source of problems. This minimization starts with appropriate tuning of main rotor characteristics. The dynamic response of a rotor blade to the aerodynamic excitation depends on blade natural frequencies, generalized masses and modal damping which gives the amplification or reduction of blade roots, vibratory loads to the aerodynamic excitation.

The aerodynamic parameters are selected mainly to improve helicopter performance in hover and in forward flight. The main parameters are :

- induced velocities
- planform shape : rectangular or tapered
- tip shape : swept, anhedral
- twist.

Induced velocities due to the fuselage or the blade vortex interactions are a significant parameter. Fuselage optimization to reduce aerodynamic drag leads to designing compact rotor heads. In these conditions, the blades are close to the body, which amplifies the interactions in terms of fuselage induced velocities exciting the blade and gives high rotor head vibratory loads.

The number of blades is thus a highly significant factor as far as vibrations are concerned. A general argument in the helicopter community is the higher the number of blades the lower the dynamic loads at the rotor head. The choice of the number of blades is strongly influenced by other criteria like performance, price and autorotation capability.

The latest aerodynamic studies are producing new blades which are no longer rectangular but tapered with evolving tips. Their twist can be modified and an anhedral added to improve their performances in hover or at high speed. The planform influences the spanwise distribution of the aerodynamic loads as well as the dynamic properties of the blades. Tapering leads, for example, to low generalized masses for those modal shapes where dynamic response and vibration level are increased.

The different results available in the literature confirm that high twist is favourable for hover and low speed performance. The linear aerodynamic theory shows that higher harmonics blade flatwise loads are proportional to twist. The current blade design methodology is an optimization of aerodynamic performances as well as a change in internal structure to improve dynamic behaviour. The simplest methodology involves retaining a margin between blade modal frequencies and hub excitation frequencies. It is possible to increase the generalized mass or shift the frequency of the modes most critical for vibrations with tuning masses. Optimization techniques involve local stiffness and mass adjustments to globally reduce aerodynamic excitations and blade response to obtain low N-per-rev hub loads (moment, vertical and lateral shears).

FUSELAGE DYNAMICS

The fuselage response to rotor excitations must carefully be considered to enable high comfort aircraft to be obtained. The fuselage response varies extensively with the excitation frequency. An example is shown in Figure 3. The fuselage transfer is more than three times lower at 30 than at 24 Hz for experimental DTP X380 helicopter. This example shows that very significant efforts in rotor optimization can easily be wasted with non appropriate fuselage dynamics. The helicopter structure is composed of elements which considerably differ in structural arrangement.

The structure design must be supported by finite element airframe analysis. In the design phase, every main architecture choice like implementation of frames, installation of heavy parts (engines, gearbox, ...) and interface between mechanical parts and fuselage must be validated by dynamics considerations. As far as new structures are concerned the effects of composites make the prediction of natural frequencies and mode shapes more difficult.

The difficulty comes from different new elastic coupling terms and the structural design concept of the composite fuselage is different from that of metals. Another problem is the structure identification methodology to ensure proper fuselage mode placement : finite element analysis as well as correlative ground shake tests are needed. The global optimization of the structural models is impractical. This is why every company is looking for simplified models which are much easier to use for parametric studies and optimization techniques.

ANTIVIBRATION DEVICES

Upgrading of performance, mission duration and versatibility looking into increased level of comfort, imperfect control of forced vibration dynamic and aerodynamic problems at design level, require the necessity for developing antivibration devices. The problem proves difficult since the vibration technology has to meet the following requirements :

- system with an unlimited service life
- reliability
- reduced maintenance
- minimum weight
- minimum dimensions.

The antivibration devices are broken down into 3 classes :

- at the rotor hub
- at the rotor-to-fuselage interface - upper deck
- in the fuselage.

In these three classes, we can distinguish three categories : passive, semi-active and active systems. These different technologies are described in Figure 4.

Blade-Mounted or hub-mounted dynamic absorbers (pendulum mass, bifilar, hub absorber or roller absorber) are the most popular antivibration devices used on helicopters. The resonance tuning of these dynamic absorbers must be close to the rotor harmonic to be reduced. The example of pendulum masses applied to Super Puma MK2 is given in Figure 5. Another example of absorber is given in Figure 6.

This absorber is secured to the rotor hub. The advantage of this technology is the ability to filter 2 frequencies in the rotating system. The example of efficiency is given in Figure 7. The reduction of vibrations in the cabin is directly correlated with the decrease of in-plane vibrations on the hub.

UPPER DECK SUSPENSIONS

Eurocopter has been one of the first helicopter manufacturers to propose, on the market, a focal point suspension system so-called "barbecue". This system was applied to SA 330 Puma. The principle was to use soft elements at the bottom of the gearbox to filter the vibrations. The satisfactory results obtained on SA 330 Puma boosted the development of several derivatives on AS 332, Dauphin and Ecureuil. The simplification comes from use of laminated elastomer mounts and flexible composite bars. The three systems shown in Figure 8, are still in operation fulfil their functions very well and provide the products with high competitiveness as concerns vibration comfort.

In the recent years, a new generation of suspension system so-called "SARIB®" was developed.
SARIB® is an anti-resonance isolation system, which consists, as shown in Figure 9, of 4 individual units equispaced around the gearbox. One unit consists of a leaf spring, the flapper arm and the flapper mass. The leaf spring is designed with two parallel flanges at the stiff end. One bolt connects, through the outer bearing, the leaf spring, to a bracket on the gearbox deck and another bolt connects the leaf spring to a gearbox strut. Elastomeric bearings are provided at both connections.

The elastic side of the leaf spring is supported at the bottom of the gearbox. The amplification needed for flapper mass oscillation is realised through the flapper arms and their connections to the stiff part of the leaf spring, close to the gearbox strut. A membrane provided between the bottom of the gearbox and the fuselage transmits the rotor torque. Excellent vibration levels were achieved with SARIB® isolation for different missions and weapon configurations.

HHC, IBC

The b/rev (where b is the number of blades) vibration of helicopters comes from higher harmonic air loads (b-1rev, brev, b+1rev) acting on the rotor blades. The HHC principle (Higher Harmonic Control) is a generation of controls on rotating swashplate at frequency (b-1) rev, b/rev, (b+1) rev through non-rotating swashplate control frequency equal to b-rev. These higher harmonic control inputs give opposing loads, which allows the reduction of vibrations in the fuselage - Figure 10.

The optimum higher harmonic controls are calculated at any time by the digital computer in which the control law identification and computation algorithm is programmed. The observation parameters are the fuselage accelerations. The efficiency of the system is shown in Figure 11 for an experimental SA 349. Very similar results can be achieved with IBC (Individual Blade Control) using the actuators directly in the rotating system.

ACSR

The ACSR (Active Control of Structural Response) principle is the superimposition of the primary vibration response given by the main rotor and the secondary imposed vibrations which are controlled to minimize the global vibrations in the cabin. The secondary imposed vibrations are applied to the structure with hydraulic actuators - Figure 12. The flight test results obtained on experimental Dauphin DTV2 are shown in Figure 13.

The control algorithms are very similar to those used for HHC and IBC. The observation parameters are the accelerometer in the fuselage or vibratory loads measured on the main load paths betweeen the mechanical parts and the airframe.

4 NEW GENERATION OF SEMI-ACTIVE ISOLATION SYSTEMS

The idea for the new generation of semi-active isolation systems comes from the need to combine the simplicity of passive isolation systems with the tuning capability of pure active systems. Figure 4 classifies three kinds of semi-active systems which have been developed : auto-tuned resonators, auto-tuned SARIB® and auto-tuned hub absorber.

AUTO-TUNED RESONATOR

The simple dynamic absorber was used in many applications where the vibration piece of machinery must be reduced - Figure 14. It is well known in the theory of dynamics that the antiresonance frequency is the natural frequency of the isolated absorber considered as an uncoupled single mass-spring system. In this condition the appropriate tuning is obtained when the phase shift between absorber and its fitting is 90° - Figure 15. The tuning of simple mass spring system is quite difficult due to the very low damping.

The main drawbacks when used on a helicopter can be listed :

- necessity for tuning the resonator on each aircraft at the time of delivery
- necessity for modifying the tuning in case of a major operation on the resonator
- no adaptation to rotor rpm variations
- no adaptation to excitation level variations
- detuning in case of modification of structure impedance or embedding.

All these problems can be avoided with auto-tuned absorber. Tuning proceeds as the stepper motor pushes the small mass secured to the flexible leaf. The controller moves the mass to keep the 90° phase shift between the absorber and its fitting on the structure. The absorber was extensively tested on the rig in laboratory, showing satisfactory resistance to the change of excitation frequency and vibration input levels. The final validation step was completed during the flight tests. The flight version was designed and tested. Excellent vibration characteristics were achieved in flight with the auto-tuned absorber.

AUTO-TUNED SARIB®

Tuning of SARIB® isolation system is a compromise between the diffferent excitation components. Figure 16 gives the vibrations level in the cabin as a function of the flapping mass for 5/6t helicopters for different rotor hub excitations. It was noted that the optimum mass is 7 kg for in-plane force up to 9 kg for pitch moment excitation. SARIB® system can be tuned with a modification of amplification ratio. This tuning is achieved with a small electric motor pushing the masses on two sliding bars. This tuning allows adjusting the isolation system to different vibratory loads which change as a function of forward speed and flight case and to different modal fuselage characteristics which can be modified with a change in aircraft loadings.

We have to cope with nonlinear control because the modification to the SARIB® amplification ratio changes the natural frequencies of the system. The gradient algorithm was applied to control the masses and movement of the autotuned SARIB®. The validation programme started with rig tests. A mock-up of the autotuned SARIB® was built - Figure 17. The analytical modes were derived for the theoretical simulations. Very good efficiency of the autotuned SARIB® was demonstrated during the rig tests. The examples of autotuning capabilities are shown in Figure 18 in the frequency change case. The good rig test results urged us to test the autotuned SARIB® in flight on a 10-ton helicopter. The flight test results - Figure 19 showed 40% vibration level reduction compared to passive SARIB®. The vibration level was very good with passive SARIB® and excellent with the autotuned version. The controller demonstrated a very stable behaviour throughout the flight envelope.

5. CONCLUSIONS

The new generation of autotuned isolation systems was presented and compared to existing passive and active systems. The full validation of autotuned SARIB® and auto-tuned resonator was achieved through rig tests and flight tests. The main advantages of this new technology are :

a/ the efficiency compared to passive systems
b/ the low-price compared to existing active systems.

The serial version studies are underway for applications on future products.

6. REFERENCES

1. Y.A. MYAGKOW
Vibrations of helicopters of « MI Family, Investigation vibrations absorbers application, buffet » European Forum, 14-16 XI 1993

2. W. HOWRYLECKI, J. KLIMKOWSKI
« Roller vibration absorber for helicopter main rotor hub » European Forum, 14-16 XI 1993

3. **P. RICHTER, A. BLAQS**
« Full scale wind tunnel investigation of an individual blade control system for the BO 105 hingeless rotor », European Forum, 14-16 XI 1993

4. **R.B. TAYLOR**
« Helicopter vibrations reduction by rotor blade modal shaping » 38th Annual Forum of the AHS, Anoheim, CA ; May 1982

5. **S.P. KING**
« Fuselage response as a factor in vibrations. Structural Dynamic and Response & its Control »
The Royal Aeronautical Society, 10 I 1990

6. **S.P. VISWANATHAN - A.W. MYERS**
« Reducing of helicopter vibrations through control of hub impedance », Journal of the American Helicopter Society - Volume 25 - Number 4, October 1990

7. **F. BEROUL - L. GIRARD - E. ZOPPITELLI, T. KRYSINSKI**
« Current state-of-the-art regarding helicopter vibration reduction and aeroelastic stability augmentation »
18th European rotorcraft Forum, 15-18 IX 1992 - Avignon

8. **M. POLYCHRONIADIS, M. ACHACHE**
« Higher harmonic control : Flight tests of an experimental system on SA 349 Research Gazelle »
42nd Annual Forum of the AHC, June 1986, Washington

9. **Richard L. BIELAWA**
« Rotary wing structure dynamics and aeroelasticity », Education Series

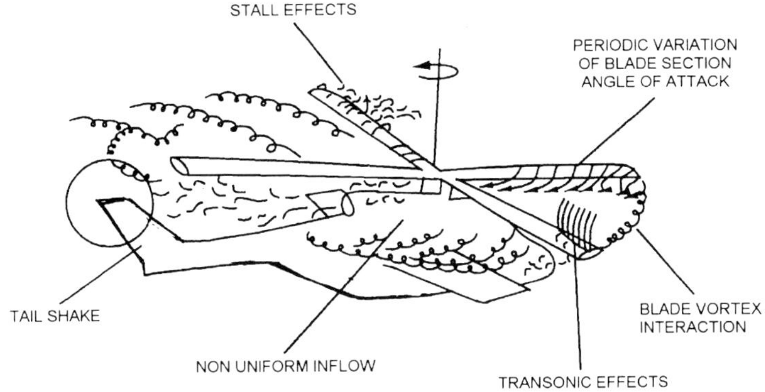

FIGURE 1 - SOURCES OF VIBRATIONS ON HELICOPTER

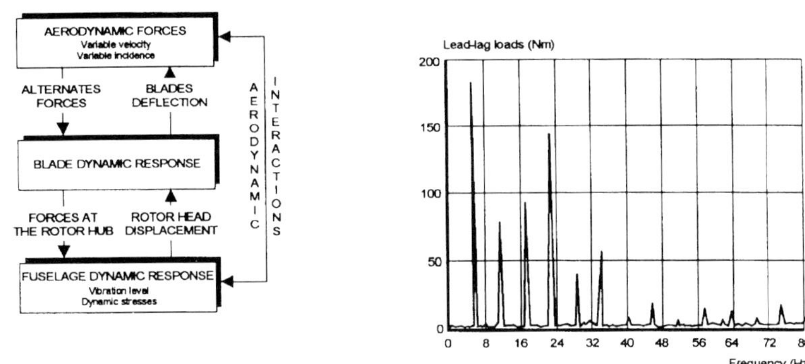

ROTOR AS A FILTER

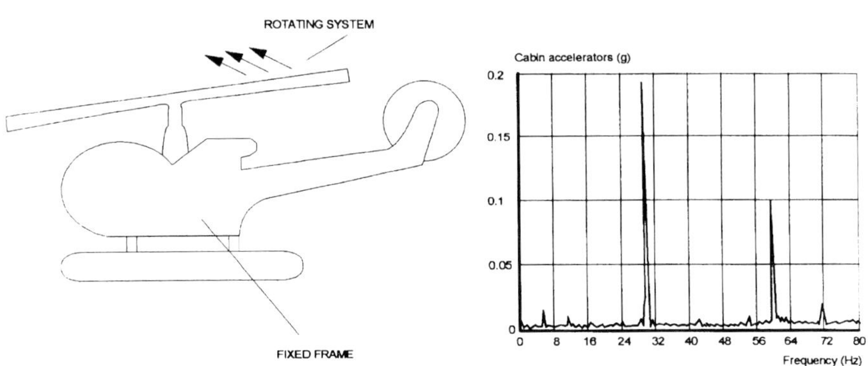

FIGURE 2 - ORIGIN OF VIBRATIONS

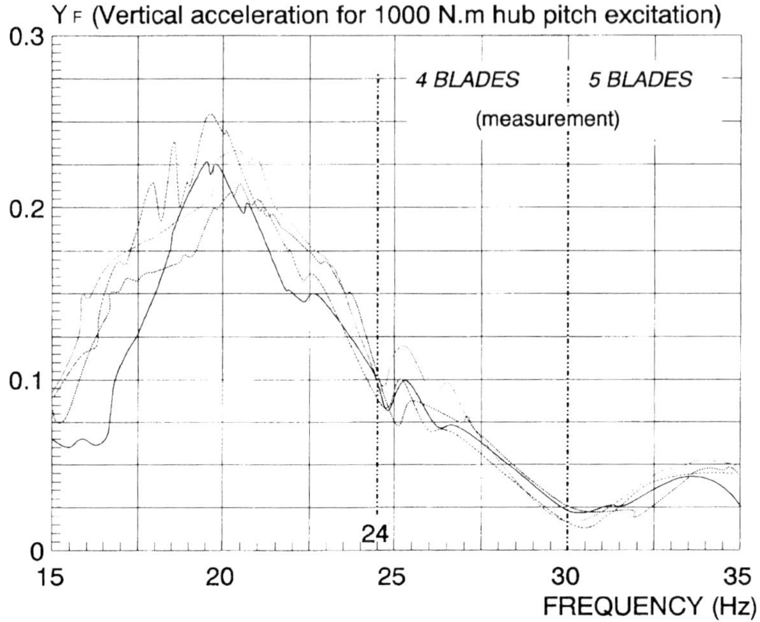

FIGURE 3 - DTPX380 FUSELAGE RESPONSE

	PASSIVE	SEMI-ACTIVE	ACTIVE
ROTOR HUB	RESONATORS - PENDULUM MASS - BIFILAR - HUB ABSORBER - ROLLER ABSORBER	Auto-tuned hub absorber	HHC IBC
UPPER DECK	B B Q SARIB RESONATORS : - HYDROMECHANIC - FLUID INERTIA	Auto-tuned SARIB	ACSR
CABIN	Mechanic RESONATORS	Auto-tuned RESONATORS	Cabin actuators

FIGURE 4 - ANTIVIBRATION DEVICES

FIGURE 5 - PENDULUM MASSES : SUPER PUMA APPLICATION

FIGURE 6 - ROTOR HUB ABSORBER : 350 APPLICATION

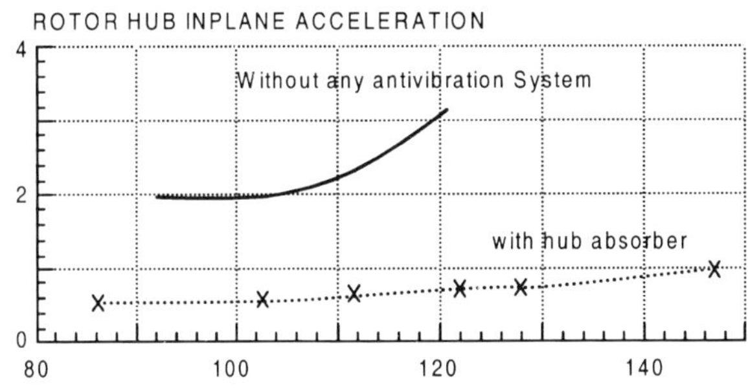

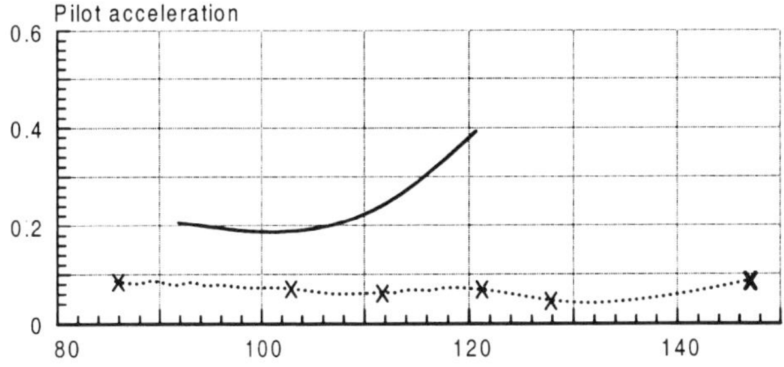

FIGURE 7 - ROTOR HUB ABSORBER : EFFICIENCY ON VIBRATIONS IN THE CABIN

SUSPENSION TECHNOLOGIES ON EC HELICOPTERS

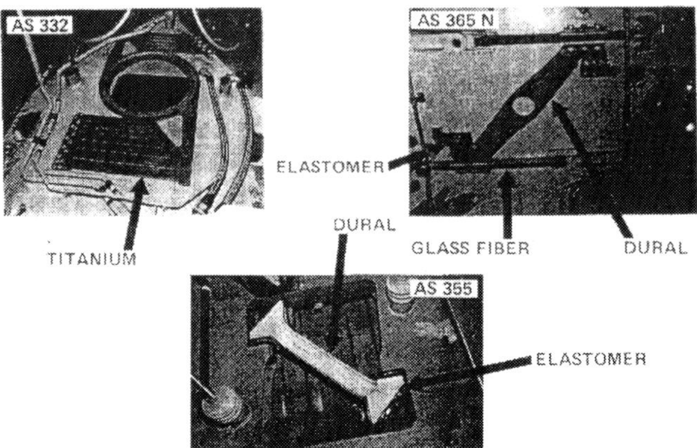

FIGURE 8 - VARIOUS PRINCIPLES OF BARBECUE SUSPENSIONS

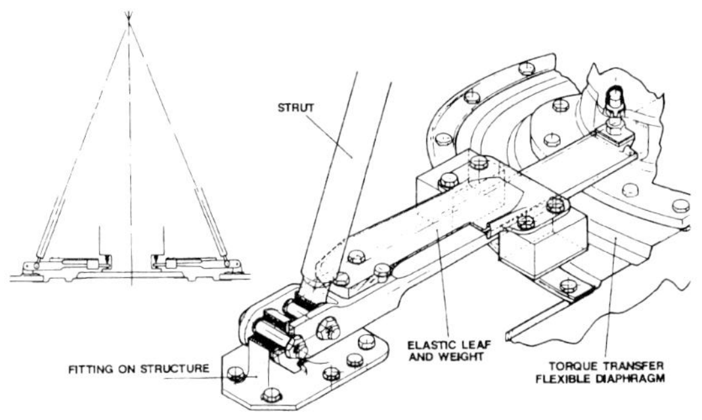

FIGURE 9 - SARIB® ANTI-RESONANCE ISOLATION SYSTEM

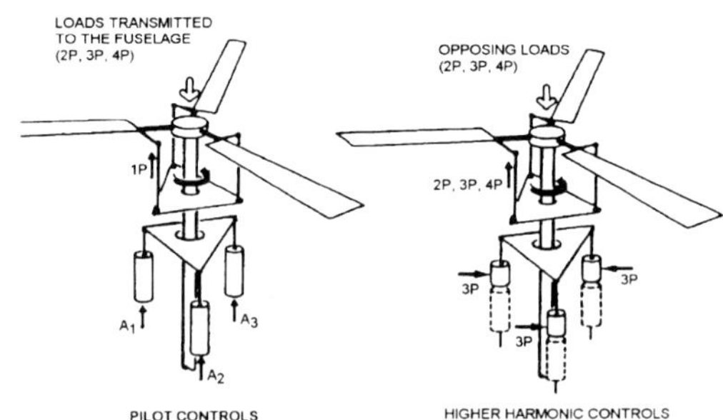

FIGURE 10 - PRINCIPLE OF HIGHER HARMONIC CONTROL ON NON-ROTATING SWASHPLATE (THREE-BLADE ROTOR)

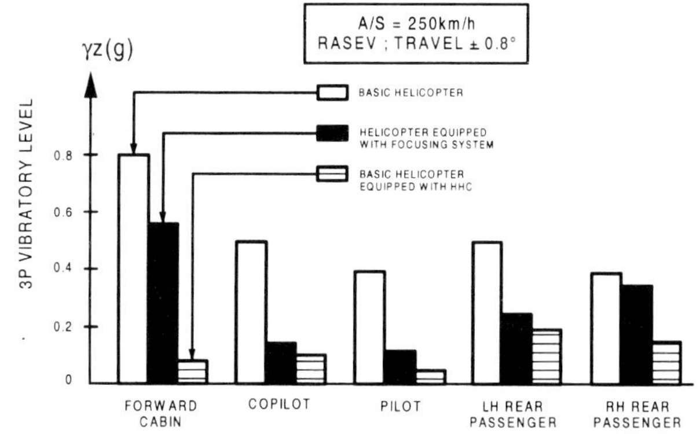

FIGURE 11 - COMPARISON WITH PASSIVE-TYPE SYSTEM

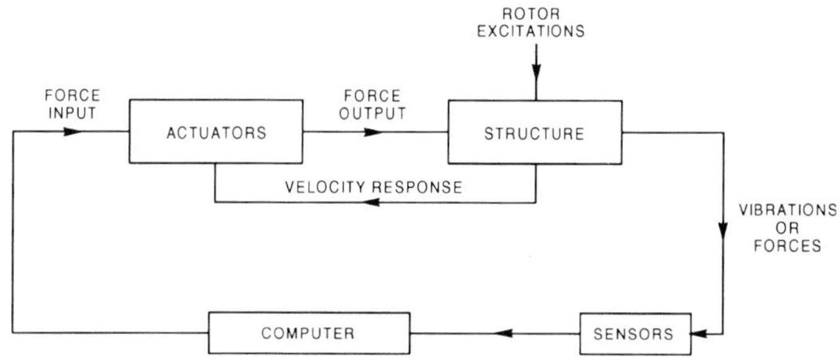

FIGURE 12 - ACSR PRINCIPLE

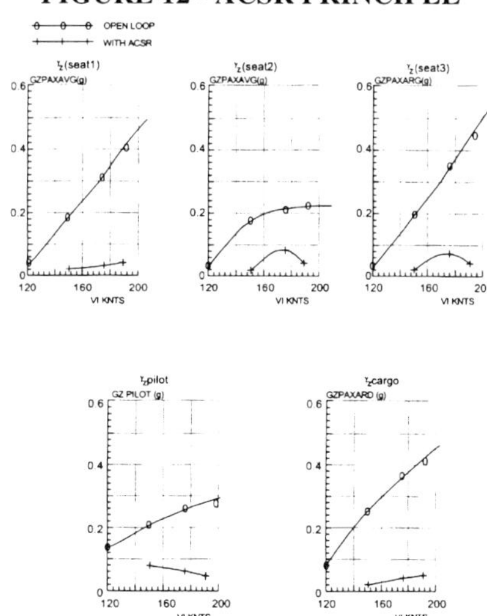

FIGURE 13 - ACSR RESULTS ON DTV2

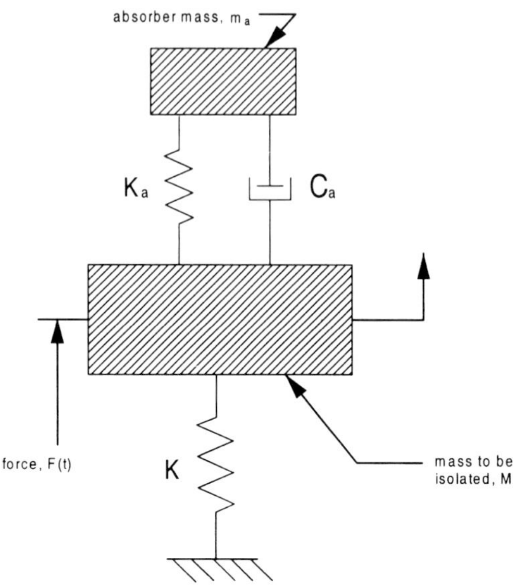

FIGURE 14 - DYNAMIC ABSORBER

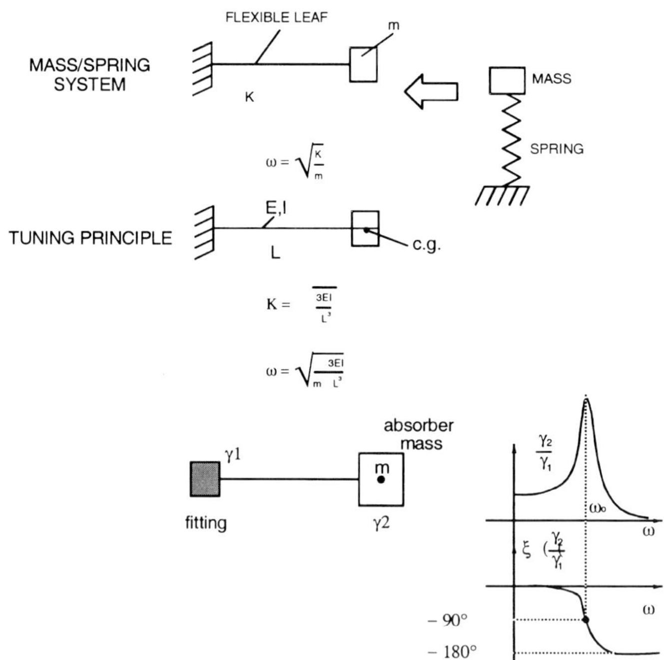

FIGURE 15 - DYNAMIC ABSORBER : TUNING PRINCIPLE

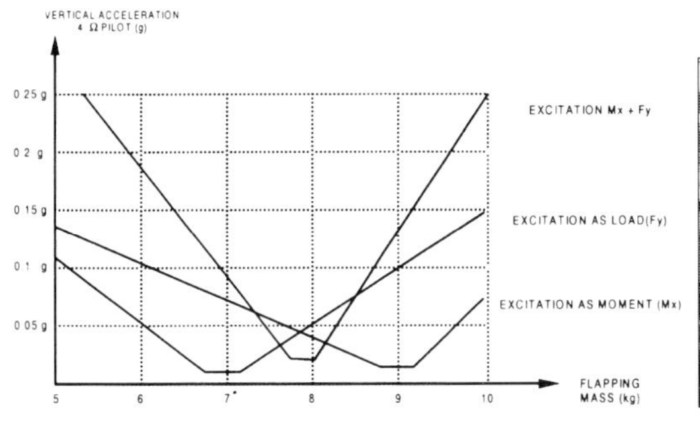

FIGURE 16 - CHANGES IN PILOT VERTICAL ACCELERATION VERSUS BEATING MASS

FIGURE 17 - AUTO-TUNED ABSORBER RIG TESTS

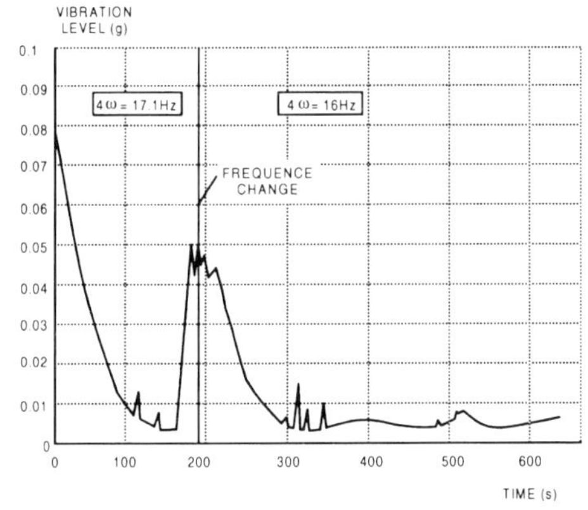

FIGURE 18 - AUTO-TUNED SARIB® - RIG TESTS RESULTS

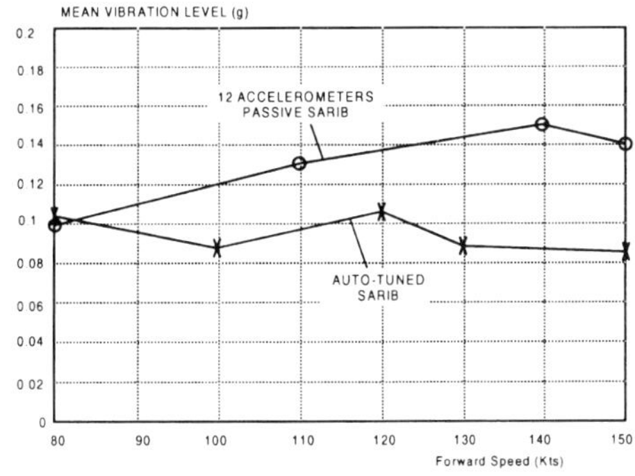

FIGURE 19 - FLIGHT TEST RESULTS WITH AUTO-TUNED SARIB®

C545/037/98

FUMS™ – An emerging technology for improved safety, reduced costs, and increased availability of aircraft

H AZZAM BSc, MSc, PhD, MRAeS
MJA Dynamics Limited, Southampton, UK

Abstract: This paper demonstrated the feasibility of an advanced Fatigue and Usage Monitoring System (FUMS). FUMS is a data management system that uses design limits and information specified by design authorities, and processes large volumes of strain gauge measurements and/or Flight Data Recorder parameters, in real time, to extract information that can improve safety, reduce costs of ownership and increase availability of aircraft.

Keywords: helicopter mathematical models, fatigue, engines fatigue, fatigue management systems, neural networks, mathematical networks.

GLOSSARY

AUM	All Up Mass
AI	Artificial Intelligence
CofG	Centre of Gravity
DA	Design Authority
FDR	Flight Data Recorder
FUMS	Fatigue and Usage Monitoring System
HUMS	Health and Usage Monitoring System
IAS	Indicated Air Speed
MoD	Ministry of Defence
PCA	Principal Component Analysis
RAF	The Royal Air Force
SI	Structural Integrity
WHL	GKN Westland Helicopters Limited
OAT	Outside Air Temperature

1 INTRODUCTION

Market forces place on aircraft manufacturers requirements for improved seat-mile revenue from civil aircraft, and for highly manoeuvrable and multi-role military aeroplanes. Aircraft designers are addressing these requirements by introducing new materials, implementing better manufacturing techniques and exploring advanced technologies. Nevertheless, the everlasting challenge the designers face is to reduce the weight of aircraft and, at the same time, maintain sufficient strength to withstand varying operational stresses with safety margins.

The varying operational stresses can cause fatigue damage and growth of microscopic defects to damaging cracks. Aircraft critical components are therefore designed with safety factors to eliminate the probability of the fatigue being accumulated (or the cracks being grown) to high values leading to failure during promulgated lives. The aircraft manufacturers use the safety factors to compensate for factors such as: possible aircraft usage outside the usage spectra as contemplated during a substantiation phase, possible operations under stress levels higher than the expected levels and possible scatter in material data. The low weight requirements prohibit the inclusion of very large safety factors during the design phase and therefore, the lives of some critical components are relatively short. Perhaps, such short lives mean that inaccuracies in operational loads or usage cannot be tolerated even if the safety factors are high, and despite every care, fatigue related accidents have occurred. On the other hand, the safety factors are often so high that a component retired after consuming its designated life could have operated for a significant extended period. Enhancing the safety standards by being more conservative, and by increasing the safety factors, will increase the costs of ownership and reduce the capability of aircraft without the conclusive elimination of the probability of failure. A new technology is therefore required to remove the artificial conflict between safety, utilisation capabilities and costs of ownership by exploiting the rapid advances made in avionics, instruments and monitoring equipment. Such a technology will improve safety, reduce costs and increase availability not only by application to future aircraft types, but also by retrospective application to those currently in service.

Whilst the literature has witnessed investigations into techniques that can lead to this technology, it was the UK studies sponsored by the MoD which formed the foundation of this technology and demonstrated the feasibility of advanced Fatigue and Usage Monitoring Systems (FUMS). The design limits and procedures as used by the manufacturers to build an aircraft would be embedded into FUMS. FUMS would utilise a model-based framework combining mathematical models, Artificial Intelligence (AI) methods and expert knowledge to automatically check that each aircraft was operating within the design limits. In other words, FUMS would automatically check the aircraft compliance with the design assumptions during its life time by evaluating loads, advanced usage information and fatigue **(1,2)**.

FUMS can be based on direct and/or indirect monitoring methods. The former requires converting data from a large number of strain gauges to loads, usage and fatigue. Maintaining a large number of strain gauges entails high operational costs especially when helicopter rotating components are considered. Indirect monitoring requires transforming flight parameters such as airspeed and accelerations into loads, usage and fatigue. These parameters are often stored by Flight Data Recorder (FDR). Over the last three decades, FDR data have been extensively used to analyse aircraft incidents and accidents. The use of readily available FDR data makes the FUMS indirect methods more attractive and cost effective.

Various research programmes in the USA and the UK have used techniques for indirect fatigue and usage evaluation. Reference **(1)** has proposed a model-based framework combining mathematical models, AI methods and expert knowledge. Haas et al **(3,4)** have trained an artificial neural network and used multivariate regression to predict helicopter rotor system loads during high speed manoeuvres. Tang et al **(5)** have ignored the mean load effects and developed an approach for fatigue monitoring based on pattern recognition. Reference **(6)** has extended Tang's method by including the mean load effects. Gunsallus et al and Gustavson et al **(7,8)** have described a system called "Holometrics" that uses a minimum number of airframe fixed sensors, FDR parameters and transfer matrices as a means to predict the loads of rotating (and non-rotating) components. The indirect methods to date have been evaluated in **(1,2)**, and it has been concluded that none of them are of sufficient maturity to warrant installing permanent FUMS into helicopters. These indirect load methods rely on data statistics and, hence, do not possess generic generalisation capability. The usage methods rely on the fragile concept that fatigue completely relates to flight regimes. References **(1,2)** have therefore introduced a model-based approach for FUMS. The approach has used the helicopter mathematical models reported in **(9,10)**, which have been developed over the last 17 years and included a wide range of helicopter fault simulations. The literature published to date has not indicated significant appreciation of the importance of such mathematical models to the rapidly evolving Health and Usage Monitoring Systems (HUMS). Perhaps the only exception to this observation is the work of Ganguli, Chopra and Haas **(11,12)**, which has started the development of mathematical models and neural networks for rotor system faults.

As mentioned above, it was the UK studies sponsored by the MoD which shaped the FUMS technology as described in the following section **(1,2,10)**.

2 FATIGUE AND USAGE MONITORING SYSTEM (FUMS)

FUMS is an advanced data management system that uses design limits and information specified by the Design Authority (DA), and processes large volumes of strain gauge measurements and/or FDR data, in real time, to extract Structural Integrity (SI) Information that can improve safety, reduce costs of ownership and increase availability of aircraft.

The DA information can include: elastic limits, proof stresses, fatigue strengths/endurance strengths, stress concentration factors, S-N curves/relationships, stress intensity factors, flight envelops, etc. Whilst each S-N curve shows the number of stress cycles in terms of amplitude and mean to cause failure, fracture mechanics can be used to predict the time to failure from some detectable crack size by calculating the rate of crack growth, which is a function of the stress intensity factor. The information used by FUMS should be compatible with the design methodology used to built the aircraft, i.e. safe life, damage tolerant or combination of both.

FUMS can be thought of as a system that integrates a set of FDR-based devices and/or direct devices. The FDR-based devices extract SI information from FDR data. The direct devices extract SI information from strain gauge measurements (or load sensor data). The SI information includes: loads, fatigue, usage and operational information such as All Up Mass (AUM) and Centre of Gravity (CofG). The usage information is expressed in terms of flight conditions and severity of conditions. The severity is determined from the induced damage. For example, the severity of a steady forward flight condition does not necessarily indicate low or high speeds, but indicates the possible induced damage associated with the flight condition. The SI information covers airframe components and engine components as well as rotor system components in the case of helicopters.

The FUMS devices generate SI information for management purposes, and can be onboard and/or ground-based devices. The definition of FUMS includes the word management to signify this important requirement. FUMS should reduce work load for pilots and fleet managers. The FUMS data management system tracks the fatigue and the other SI information of components. It can be a ground based system including information such as: aircraft type, acquisition equipment identifier, aircraft tail number, flight number, flight date, flight time, mission type, pilot name, location, squadron, etc. FUMS can improve aircraft maintenance by automating the decision making process, computerising the if-then planning process and using advanced communication technologies. The system should generate management reports including graphical outputs and statistics in unambiguous, readily clear, format. FUMS should also provide specialised reports for engineers, scientists, investigators and evaluators.

The main objectives of FUMS are to improve safety, reduce costs of ownership and maximise availability of aircraft. A matured and qualified FUMS can be used by pilots and fleet managers to explore the extremes of the flight envelopes with the design limits being the only restraints. In this way the capability of aircraft can be maximised in emergency. Under normal conditions however, FUMS information regarding flight conditions and associated damage costs can be used to define the practices that preserve the lives of high cost components. The costs of ownership are also reduced by the management system that ensures optimum maintenance scheduling and maximise aircraft utilisation. The safety standards are enhanced by observing the design limits in flight, identifying operations outside the flying envelopes and evaluating the costs of these operations. The FUMS objectives are achieved by ensuring that the system is designed using a bottom-up approach starting from the requirements of users and resolving the difficulties they face.

HUMS utilises incipient fault symptoms to trigger maintenance actions including component replacements. HUMS is therefore a diagnostic system that infers presence of defects from symptoms. It is applicable to defects that produce measurable symptoms and have low growth rates that provide adequate warning before severe failure. FUMS is mainly a prognostic system that provides advanced probabilistic indications of failures. Unlike HUMS, FUMS can cope with defects growing under a low usage rate with undetectable symptoms. HUMS and FUMS complement each other. For example, accidental damage and maintenance induced faults can be detected by HUMS, and the effects of defects growing under normal usage conditions can be evaluated by FUMS.

FUMS can process a large volume of strain gauge measurements and/or FDR data in real time. For example, FUMS was found to be capable of real time, on-line management of a large number of helicopter load sensors; more than 50 load sensors with strain measurements sampled at 840 samples/second using a 486DX 66 MHz PC computer **(6)**. It was also found that by using less than 53% and 20% of the computational power of the 486DX 66 MHz, FUMS could manage in real time the fatigue of 100 engine components and 100 aeroplane load sensors with measurements sampled at 32 samples/second **(15,16)**. It is worth mentioning that the recent advances in avionics and computational hardware can offer FUMS equipment that performs an order of magnitude better than the 486DX 66 MHz PC, in terms of computational speed and data storage capacity. FUMS therefore can be qualified/certified as onboard and/or ground-based system.

3 FUMS DATA

The FUMS direct devices require data from a large number of strain gauges/load sensors fitted on all critical locations of each aircraft. Maintaining a large number of strain gauges fleet-wise entails high operational costs. The strain gauges must be accurately bonded and calibrated, and their performance monitored to detect undesirable degradation. The strain gauges must provide information about the mean stresses and the alternating stresses in all weather conditions without loss of accuracy. A 5% inaccuracy in strain gauge measurements was found to cause more than 500% error in the fatigue induced by low stress levels close to the endurance strength. For typical operational stress levels, a 5% inaccuracy in strain gauge measurements was found to cause about 20% error in fatigue. The FDR-based devices require training on aircraft measurements. In other words, they require calibration coefficients so that their results agree with strain gauge measurements. The data required to calibrate the FDR devices can be gathered from one or two aircraft: Maintaining the accuracy and integrity of the strain gauges of two aircraft is much more manageable than maintaining those of the entire fleet under various operational conditions including emergencies.

Whilst operators and DAs are required to specify the types of the FUMS devices to be used, a policy similar to that adopted by the Royal Air Force (RAF) is highly recommended **(13)**. For each aircraft type, one or two aircraft are fully equipped with strain gauges/load sensors. The strain/load measurements are used to calibrate the FUMS FDR-devices which are fitted fleet-wise on each aircraft. The integrity and accuracy of the strain gauges of the two aircraft are maintained and used as required (e.g. once a year) to check the effects of factors such as age or aircraft modifications on the calibration coefficients of the indirect FDR devices.

In theory, uncalibrated FDR devices can be derived from a comprehensive aircraft simulation. Generally, the accuracy of the uncalibrated devices can be very poor. An extensive research programme is usually required to improve the accuracy. The target of the programme is to refine the comprehensive simulation and validate the refinements by measured data. Thus, both the calibrated and the uncalibrated FDR devices require measured data. The former uses the data for training and the latter to refine the comprehensive simulation upon which it is based. The uncalibrated FDR devices require more development effort than their calibrated counterparts (in terms of costs and time).

As indicated above, central to the success of the FDR-based devices is the availability of consistent training examples. Each training example consists of input parameters and desired outputs; the outputs can be loads, flight conditions or fatigue. The input parameters have been classified as direct and indirect parameters **(1,2)**. The direct parameters include aircraft controls, and sensor measurements that indicate undercarriage reactionary loads, brake loads, impact reactionary loads and environmental inputs. The indirect parameters include configuration parameters such as AUM and CofG, and response parameters such as speeds, accelerations and orientation angles.

On board ship, an aircraft critical component could have started to corrode, but the corrosion might have not been severe enough to cause concern during periodic inspection. Afterwards, a relatively small number of flying hours could cause an unexpected failure even if operating in a non-corrosive environment. Evaluating the effects of factors such as sand storms and humidity on life consumption of components should be therefore considered. Generally, a component life is consumed as a result of the component being stressed mechanically, thermally and/or chemically. For example, under the simultaneous action of corrosion and repeated stresses, the fatigue strengths of most metals are reduced. Corrosion can occur in components exposed to salt water. Equally, metallic surfaces rubbing together that release sufficient energy for chemical formation can cause corrosive (oxidative) wear. The fatigue fretting associated with surface rubbing can be aggravated by humidity. FUMS should therefore consider simple environmental parameters such as humidity, salt contents, sand contents and wind speed.

The energy imparted to an aircraft from turbulent air can be transformed to severe vibration and induce fatigue. State sensors operating at high sampling rate can capture global effects, which are induced by factors such as air turbulence and shock landing loads. A global effect can be considered as an external input to the entire helicopter. It is felt almost everywhere by the components of interest, and it can be captured by one to three state sensors. It has been recommended to fit the global sensors in the fixed frame away from the parts where severe local effects can be encountered **(2)**. If the global sensors are not available, the effects of gusts and turbulence on fatigue will require input parameters describing, for example, the weather condition.

The above discussion indicates that a considerable part of fatigue damage relates to the control and response parameters, which can be sampled at low rates. HUMS can be used to track local effects. The remaining effects span a relatively wide frequency band, and considerable part of them can be captured by a global sensor. The extent of a system capability is a function of the chosen parameters from which the fatigue of components will be evaluated.

3.1 Fixed wing FDR data

The control parameters include positions of control surfaces and engine control parameters. For a military aeroplane, the parameters can include: wing sweep angle, flap position, slat position, spoiler angles, taileron positions, rudder position, engine throttle positions and nozzle areas. The response parameters include: speed, roll acceleration, pitch acceleration, yaw acceleration, normal acceleration, longitudinal acceleration and lateral acceleration. The environmental parameters can include outside air temperature and altitude.

3.2 Helicopter FDR data

The direct parameters include main rotor, tail rotor and engine control inputs. These control inputs are transformed into high frequency loads through the rotor system. Therefore, mapping control parameters of low frequency contents into desired output of high frequency contents can not be achieved without a mathematical understanding of the dynamics and aerodynamics of the rotor system. A model-based approach can relate control parameters sampled at adequate rates (2 to 20 samples/second) to nR vibration (n is an integer and R is the rotor frequency). It is worth mentioning that the vibration measured in the rotating frame is dominated by 1R to bR (b is the number of blades). In practice, the vibration is measured by sensors fixed in the fuselage, and is dominated by bR vibration. High sampling rates are required to capture the damaging effects of the reactionary loads. The direct input parameters can yield a reasonable prediction of average fatigue damage that relates more to a helicopter type than to an individual helicopter.

A considerable part of the bR vibration, and the associated high frequency loads and fatigue, depends on indirect parameters such as All Up Mass (AUM), helicopter speed and Centre of Gravity (CofG), and can be modelled. The response, configuration and local parameters are indirect parameters. The response parameters include helicopter speeds, accelerations, and orientation angles. The dynamic characteristics of a component can be fully determined if all inputs and independent responses are known. If the response parameters are sampled at a rate of about 15 Hz, 1R vibration and rigid body motion effects will be captured by the training examples for all helicopter types. The 1R vibration is induced, in the fixed frame, by a variety of main rotor system faults and, hence, the relationship between the control parameters and this 1R vibration is complex and, difficult to model; FUMS can only learn about the fixed frame 1R effects from training examples. Configuration parameters such as AUM and CofG can significantly influence the vibration levels. Monitoring these parameters is far from being straightforward. It is however possible to estimate these parameters from other parameters. The local parameters are produced by sensors targeted at components such as gearboxes or shafts. The purpose is to capture local effects such as shaft misalignments, which can induce excessive internal stresses and considerable damage. HUMS monitors the health of local components and produces local parameters. Solving the fatigue problem associated with local effects requires a different strategy: FUMS can classify local parameters. If the classified parameters are dissimilar to the typical cases, which are covered by the training data, the predicted fatigue damage can be modified according to some dissimilarity measure.

4 THE FUMS MODEL-BASED FRAMEWORK

As mentioned previously, FUMS transforms FDR data into fatigue, loads and operational information by FDR-based devices. The heart of each device is a mathematical network. The mathematical network has been proposed to provide a framework that allows interactions between Artificial Intelligence (AI) methods and mathematical formulae **(1,2,10)**. The mathematical network is in fact a number of networks, which can work in parallel to each other. Whilst the details of such a network are application dependent, Fig 1 shows a simplified schematic that illustrates the generic functions of the network. The network can check the integrity of parameters and correct suspect values through interpolation. Lost or corrupt signals can be also reconstructed from redundant measurements. A set of merging functions is used to combine the input parameters. These functions should be derived from mathematical models or engineering relationships. The time trace of each merging function is divided into a number of time blocks. Each block contains a number of points between a minimum of one point and a maximum of all points. For each time block, features such as average, standard deviation, etc., are extracted from the values of the merging function. A set of compressors can be used to compress a large number of features to a smaller number. The compressors can use Principal Component Analysis (PCA) or can be auto-associative neural networks. The compressors can also compress the features such that the contribution of the features which relate significantly to the desired output is rewarded and the signal noise attenuated.

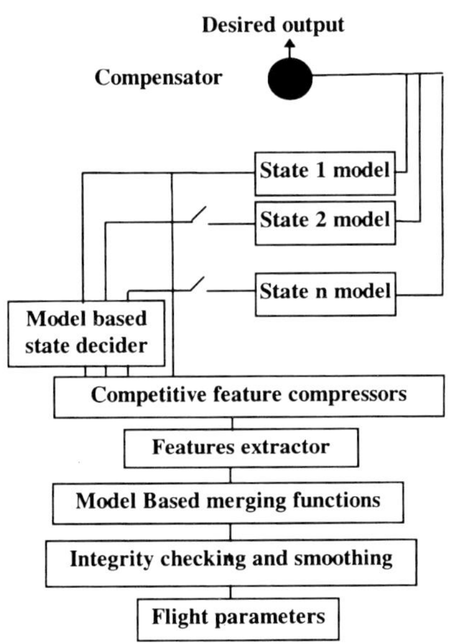

Fig 1 The mathematical network

A set of features can be classified by a module called the state decider. The state decider can be driven by mathematical or engineering relationships. For example, in some applications, the Mach number is used to identify states such as subsonic, transient and supersonic. Alternatively, the state decider can be a network that learns the relevant states from a set of examples. Generally, the state decider can learn how to identify states from a set of features through supervised learning and/or unsupervised learning. The output of the state decider can be used to select an appropriate state model. Each state model can be embedded into a network that receives a set of compressed (and non-compressed) features and delivers an output. The differences between the output values of a state model and the desired values can be mitigated through a module called the compensator. The compensator can be based, for example, on expert rules, statistical processes or engineering relationships. It can be also a network. The process of calculating the weights of the modules of the mathematical network is called training.

5 EVOLUTION OF FUMS

FUMS is a complex system that operates on inputs from a large number of sensors, and manage a large number of desired outputs. The outputs, for example, can include: synthetic load information at 20 critical locations, fatigue damage of components at these locations, information about flight conditions which can be linked to induced damage, synthetic AUM and CofG. Training a large number of FDR-based devices to produce these outputs requires a dedicated flight test programme. If the potential benefits of FUMS are demonstrated, the high costs of such a programme can be justified. Demonstrating the feasibility of FUMS can be based on available infield data and synthetic data. In this way, the mathematical models are not only used to derive the mathematical attributes/equations of FUMS, but also to generate a large training database. FUMS can therefore evolve from available flight data and mathematical models. A second evolution phase should involve a comprehensive flight test programme. A final phase should mature the FUMS technology and should consider data gathered from individual helicopters/aeroplanes.

In the following sections, infield data are used to demonstrate the feasibility of a number of FDR-based devices. The sections concentrate on demonstrating the capabilities of the mathematical networks. The sections do not describe the details of each mathematical network, since these are application dependent. Nevertheless, the last section describes the process of generating synthetic data and gives more details about the design of an FDR-based device.

5.1 The infield data

The results presented in this paper required the analysis of 16.4 flying hours of helicopter data, 766.8 logged flying hours of engine data and about 20 flying hours of military aircraft data. This is the largest data set known to the author, to have been used world wide for similar research programmes.

The helicopter data covered a variety of manoeuvres. The data were acquired from a Lynx helicopter by GKN Westland Helicopters Limited (WHL) over two and a half years, and involved four pilots and different helicopter configurations. The WHL data gathering objective was the substantiation of component fatigue lives against standard flight spectra. The objective was not to establish relationships between flight parameters and fatigue. In other words, WHL's attention was primarily focused on the accuracy and integrity of strain measurements. Rotor system faults were not seeded, and FDR parameters such as the rate of climb/descent, fore/aft acceleration, angular accelerations and sideways speed were not available. The Lynx FDR data had been all sampled at a rate of 26 samples/second. The response parameters (accelerations) had been low pass filtered at a rate of 1.6 Hz before they were acquired at the 26 samples/second rate. The inputs to each network were selected from the following FDR parameters: Outside Air Temperature (OAT), Indicated Air Speed (IAS), collective lever position, tail rotor pitch angle, normal G, lateral G, roll rate, pitch rate, yaw rate, main rotor rpm, engine 1 torque, engine 2 torque, F/A cyclic stick position, lateral cyclic stick position, pitch attitude and roll attitude. The desired outputs were prepared from strain gauge measurements. Accumulative fatigue damage of a rotating blade lug section was evaluated from two strain gauges. Tail and main rotor torque were measured by strain gauge systems. The tail drive torque had been measured using a strain-gauge bridge installed on the drive shaft. The bridge had been powered via a short range radio frequency telemetry link from the static shroud around the shaft **(14)**. The output signals were transmitted from the shaft to the shroud in a similar way. The main rotor torque had been measured using strain gauges mechanically calibrated to read torque applied to the rotor by the main rotor drive shaft.

In order to demonstrate the applicability of FUMS to fixed wing aeroplanes, a mathematical network was trained to synthesise wing pivot fatigue . The training data covered a period of 4 months, and involved 14 sorties, about 17 flying hours and 7 missions.

Rolls-Royce RB199 engine data covering a period of about two and a half years were used. The data were acquired from two aeroplanes that had been fitted with a total of thirteen engines. The aeroplanes were flown from 12 locations. The data involved 250 sorties and 766.8 logged flying hours. The sorties covered a wide range of missions. The engine parameters included, for example, low pressure spool speed (Nl), high pressure spool speed (Nh), total engine inlet air temperature (T_{t1}), Indicated Air Speed (IAS) and altitude (H). The desired output fatigue was calculated by a progressive damage algorithm **(15)**.

5.2 Blade lug fatigue

In 1994, a mathematical network was trained to synthesise the fatigue usage of a Lynx blade lug section based on data from 10.5 flying hours. After training, the network was blind tested (that is, without strain gauge data) using other FDR data (about 5.6 flying hours). The network synthesised the fatigue from the following FDR parameters: OAT, IAS, collective lever position, F/A cyclic stick position, lateral stick position, tail rotor pitch angle, normal G, lateral G, roll rate, pitch rate, yaw rate, main rotor rpm, pitch attitude, roll attitude and torque of engines (Fig 2). The control stick positions were adjusted three times during data gathering. The use of calibrated control angles would further enhance the results of the network.

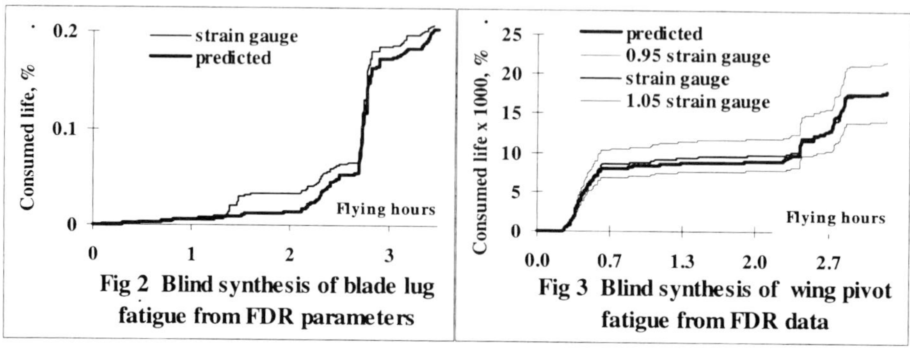

Fig 2 Blind synthesis of blade lug fatigue from FDR parameters

Fig 3 Blind synthesis of wing pivot fatigue from FDR data

5.3 Wing pivot fatigue

In 1995, a mathematical network was trained to synthesise wing pivot fatigue of a military aeroplane. After training, the network successfully synthesised fatigue from FDR data (Fig 3). The errors in the synthesised accumulative fatigue after about 3 flying hours was 1.4%. Error bands from ±5% strain gauge measurement errors are shown; the fatigue was calculated from strain gauge measurements multiplied by 0.95 and 1.05. Fig 3 suggests that the network would be better than a strain gauge system with 5% measurement error for both the blade lug and the wing pivot cases. The blind test sorties included two cases where the configuration and mission were significantly dissimilar to those seen by the network during training **(16)**. This would suggest that the network possess good generalisation capability.

5.4 Lynx tail rotor torque

The helicopter tail rotor and associated drive shafts are vulnerable components. Load excursions caused by factors such as adverse wind induced loads and blade-vortex interaction could cause serious over-load and tail rotor damage. Accurate synthesis of tail rotor torque could procure safety and operational benefits. A mathematical network was therefore trained to synthesise the Lynx tail rotor torque at rates of 0.82 and 6.5 samples/second. At these rates, the quasi-steady (low cycle) torque would be reasonably portrayed. The AUM and fuel consumption were not required. This would eliminate any additional requirements for a means of announcing crew size, stores and/or weapons upload and download. Data from about 3.6 flying hours were used to train the network, and 12.8 flying hours were used for blind tests. The blind test results indicated that the network could synthesise the tail rotor torque within an average error less than 5%. The network synthesised the tail rotor torque accurately after about 1.5 years from the date of training (Fig 4). The control stick positions were adjusted three times and FDR parameters such as the rate of climb/descent and the fore/aft acceleration were not used. The use of calibrated control angles and additional FDR parameters would have improved the results of the network.

The blind test results also suggested that an FDR-based measuring device could have been more durable than a strain-gauge device. For four flights, the mathematical networks indicated zero tail rotor torque at zero rpm, and the strain-gauge device indicated about 1800 lb ins (Fig 5). These results suggested that the tail rotor torque measuring device could have been faulty, and WHL were requested to check the integrity of the device during these flights. WHL indicated that there had been no requirements for use of tail rotor signals from these flights, so the system had not been maintained during this period **(14)**. They confirmed that the mean values of tail rotor torque on these flights should be considered suspect. Fig 5 suggests that the mathematical networks can be used to reconstruct lost or corrupt signals.

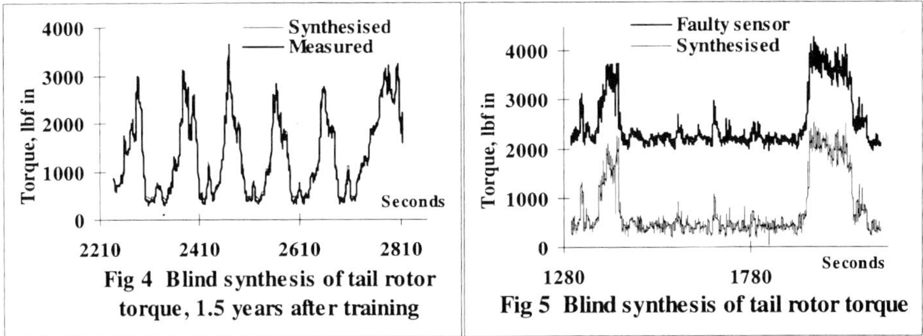

Fig 4 Blind synthesis of tail rotor torque, 1.5 years after training

Fig 5 Blind synthesis of tail rotor torque

5.5 Lynx main rotor torque

Data from about 3.6 flying hours were used to train a mathematical network, and 12.8 flying hours were used for blind tests. The blind test results indicated that the network synthesised main rotor torque adequately at rates of 0.82 and 6.5 samples/second within average errors of less than 2.5% (Fig 6). The network synthesised torque for manoeuvres that were not seen by the network during training. The AUM and fuel consumption were not used. The low cycle fatigue calculated from synthetic main rotor torque was in excellent agreement with the low cycle fatigue calculated from measured torque (Fig 7). The error in low cycle fatigue usage of flights covering a period of two and a half years was 2.54%.

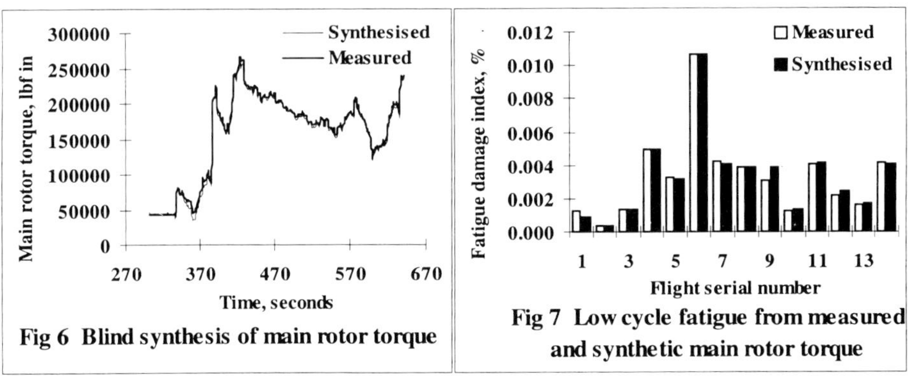

Fig 6 Blind synthesis of main rotor torque

Fig 7 Low cycle fatigue from measured and synthetic main rotor torque

5.6 Engine fatigue usage

Mathematical networks were trained to synthesise non-thermal transient fatigue of an engine component and thermal transient fatigue of two engine components **(15)**. These components are referred to as f1, f2 and f6. The core of these networks is a 30-5-1 neural network with 30 input features. The engine data involved 250 sorties and 766.8 logged flying hours. The sorties covered a wide range of missions. Fifty sorties were randomly selected and reserved for the blind tests. The desired output fatigue has been normalised by the fatigue damage of a reference stress cycle **(17)**. The fatigue damage presented in this paper is therefore expressed as the number of the equivalent reference damaging cycles to which the component would have been exposed.

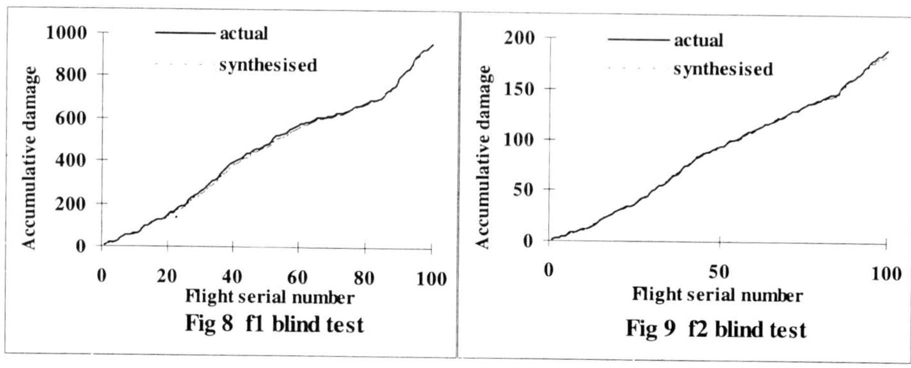

Fig 8 f1 blind test

Fig 9 f2 blind test

In training, the average errors of the synthesised accumulative fatigue were 0.66%, 2.03% and 1.2% for f1, f2 and f6 respectively. The blind tests indicated average accumulative fatigue errors of 1.98%, 1.27% and 4.84% for f1, f2 and f6, and the accumulative fatigue errors at the end of 50 sorties were -0.22%, -2.03% and -4.5% (Figs 8-10). Figs 11-13 demonstrate the quality of synthesising fatigue for individual sorties. Other blind tests indicated that the accumulative damage errors at the end of 50 sorties were -2.43%, -5.16% and 4.4% for f1, f2 and f6 respectively **(15)**.

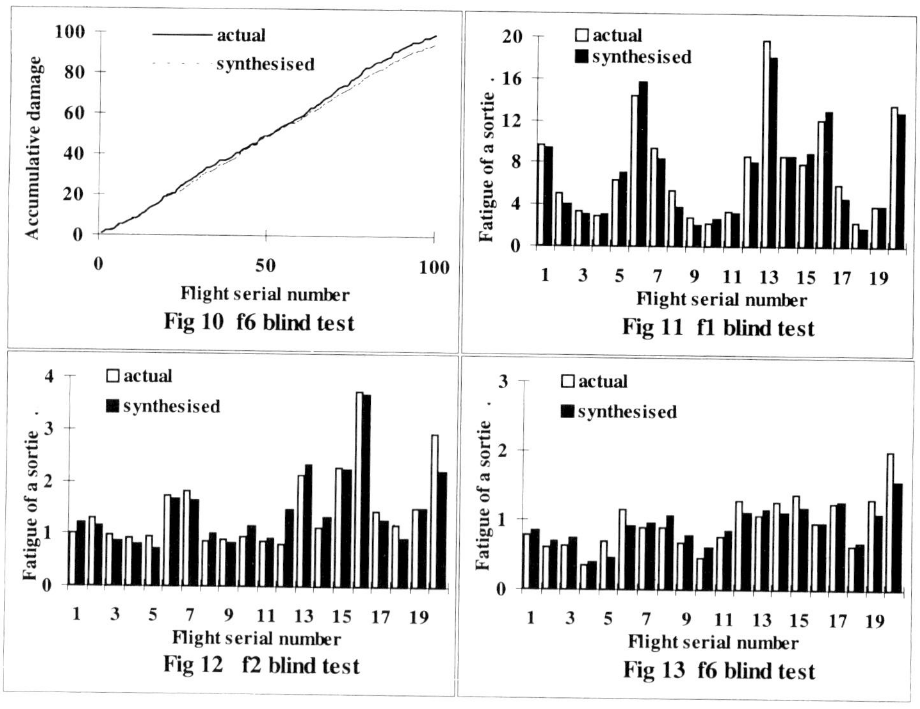

Fig 10 f6 blind test

Fig 11 f1 blind test

Fig 12 f2 blind test

Fig 13 f6 blind test

5.7 Synthetic AUM and CofG

In 1993, a quick-look theoretical assessment on the feasibility of synthesising AUM and CofG was encouraging, but historical HUMS data suggested possible difficulties. The historical data highlighted the problems that would arise from data variability. For helicopters having the same AUM and CofG, the term variability is used to indicate that FDR parameters such as controls can exhibit different values between flights at the same test point. Factors that can induce variability includes: operational conditions, gusts, recording equipment noise, individual helicopter factors such as age, faults and play in control systems. In 1996, the following steps were proposed to investigate the feasibility of synthesising AUM and CofG:

- **a.** Generate synthetic data using the mathematical models of **(10)**, contaminate the data with noise and introduce variability effects.
- **b.** Train mathematical networks how to synthesising AUM and CofG using the simulated data.

The feasibility study covered a conventional helicopter with a single main rotor (S61) and a tandem helicopter (Chinook).

The S61 synthetic HUMS database was generated using a mathematical model for the S61 **(10)**. The database contained about 480,000 records. Each record consisted of FDR parameters and associated AUM and CofG values **(18)**. About 280,000 records of the database covered a forward speed range of 95 to 105 knots, an altitude range of 500 to 5000 feet and an airfield temperature range of 0 to 30 degrees. About 200,000 records covered a speed range of -5 to +5 knots (hover), an altitude of 250 feet and an airfield temperature range of 0 to 30 degrees. The database was contaminated with random variability (uniform deviates of ±10%), this would be a worst scenario for the observed variability within the historical HUMS data. By scrutinising the synthetic data, it was possible to identify the causes of the variability, and contemplate methods that remove its adverse effects on synthetic AUM and CofG.

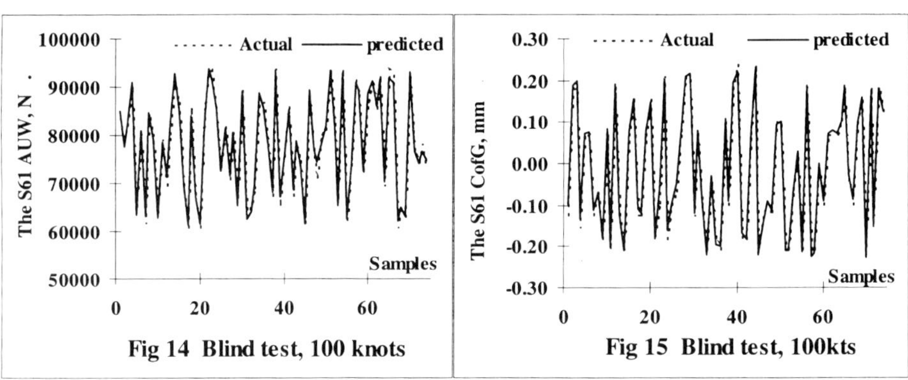

Fig 14 Blind test, 100 knots **Fig 15 Blind test, 100kts**

The Chinook synthetic HUMS database was generated using a mathematical model for the Chinook **(10)**. The database was contaminated with random variability (uniform deviates of ±10%). The database contained over 2,000,000 records **(19)**. About 244,900 records covered a forward speed range of 97 to 103 knots, a climb/descent range of ±3 knots, an altitude range of 900 to 10000 feet and an airfield temperature range of -30 to 15 degrees. About 241,200 records covered a forward speed range of ±2 knots (hover), a climb/descent range of ±2 knots, an altitude range of 900 to 10000 feet and an airfield temperature range of -30 to 15 degrees.

About 1,600,000 records covered a forward speed range of - 6 to +120 knots, a climb/descent range of ±42 knots, an altitude range of 0 to 10800 feet and an airfield temperature range of -20 to 15 degrees; these records simulated a wide range of flight conditions including: climb, descent, forward speed, acceleration, deceleration and climb/descent acceleration/deceleration.

By operating on the synthetic databases, the optimum set of FDR parameters was identified, and the modules of the mathematical networks shown in Fig 1 were investigated as indicated below:

a. Integrity checking: Correcting corrupt data and reconstructing lost signals would enhance system reliability.

b. Smoothing: Filtering the data reduced the adverse effects of random data variability.

c. Merging functions: The Mach number would be useful. Other merging functions including density reduced operational variability effects.

d. Features: The features were the instantaneous values of filtered FDR parameters and merging functions.

e. Compressors: The number of features was small, so compressors were not used.

f. State decider: There was no evidence in the synthetic database to suggest the need for more than one state model, but real systems could benefit from more than one. The state model consisted of two neural networks (one for AUM and the other for CofG). One hidden layer was sufficient (4 to 8 neurons). The input layer contained 7 to 21 neurons. The output layer produced synthetic AUM/CofG. The small number of hidden neurons was chosen to synthesise AUM and CofG at one flight condition such as hover. Eight hidden neurons were used for instantaneous synthesis across flight conditions.

g. Compensator: The compensator was essential. The compensator would include delay units and a median/average process applied on the delayed outputs + the present output. In this way the compensator function was the evaluation of the Median of Repeated Predictions as proposed in **(10)**.

h. Collective Synthesis: The reliability of real systems would benefit from Collective Synthesis: several networks can be designed to look upon a problem from different perspectives (by using different input data, merging functions, etc.). The synthesised output can be the median (or the union) of the outputs of the networks **(10)**.

i. The large volume of synthetic data available would be extremely useful to qualify the network and determine the overall reliability of the mathematical network.

The core neural networks were trained using a feed-back error propagation technique with error relief as described in **(10)**: If the average of an output, as synthesised by a core neural network, is not the same as the average of the desired output, the error in the average is relieved at the final stage of training. This is achieved by reducing the learning and momentum rates of the network gradually over a number of training epochs.

Figs 14-15 indicate the quality of the synthetic AUM and CofG. Figs 16-17 show the favourable effects of a compensator on the mathematical network capabilities.

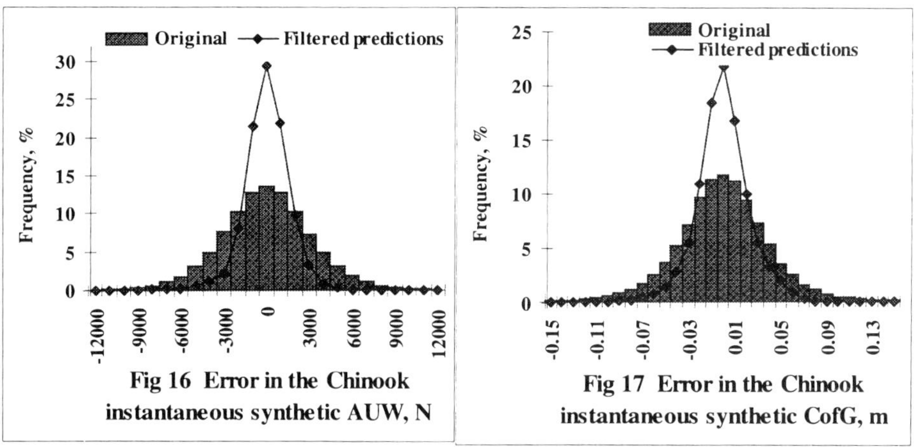

Fig 16 Error in the Chinook instantaneous synthetic AUW, N

Fig 17 Error in the Chinook instantaneous synthetic CofG, m

The average errors of the mathematical network reduced by increasing the number of filtered input records and/or outputs, and by increasing the size of the training data. The average errors in the S61 AUM were 0.25% to 1.2% at 100 knots, and 0.1% to 0.6% in hover. The average errors in the CofG were 2 to 12 mm at 100 knots, and 2 to 16 mm in hover. The Chinook study indicated that the AUM and CofG could be synthesised instantaneously from FDR parameters. The mathematical networks predicted the AUM within an average error of 1.15%; error less than 360 kg in 96% of 91000 blind test cases. The error in the synthetic CofG was less than 25mm in 77% of the cases, and less than 50 mm in 96% of the cases.

CONCLUSIONS

- By embedding design limits and Design Authority (DA) information into a Fatigue and Usage Monitoring system (FUMS), and by processing a large volume of Flight Data Recorder parameters and/or strain gauge measurements, FUMS will automatically check the compliance of aircraft with design assumptions. In this way FUMS will improve safety, reduce costs of ownership and increase availability of aircraft.
- FUMS was conceived as a set of devices, which would synthesise fatigue, usage and other Structural Integrity (SI) information. The FDR-based devices are FUMS indirect devices that synthesise the SI information from FDR data. The heart of the indirect devices is mathematical networks, which provide a framework combining artificial intelligence methods and mathematical formulae.
- The model-based approach presented in this paper used mathematical models not only to determine the mathematical attributes of the networks, but also to create a large database of synthetic data. The synthetic data and available infield data can be used to demonstrate the feasibility of FUMS, and qualify its neural networks and assess their reliability.
- The feasibility of the FUMS technology was demonstrated by synthesising the following: the fatigue of a helicopter rotating component, the fatigue of an aeroplane component, the main and tail rotor torque, the AUM and CofG of helicopters.

ACKNOWLEDGEMENTS

The work presented in this paper would not have been realised without the support of the Ministry of Defence. Acknowledgements are expressed to Cdr Trevor Pritchard (RN), Wg Cdr Steve Welburn, Sqn Ldr David Aunger, Mr Fred Tufnell and Mr John Cochrane for their interest in this investigation, and their discussions and valuable comments. The author would like also to acknowledge the co-operation and support of Rolls-Royce and GKN Westland Helicopters.

REFERENCES

1 **Azzam, H,** The use of helicopter flight parameters for fatigue damage prediction, MJAD/R/125/93, MJA Dynamics, April, 1993.
2 **Azzam, H,** A practical approach for the indirect prediction of structural fatigue from measured flight parameters, Journal of Aerospace Engineering, Proc Instn Mech Engrs, Part G, Vol. 211 No G1, pp 29-38, 1997.
3 **Haas, D J, Milano, J and Flitter, L,** Prediction of helicopter component loads using neural networks, AIAA paper 93-1301, AIAA 34th Structures, Structural Dynamics, and Materials Conference April 19-22, 1993.
4 **Haas, D J, Flitter, L and Milano, J,** Helicopter flight data feature extraction for component load monitoring AIAA paper 94-1308, AIAA 35th Structures, Structural Dynamics, and Materials Conference April 18-20, 1994.
5. **Tang, S S and O'Brian, L J,** A novel method for fatigue life monitoring of non-airframe components, AIAA paper 91-1088, the American Institute of Aeronautics and Astronautics, 1357-1370, 1991.
6. **Azzam, H,** Indirect prediction of helicopter structural fatigue using measured aircraft parameters - training and test, MJAD/R/142/93, MJA Dynamics, December 1993.
7. **Gunsallus, C T, Pellum, W G, and Flannelly, W G,** Holometrics: An information transformation methodology, 44th Annual National Forum of the American Helicopter Society, Washington, 879-885, June 1988.
8 **Gustavson, B, and Pellum, W L,** Systematic application of Holometric synthesis for cost-effective flight load monitoring, the American Helicopter Society 49th Annual Forum, St. Louis, Missouri, May 19-21, 1993.
9 **Azzam, H and Andrew, M,** The use of math-dynamic models to aid the development of Integrated Health and Usage Monitoring Systems, Journal of Aerospace Engineering, Proc Instn Mech Engrs Vol 206 No G1, pp 71-76, IMechE 1992.
10 **Azzam, H,** The use of mathematical models and artificial intelligence techniques to improve HUMS prediction capabilities, Innovation in Rotorcraft Technology Proceedings, The Royal Aeronautical Society, 24-25 June 1997.
11 **Ganguli, R, Chopra, I and Haas, D J,** Formulation of a helicopter rotor system damage detection methodology, Journal of the AHS, October 1996.
12 **Ganguli, R, Chopra, I and Haas, D J,** Detection of helicopter rotor system simulated faults using neural networks, Journal of the AHS, April 1997.
13 **Brindley, J,** Report on an investment appraisal into fitting fatigue usage monitoring systems (FUMS) to in-service helicopters, MJAD/R/168/95, May 1995
14 **Venn, G,** GKN Westland Helicopters Limited, Private communications, 1997.

15 **Azzam, H,** Mathematical networks for thermal transient and non-transient progressive fatigue of engine components, Journal of Aerospace Engineering, Proc Instn Mech Engrs, Part G, Vol. 212 No G2, 1998.
16 **Azzam, H,** Indirect prediction of military aeroplanes fatigue using mathematical networks and measured flight parameters, MJAD/R/177/95, June 1995.
17 **Bryant, M,** Rolls-Royce, Notes and private communications, June 1996- April 1997.
18 **Hazell, J,** An investigation into the feasibility of heliucopter CofG/AUM calculation using HUMS/FDR parameters, MJA Dynamics, MJAD/R/216/97, June 1997.
19 **Wallace, M,** An Investigating into the feasibility of helicopter AUW/CofG calculations using HUMS/FDR parameters, MJAD/R/231/97, December 1997.

Simulation and Light Aircraft

C545/099/98

International regulatory harmonization of simulator qualification evaluations

P A RAY
National Simulation Program, US Federal Aviation Administration, USA

ABSTRACT

In November 1997, at the Royal Aeronautical Society, representatives of the United Kingdom Civil Aviation Authority (CAA) and United States of America Federal Aviation Administration (FAA) signed a document which, for the first time in aviation regulatory history, allowed for simulator qualification evaluations performed by one authority to be recognized by another sovereign authority. The document, Simulator Implementation Procedures (SIP), was not the overnight creation of the respective authorities. Rather, it was the culmination of many years experience (yes, fraught with some degree of trial and error) between organizations committed to a common goal of internationally recognized flight simulator standards.

This paper provides a historical overview of issues and efforts toward international standards, provides in practical language, content of a SIP and provides a summary of continuing efforts towards expanding regulatory harmonization through the pursuit of potential SIP agreements between the FAA and other regulatory authorities. I will touch briefly on a few areas related to the development of the a basic SIP agreed upon between the Federal Aviation Administration (FAA) and the European Joint Aviation Authorities (JAA). After sharing some background on harmonization in this area, I will provide some background on the development of the regulatory framework for our cooperation and how, in collaboration with our European colleagues, we are making a more efficient regulatory process a reality in Europe.

INTRODUCTION

To an ever increasing degree, airlines and the aviation industry have departed from their traditional role as a national enterprise by their expansion into what is rapidly becoming a global industry. National aviation authorities, if they are to fulfill their mandate to uphold the highest levels of aviation safety, must adapt to the impact of the global nature of the aviation industry by increasing their partnerships with other regulatory authorities. This will provide the only effective and efficient global means to accomplish surveillance and foster safety in an era of diminishing resources.

No where is this clearer than in the area of simulator qualification evaluations, where not only is the simulator required to undergo an evaluation by the regulatory authority of their own country, they must also endure such an evaluation by each of the regulatory authorities from each of the foreign countries whose airlines purchase time on the subject simulator. This makes for neither an efficient nor effective operation from either an operator's or regulator's standpoint. If even a portion of these multiple evaluations are able to be eliminated while safety is maintained, all will benefit.

FUNDAMENTAL ISSUES

The benefits from harmonization of international flight simulator standards was clearly recognized during the process of their formation in the forum graciously provided by the Royal Aeronautical Society (RAeS). Their adoption at the RAeS in 1992 and later acceptance by the International Civil Aviation Organization in 1994 have provide a common standard accepted by regulatory authorities around the world. During the formal and informal discussions leading up to the acceptance and publication of international standards, many thoughts were exchanged by regulatory authorities and users regarding the follow on benefits of a common standard. The central question raised was - Would regulatory authorities be able to accept each other's work on evaluation of simulators to a common standard? If successful, such a program would provide benefits to regulators, users and manufacturers.

In one aspect, the issue of flight simulator standards is miles, if not light years, ahead of the more classic regulatory issue of "certification" such as aircraft certification, repair station certification, environmental certification, etc. With regard to classic certification issues, few, if any pair of regulatory authorities possess the exact same rules (standards). With the adoption of the international simulator standards, those of us associated with flight simulation find ourselves in a rather unique position when we participate in discussions with our colleagues dealing with the development of implementation procedures regarding classical certification issues.

It could easily be concluded by some that, recognizing the international acceptance of common simulator standards, reciprocal acceptance should be routine. However, one must recognize a fundamental aspect of flight simulator evaluation and qualification. A flight simulator is qualified by a regulatory authority for use in a specific airline or training center's training program. Some may believe a simulator acceptable for one training program is automatically acceptable for any training program. Such a belief is not accurate. Differences between training programs regarding flight simulators clearly begin with the basic cockpit configuration and system(s) configuration and extends into a significant number of training issues. This list of differences, although not endless by any means, must clearly be addressed by the users and regulatory authorities.

Additionally, and likely more fundamental, is the issue of how do we as independent regulatory authorities apply our common flight simulator standards? Does an inspector responsible for upholding those standards require any prerequisite knowledge or training? Should a specific regulatory authority's record keeping be a subject of assessment in determining the acceptability of developing an agreement? In summary, does one regulatory authority have the confidence to make a regulatory decision based upon the findings and recommendations of another authority's findings and recommendations? These and other issues must be thoroughly addressed by each regulatory authority prior to coming to a conclusion that reaching an agreement to cover simulator qualification is appropriate.

A REGULATORY FRAMEWORK FOR INTERNATIONAL PARTNERSHIPS

In September 1995, the United States signed the world's first Bilateral Aviation Safety Agreement (BASA) with The Netherlands in The Hague. Since that inaugural signing, BASAs have been signed between the US and a growing list of countries within the European community including Austria, France, Germany, Ireland, Switzerland and the United Kingdom.

The development of BASAs and their implementation procedures between the U.S. and JAA countries affords us a means of realizing this hope of improving our efficiency in simulators without compromising aviation safety in just such a manner. The intention of these agreements is to help promote aviation safety while at the same time improving the use of regulatory oversight resources. After conclusion of an umbrella bilateral aviation safety agreement between governments, aviation authorities develop implementation procedures (IPs) concerning specific areas of required aviation regulatory oversight. Unlike previous international aviation agreements which required foreign ministry level approval to amend a document, BASAs afford the FAA and the respective National Aviation Authority (NAA) the ability to make required amendments without having to move through diplomatic channels for each new change to the IP. In the case of simulators, upon mutual satisfaction with each other's systems and standards, Simulator Implementation Procedures (SIPs) may be executed for reciprocal acceptance of the other authority's work on simulator qualification evaluations.

BASAs offer a significant level of accountability across national borders, such that one authority's expectations of another are absolutely clear. A BASA and the IPs establish a clear system of findings and acceptances which also provides for accountability at each step of the regulatory process. It forms a system whereby regulatory authorities recognize each other's abilities to collect data and accomplish surveillance, such that the authority receiving the information may consider awarding an approval consistent with its legal responsibilities.

In progressing toward an agreement in simulators or any other aviation regulatory discipline, it is vital that safety remain at the heart of any regulatory approach. Inspectors on both sides of the Atlantic vested with the responsibility for insuring compliance with regulatory requirements are essential participants in any agreement involving reciprocity. As Anthony Broderick, former FAA Associate Administrator for Regulation and Certification, and a strong supporter of BASAs and IPs stated at the 12th Annual FAA/JAA Harmonization Meeting, "...I want any of our inspectors to be able to sit before a congressional committee and testify that a similar level of safety is achieved under an agreement with another civil aviation authority as we have under our own direct oversight. For this to happen, the agreements we enter into have to be with technically competent and legally accountable entities. Such agreements also must have a sole focus -- SAFETY. Even the appearance of any other objective beyond the safety of the aviation system would be unacceptable." Mr. Broderick went on to state "...And when something goes wrong or an accident occurs, and as we in this room know it will, it is crucial that the relationship we have structured with each other has no other competing objectives or goals than the safety of our citizens." Inspector familiarization, confidence and acceptance of the agreements is clearly a necessity.

MAKING IT A REALITY IN EUROPE

Recognizing the FAA's commitment to work as a partner with the Joint Aviation Authorities (JAA) in the development of JAA rules and policies, the FAA determined the most effective course of action regarding a SIP under a BASA with our European regulatory colleagues would be JAA acceptance of a model SIP prior to signing a SIP with a JAA member state. In June 1997, the FAA committed to provide an initial generic draft SIP and a work plan for the development of a SIP with any JAA member state. Negotiations on the draft SIP and work plan progressed rapidly and concluded with the announcement in Washington, DC in September 1997 that the FAA and JAA had agreed upon the SIP and work plan. Formal JAA Committee approval occurred in October 1997followed by the FAA and U.K. CAA SIP signing in London in November 1997.

The FAA has designated the National Simulator Program (NSP) as the FAA office accountable for implementation and administration of the FAA/CAA SIP and future SIPs adopted with any NAA. The NSP will maintain records of all SIPs and any evaluations conducted by an NAA in accordance with a SIP and recognized by the FAA as meeting a specific evaluation requirement.

DEVELOPMENT OF SIMULATOR IMPLEMENTATION PROCEDURES

The work in the early 1990s to developing a SIP by the FAA with the Canadian Ministry of Transport (MOT) and UK Civil Aviation Authority (CAA) and by the MOT and CAA with the FAA was not an efficient process. That is not to say any of the parties involved would apologize for those efforts or our inefficiencies. Like any new effort, hindsight and experience typically reveals inefficiencies of newly developed products or processes. However, we did have the foresight to recognize the potential benefits of our work and thereby documented the issues addressed during our efforts. Our documentation forms the heart of what has been entitled SIMULATOR IMPLEMENTATION PROCEDURES WORKING PLAN. The work plan will be jointly followed by the FAA and NAA to determine the acceptability of signing a SIP.

Development of the level of understanding required to enter into acceptance of another authority's work on flight simulator evaluations is not an overnight exercise. As previously noted, a number of significant issues must be addressed by each regulatory authority before both parties can be assured all regulatory concerns are met.

Formal work plan

As stated in the FAA/JAA SIP Work Plan, this process lays out a clear and flexible process which could "lead to the establishment of Simulator Implementation Procedures (SIP) addressing simulator evaluations between the FAA and NAA of JAA member States where the FAA conducts simulator evaluations for FAA certificated airlines or training centers; and the NAAs of JAA Member States conduct simulator evaluations for NAA certificated airlines, operators or training centers."

Three phase SIP development and evaluation process

The process for developing SIPs and the associated evaluation activities were derived from the documentation of the FAA, MOT and CAA efforts of the late 1980s and early 1990s. Based upon lessons learned in the pursuit of implementation procedures in other regulatory areas, specific time frames associated with this process are not listed but are dependent upon the alignment of FAA and JAA NAA simulator evaluation policy, regulatory resources available and the number of concurrent (joint) evaluations that can be satisfactorily accomplished. Determination of the acceptability of FAA or NAA policies and practices is based upon a three phase process.

Phase I – Authority system familiarization

An initial series of meetings will be conducted with the principal regulatory authority representatives of FAA and the JAA NAA involved in evaluating and qualifying aircraft flight simulators to review and discuss: the regulatory basis under which each country conducts its simulator evaluation/qualification process; the establishment of those principles that are similar and those that are different in the accomplishment of simulator evaluations and the criteria necessary for qualification or denial of qualification status; the practices involved when qualifying a simulator, including appropriate restrictions due to particular problems with missing, malfunctioning, or inoperative equipment or components; the background and experience of persons actually conducting the evaluation and recommending the qualification of simulators; the training required/conducted for persons actually conducting the evaluation and recommending the qualification of simulators; the reporting and documenting of evaluations conducted, qualifications granted, discrepancies noted, corrective action taken, and any limitations recommended for imposition or removal from the simulator; the associated documents which will form the basis of the objective and subjective evaluation of flight simulators; the means of accounting for and tracking of simulator faults and discrepancies, including the downgrading of qualification levels; and whether or not the simulators under consideration are to be used for checking activities and whether or not "zero flight time" training and checking is to be permitted.

Phase II – Joint evaluations

Scheduling of joint (concurrent) evaluations -- both initial and recurrent -- of simulators for which each country has an operational need to evaluate or inspect. The number of such concurrent evaluations must be sufficient for each participating country's representative(s) to gain a full understanding of the way in which each of the items listed in Phase I is addressed.

A follow-on series of meetings between the principals to review again the items in Phase I, and how they might be affected by the concurrent evaluations. These follow-on meetings should focus on any additional requirements that might be necessary to ensure that both FAA and JAA simulator criteria are met. The FAA and the JAA NAA will have to determine how any such adjustments might be incorporated into their own processes to accommodate specific requirements of the other authority.

Phase III – Follow-up
Joint (concurrent) evaluations incorporating any required adjustments to individual countries process of simulator evaluation, qualification, record keeping, or reporting will continue as necessary.
The Simulator Implementation Procedures work plan will be complete when all questions regarding the way in which the FAA and an individual JAA NAA accomplish simulator evaluations have been resolved including: (i) a full understanding and acceptance of each other's practices; and (ii) a level of confidence has been achieved such that each authority's practices and policies are mutually acceptable so as to allow both the FAA and the JAA NAA to conduct a simulator evaluation on each other's behalf.

Key to the success of this work plan will clearly be the authorities mutual acceptance of how each authority applies the agreed upon standards. This three phase process should readily identify differences. Resolution of differences has been made easier through publication by the Royal Aeronautical Society of the Airplane Flight Simulator Evaluation Handbook, Volumes I and II, which provide supplementary information regarding the evaluation of flight simulators. Malcolm Blackwood and Simon Wood, principal authors of the publications, worked in collaboration with regulatory authorities in the development of the documents which should greatly reduce the time involved to resolve differences through the detailed information contained in each volume regarding specific objective and functional testing requirements.

SIMULATOR IMPLEMENTATION PROCEDURES

Any international agreement between authorities is only as good as the level of understanding and cooperation at its foundation. For a SIP that foundation is built upon completion of the work plan listed in the preceding discussion to the complete satisfaction of both authorities and the acceptance and confidence by both authority's inspectors of the competency of another regulatory system.

Based the international cooperation of the flight simulation industry including users, manufacturers and regulators, the model SIP signed by the FAA and JAA embodies the simplicity of purpose and application we have long felt was achievable. As noted in the following short discussions of several key points within the SIP, its simplicity should be readily apparent. Only a moderate amount of detail is included in the twelve page document, mostly to insure regulatory requirements are maintained by the FAA and NAA. The two key points for non-regulatory readers should be those items addressing Applicability and Standards Used.

Objective
The objective of the model SIP developed by the FAA and JAA is to outline the terms and conditions under which the FAA and an NAA can accept each other's evaluations of flight simulators for findings of compliance with FAA AC 120-40, as amended, and/or Joint Aviation Requirement-Synthetic Training Devices (JAR-STD)-1A, as amended, for the

purpose of Simulator Qualification only. The FAA and NAA will agree, subject to the terms of the SIP, to accept each other's evaluations for findings of compliance with their respective requirements as the basis for its own qualification of a simulator.

Applicability
The SIP is applicable to those Level B, C and D Simulators that are required to have FAA and NAA evaluation leading to a Qualification Level because they are used by an FAA-certificated entity (U.S. air carrier or FAR part 142 training center) and an NAA-approved entity (Air Operator Certificate holder or Flying Training Organization).

Notably absent from the list of simulators eligible under the SIP is a Level A simulator. This exclusion is necessitated due to the differing standards for Level A simulators between the FAA and JAA. This point should not be lost because it again focuses us back to the benefits of the adoption of internationally accepted standards.

Standards used
Qualification evaluations will be conducted to the standards defined in AC120-40, as amended, or JAR-STD-1A, as amended. These standards incorporate the content of the International Civil Aviation Organization (ICAO) Document 9625-AN/938, "Manual of Criteria for the Qualification of Flight Simulators.

Reference to the ICAO document was not an afterthought. It is an acknowledgment of the value of our international standards. Without the common purpose efforts of those involved with their development and adoption, the common standards adopted by the FAA and JAA would not likely have occurred, therefore greatly reducing the likelihood of an FAA/JAA SIP.

Evaluation reporting
The FAA and NAA will provide the other authority with the a blank evaluation report form to be completed during an evaluation including any special instructions or requests at least 30 days prior to an evaluation. The completed report will be returned to the originating authority within 30 days of the evaluation. The report will identify the standards used, including the appropriate revision level.

Acceptance of evaluation
In order to meet the regulatory requirements of both authorities and as stated in the BASA, the evaluation report will be reviewed by the authority requesting the evaluation for content and recommendations resulting from discrepancies discovered. If found acceptable, the evaluation will constitute an evaluation conducted by the requesting authority.

Right to make independent verification
The SIP provides that the FAA and NAA acknowledge the right of the other authority to make its own evaluation or re-evaluation of a simulator at any time. Quite correctly, neither authority should completely relinquish an inherent regulatory requirement.

Assuming the work plan leading up to the signing of an individual SIP resolved practical or philosophical differences between the FAA and NAA, we would not anticipate a need to re-accomplish a given evaluation. However, recognizing a flight simulator is a tool within a given training program, situations may clearly occur wherein simulator hardware or software changes for a specific operators training program may be required. Based on the judgment of the appropriate regulatory authority evaluation of those changes may require their assessment of specific changes. We would anticipate close coordination by both authorities to perform such an assessment during a routine evaluation which may involve both authorities. Similarly, the SIP does not preclude periodic concurrent (joint) evaluations when, in the judgment of the authorities, such evaluations would be appropriate.

Continued cooperation
The SIP formalizes the commitment of the FAA and JAA to continue our cooperative efforts in development and maintenance of common flight simulator standards and the use of flight simulators in quality training programs. The SIP specifically mandates the FAA and NAA meet at least annually to discuss the SIP, on-going projects, changes in their own organizations, any revisions to their requirements, technical assistance requests, or any other matters relating to the SIP. Additionally the SIP clearly states a requirement for both authorities to provide any technical assistance which may be requested, including but not limited to: evaluation recommendations, surveillance activities, enforcement actions and safety investigations.

PROGRESS TOWARDS A SIP

Presentlysix JAA member states would meet the prerequisites for a SIP. Simulators located in the United Kingdom, France, Germany, Switzerland, Sweden and the Netherlands are required to have FAA and the respective NAA's evaluation and approval because they are used by an FAA-certificated entity and an NAA-approved entity. Utilizing the documentation of our efforts with the CAA, the NSP initiated SIP work plans with the French DGAC in December 1996 and the Swiss FOCA in April 1998. Additionally, preliminary discussions with the German LBA regarding a SIP work plan were conducted in May 1998.

SIPS – EVERYONE WINS

Practical application for a SIP is plainly evident for those simulator operators who provide simulator time to airlines from several foreign countries. Currently, not only is the simulator required to undergo an evaluation by the regulatory authority of their own country, they must also endure such an evaluation by each of the regulatory authorities from each of the foreign countries whose airlines purchase time on the subject simulator. With the elimination of redundant evaluations, not only will the simulator operator enjoy significant increase in the time available on the simulator to sell or use as he pleases, but he will also enjoy the benefit of the additional time of his technical staff to address the normal occurrences of simulator operation instead of having to provide technical support for yet another regulatory authority evaluation. Needless to say, such a reduction in regulatory evaluations will benefit each regulatory authority as well, by not having to spend the increasingly hard to come by money to support such travel and inspection, but also will provide the benefit of the individual staff time to devote to other increasingly pressing demands on that authority.

SUMMARY

The cooperative work conducted between the FAA and JAA, principally via the JAA STD Committee and RAeS forums, has clearly shown the answer to our question of many years ago "Would regulatory authorities be able to recognize each others work on the evaluation of simulators to a common standard?" has been a resounding YES. A means to achieve this regulatory cooperation in acceptance of simulator evaluations has now been provided by BASAs and the model SIP.

We clearly look forward to working with our JAA counterparts in Europe and other parts of the world regarding the potential of establishing a SIP between our respective authorities. The FAA and NSP look with great anticipation to advancing greater regulatory cooperation among its peers. We are confident that the resulting potential for improved regulatory efficiency provided by BASAs and IPs will be viewed positively by both industry and governments as we are able to both maintain and enhance current safety levels while reducing redundant regulatory costs. We look forward to working with our European colleagues and those from other parts of the world to building strong, transparent, accountable, and efficient system of international aviation safety regulation.

ACKNOWLEDGMENTS

I would be remiss if I did not acknowledge former JAA Operations Director Richard Yates who played a central role in the model SIP and SIP Work Plan negotiations and deserves a great deal of credit for the rapid manner in which negotiations proceeded following the JAA meeting in Berlin in June 1997. Mr. Yates' efforts and those of Capt. T. (Paddy) Carver, former JARSTD committee chairman, were instrumental in attaining accord with the FAA regarding the SIP and SIP work plan. We are indeed indebted to them for their efforts.

C545/094/98

JAR STD: The european synthetic training device family

D R IRVING BSc, MBA, CEng, MRAeS
Civil Aviation Authority, Gatwick, UK

The JAR-STD Working Group has over the past seven years produced a family of documents defining standards for a family of flight crew training devices.

These devices range in capability from simple instrument procedures trainers through to full motion daylight visual zero flight-time capable simulators. This new family is based on international flight simulator standards developed by a Royal Aeronautical Society group.

The challenge now is to use these new standards to underpin the European wide harmonisation of flight crew training and testing. The standardisation of the JAA regulatory bodies is under now way to ensure the common application of the rules.

The next step will then be to extend the acceptance of common standards to include the USA.

1 SUMMARY

Within Europe there is today one new standard against which Synthetic Training Devices (STDs) can be qualified for use in pilot training and testing: JAR-STD

Under the umbrella of the Joint Aviation Authorities (JAA), the JAR-STD Working group has created a series of documents defining standards for the qualification of a family of aeroplanes synthetic training devices. The family defines standards for flight simulators, flight training devices and flight and navigation procedures trainers (FNPTs).

To ensure the widest possible acceptance of these new standards the membership of the JAR-STD Working group was drawn from 15 countries including the USA.

The Working group has just started to develop a similar family of documents applicable to helicopter training devices.

In parallel with this JAA focused effort, the FAA has started the process of laying down the foundations for a future bilateral agreement with individual JAA states mutually recognising flight simulator qualifications.

2 BACKGROUND

In the late 1970's the US FAA developed objective and subjective standards in support of their Advanced Simulation Plan. This plan paved the way for the concept of Zero Flight Time (ZFT) conversions within an airline environment. In Europe and probably in many other parts of the world flight simulators were being used as part of a range of tools in pilot training and testing with the aeroplane retaining its role as the tool for the final proficiency check.

As flight simulators became more common due to their wider acceptance by the pilot training community and due to advances in technology, industry saw the need for common technical standards throughout the world to replace the many and varying local standards. Some of these standards including those of the UK's CAA were in part based on the FAA's work. That being the case it made sense to base any common standard on the FAA's publications.

During 1989 at a Royal Aeronautical Society conference a working group was formed to define an international standard for qualifying aeroplane flight simulators. The group was fortunate to gain support from a wide range of airlines, manufacturers both aeroplane and simulator, independent training centres, data supplies and regulators. Four meetings later the group was able to publish the RAeS, "International Standards for the Qualification of Aeroplane Flight Simulators". This document was then adopted by ICAO and subsequently published as "Manual of Criteria for the Qualification of Flight Simulators". In support of this international standard two supporting hand books were published by the RAeS: Aeroplane Flight Simulator Evaluation Handbook Vols. I & II

These books enlarged on the content of the RAeS's international standard giving clear guidance on both subjective and objective testing to people intending to apply the standard.

It must be remembered that under-pinning the development of these "standards" was an assumption that high quality data packages existed for each aeroplane type. Under the auspices of IATA, a group of airlines and manufacturers have tirelessly shadowed the development of the regulatory standards. This group published a document (now in its 5th edition) specifying in great detail the data requirements to build a flight simulator capable of meeting the highest regulatory standards (Flight Simulator Design and Performance Data Requirements).

3 JAA

The Joint Aviation Authorities (JAA) started work during the early 1970s. The driving force behind the early work was the need to develop common standards for the certification of large transport aeroplane (JAR25). From this the JAA slowly grew to its current position today. It exists with a small permanent staff in Holland providing a focus for the development of a wide range of regulatory codes for certification, maintenance, licensing and operations.

During 1991, a working group (JAR-STD WG) was formed to generate a JAA code for the standards for a range of synthetic flight training devices to support pilot training, testing and operational competence as required by JAR-FCL (Flight Crew Licensing) and JAR-OPS (Air Transport Operations). The Working group drew its membership from a wide range of European Authorities, airline associations, simulator operations and manufacturers. In support of one of the key JAA objectives - that of harmonisation with the FAA, the Working group was fortunate to have representatives from both the FAA and the US Air Transport Association (ATA).

Since 1991 the JAR-STD Working group has been meeting regularly to develop the family of documents defining standards for a range of training devices to support both aeroplane and helicopter pilot training testing and checking..

4 AR-STD FAMILY : AEROPLANES

The first member of the JAR-STD Family to be developed was JAR-STD 1A which specifies standards for aeroplanes flight simulators. (The ICAO definition of a flight simulator states that the device has both motion and visual systems).

This document which defines 4 levels of flight simulation (Appendix 1) in theory should have been relatively straightforward to produce: The technical standards were already established (RAeS: International Standards for the Qualification of Aeroplane Flight Simulators) and the working group representatives were in the main familiar with all aspects of flight simulator manufacturing and use. Six years later JAR-STD 1A was finally completed and published. Language and culture is not common throughout the JAA community and thus much meeting time was devoted to carefully ensuring that all parties understood the intent of the baseline standards. In addition the standard as developed by the RAeS working group was purely technical and did not address any regulatory issues. The working group had therefore to develop the essential regulatory material from scratch.

JAR-STD 1A does not in fact represent a major change for many countries. It does however clearly define a set of common standards and procedures which will be used by all JAA countries when qualifying flight simulators.

Following on from JAR-STD 1A, the working group developed standards for Flight and Navigation Procedures Trainers (FNPTs), as required by JAR-FCL to support initial professional pilot training. JAR-STD 3A defines two levels of FNPTs, Level 1 is a basic IFR procedures trainer. Level 11 is really a tailored generic flight simulator without motion. Under the provisions of JAR-FCL an FNPT11 can be used for training and some testing (abnormal and asymmetric) for the initial issue of a professional pilots licence for the issue of an instrument rating and for the multi-crew co-operation (MCC) module. (Appendix 2).

JAR-STD 2A the standard for Flight Training Devices (FTDs) is the final part of the aeroplane STD family. FTDs are type specific devices and may be used to complement the use of flight simulators in an approved type rating course. There are two levels of FTDs, the simplest of which need only simulate a single aeroplane system. The higher level device is a complete system simulator supported by a rudimentary flight model. This model will be sufficient to enable the device to be flown via its autopilot but not of sufficient fidelity to practice manual flying tasks. (Appendix 3).

5 JAR-STD FAMILY : HELICOPTERS

Work has just commenced on developing the first member of the helicopter synthetic training device family: JAR-STD 1H. The basis for the work is the FAA's advisory circular AC120-63. Perhaps the real challenge in developing standards for helicopter STDs lies within the industry. Unlike the fixed wing world the helicopter world is much more diverse in terms of equipment and operations. Flight simulators for helicopters are relatively rare thus there is not the enormous body of knowledge and experience from which to develop JAR-STD 1H. There are also some very demanding technical issues to be debated, the main ones concerning the visual system. Because of their design and operating modes helicopters use very large fields of view particularly in the vertical plane. The conventional 150° x 40° field of view does not fulfil the needs of the helicopter pilot and the issue of chin windows looms large. Technical solutions do exist but there is very little practical experience from which to draw any conclusions. A further balance to be struck is that of operating costs vs. fidelity. Many helicopters are small and therefore cheaper to buy and probably cheaper to operate than the equivalent simulator. The dilemma faced by the working group is that of technical complexity and thus cost versus fidelity and fitness for purpose.

The helicopter STD family will in time be extended to include flight and navigation procedures trainers (FNPTs) as part 3H and should a requirement exist, FTDs as part 2H.

6 INTERNATIONAL COMPARISONS

As much of the work is based on existing standards it may be useful to illustrate some parallels:

6.1 For Aeroplane Flight Simulators

JAA	FAA	CAA	IQTG
D	D		11
C	C	4	1
B	B	B	-
A	A	2B	-

6.2 For Aeroplane FNPTs & FTDs

JAA	FAA	CAA
FTD 11	5/6	-
1	2/3	-
FNPT 11	?	-
1	?	1

At present it is not possible to draw simple parallels in the helicopter family. The flight simulators levels will mirror the current FAA's AC120-63 levels with some potential differences.

One drawback of trying to illustrate simple parallels is that of revisions. None of the standards produced can anticipate future training needs and new technologies. Whilst the JAA was developing its flight simulator standards the FAA was reviewing its own standards and updating them to accommodate the needs of their commuter airlines. JAA concentrated on lowering the cost of its Level A standard whilst the FAA revised its Level B standards. Despite this work the parallels shown are still valid with only small deviations.

7 STATUS OF FAMILY

Aeroplanes

JAR-STD	1A	4 levels of Flight Simulators	effective 1/4/98
	2A	2 levels of Flight Training Devices	NPA
	3A	2 levels of Flight & Navigation Procedures Trainers	effective 1/7/99

Helicopters

JAR-STD	1H	In preparation
	2H	??
	3H	2 levels required by JAR-FCL

8 IMPLEMENTATION PROCESS

To ensure that its codes are carefully developed and consistently applied the JAA has a standard process for implementation.

The Working group which develops the code draws its membership not only from the national regulators but from a wide cross section of the appropriate industries. When a code has been completed by a working group it is passed up to the "parent" committee for approval. (In the case of the JAR-STD family this involves both the JAR-FCL and JAR-OPS committees). Having passed this stage the codes are then checked for consistency and any possible conflict with other codes by the JAA's Regulation Director. The final in house stage of approval is by the JAAC, JAA's top level committee who's members are drawn from the leaders of the various national authorities.

The draft code can then be published for public comment, the NPA (notice of proposal amendment) stage. All comments have to be considered by the original working group who will amend their text if necessary. The final code can them be published with an adoption date set in the near future and with an implementation date typically two years away. After the adoption date the code may be used, after the implementation date the code should be used.

The authorities who participate within the JAA are subject to a standardisation process to ensure that their application and interpretation of any code is fair and consistent. Each code has a parallel set of "Procedures" which detail how a code should be implemented in practice. The "procedure" for the JAR-STD family sets out how each synthetic training device should be evaluated. The type of evaluation team and the content of the inspection are specified. The "procedures" also specify how the process should be standardised via a system of audits and introduces an appeals procedure from reconciliation on the event of a dispute.

JAR-STD procedures establish an Advisory Board (JAR-STD AB). The STD AB draws its members from the regulatory community. It's task is to give guidance and interpretation as required and to organise the standardisation process. The standardisation team (STD STs) are drawn from the national regulators. Their task is to conduct an annual audit of each regulator to ensure that JAR-STD is applied appropriately.

At present, the JAR-STD Advisory Board has just started it work. It has a number of technical issues to resolve some of which may result in amendment of the codes. The actual standardisation process has started and it is anticipated that the initial group of countries will be audited before the end of 1998. The satisfactory completion of an audit allows a country to issue simulator qualifications based on JAR-STD 1A. That country can then expect that its simulator qualifications will be mutually recognised by all other JAA states thus avoiding the need for multiple inspections of one simulator by each authority.

9 HARMONISATION WITH FAA

In parallel with its work on the JAA working group developing codes for STDs, the UK CAA has over approximately 8 years been sharing information and ideas with the FAA mainly through joint simulator evaluations. This process has now proceeded to the point where although differences in process do exist, the outcome results in a consistent and common standard of simulator qualification. Under the umbrella of the UK/USA BASA (Basic Air Safety Arrangement) the SIP (Simulator Implementation Procedures) allow each country to conduct every other evaluation on behalf of one another. The UK current by "uses" about 6 simulators within the USA. The SIP therefore save a small but useful number of visits to the USA. A number of European states are in various states of negotiation with the FAA to establish their own SIPs.

10 LOW COST

In the large aeroplane world the use of flight simulators and lesser supporting devices is well established. The cost, safety and environmental issues have been responsible for a well established ZFT training and testing strategy in the USA and here in the UK. Within the smaller aeroplane community it is a very different story: Simulators are rare because operating costs are equal too and perhaps higher than those of the aeroplane. Since there are only a few devices, crew positioning adds further to the cost imbalance therefore pilot training and testing is still mainly conducted in the aeroplane.

Recognising this dilemma, the JAR-STD Working Group specifically addressed the issue of a low cost simulator positioned at Level A standard. The Working Group relaxed the standard for data packages, motion, visual, parts and for the qualification test guide with a view to encouraging the ownership of flight simulators representing smaller commuter type aeroplanes.

The simulation industry, both constructors and owners now recognise their responsibilities towards lowering all costs. They have proposed a number of positive initiatives to lower the cost of the highest Level (D) devices:

Fewer malfunctions
Less customisation
Standard data packages

Technology is also playing its part enabling key components of a simulator to be produced at a fraction of historical costs.

It will be difficult to quantify the success of any of these changes, but whatever the outcome the use of the actual aeroplane as a basic tool for type conversions is very limited. It is widely accepted that a flight simulator is a far more effective training tool albeit that the final proficiency check may still take place in the aeroplane.

11 QUALIFICATION & APPROVAL

There exists wide spread confusion regarding the qualification of a flight simulator and the subsequent approval to use a flight simulator.

Under the provision of the JAR-STD family, specifically Part 1A all flight simulators must be qualified. The process leading to a qualification consists of:

- a qualification test guide (QTG) submitted to the local authority fulfilling the requirements of AMC030.
- A satisfactory review by the authority of the results contained within the QTG.
- a satisfactory objective and subjective evaluation of simulator by an authority team consisting of at least a type rated training inspector and a technical inspector aided by the operators staff.
- the existence of an effective Quality Management System.

A qualification is valid for a maximum period of 12 months after which this must be requalified. Normally this is an annual process but the rhythm may be disturbed by major modifications upgrades, relocation or poor reliability of the simulator.

For an airline or training centre to use a flight simulator as part of any approved training program:

the simulator must be qualified and
the user must gain approval to use the simulator.

The extent of the user approval will vary according to a number of parameters including:

experience level of trainees
experience level of training organisation
differences between simulator and aeroplane configuration

The key point is however, that the level of user approval can **never** exceed the level of simulator qualification.

12 JAR-STD 1A

Technical Requirement and Credits

Qualification Level	General Technical Requirements	Maximum Credits
LEVEL A*	The lowest level of simulator technical complexity. A full scale replica of the aeroplane cockpit/flight deck including simulation of all systems, instruments, navigational equipment, communications and caution and warning systems. An Instructors station must be provided as must be seats for the crew members and two additional seats for inspectors/observers/instructors. Control forces and displacement characteristics must correspond to that of the replicated aeroplane and they must respond in the same manner as the aeroplane under the same flight conditions. The use of class specific data tailored to the specific aeroplane type with fidelity sufficient to meet the Objective Tests. Functions and Subjective Tests are allowed Generic Ground Effect and ground handling models are permitted. Motion visual and sound systems sufficient to support the training, testing and checking credits sough are required. The visual system must provide at least 45 degrees horizontal and 30 degrees vertical field of view per pilot. A night scene is acceptable. The response to control inputs shall not be greater than 300 milliseconds more than that experienced on the aircraft. Windshear need not be simulated.	Suitable for - Crew procedures training. - Instrument Flight training. - Transition/Conversion training, testing and checking except for take off and landing manoeuvres. - Recurrent training, checking and testing (Type and Instrument Rating Renewal/Revalidation).
LEVEL B	**As for Level A plus**: Validation Flight Test Data must be used as the basis for flight and performance and systems characteristics. Additionally ground handling and aerodynamics programming to include Ground Effect reaction and handling characteristics must be derived from validation Flight Test Data.	**As for Level A plus**: - Recency of experience (three take offs and landings in 90 days). - Transition/Conversion training for take off and landing manoeuvres. - Transition/Conversion testing and checking except for take offs and landings.

Qualification Level	General Technical Requirements	Maximum Credits
LEVEL C	The second highest Level of simulator performance. **As for Level B plus**: A Dusk/Night Visual system is required with an instantaneous horizontal field of view of not less than 75 degrees per pilot. A six axis motion system shall be provided. The sound simulation must include the sounds of precipitation and other significant aeroplane noises perceptible to the pilot and must be able to reproduce the sounds of a crash landing. The response to control inputs shall not be greater than 150 milliseconds more than that experienced on the aircraft. Windshear simulation must be provided.	**As for Level A plus**: - Transition/Conversion testing and checking of take offs and landings for crew members whose minimum experience level is defined by the Authority.
LEVEL D	The highest Level of simulator performance. **As for Level C plus:** A full Daylight/Dusk/Night visual system is required and there must be complete fidelity of sounds and motion buffets.	**As for Level C plus**: - Transition/Conversion testing and checking of take off and landings for crews, who may be required to meet a minimum experience level defined by the Authority.

Qualification Level	General Technical Requirements	Maximum Credits

LEVEL 1	Type specific with at least 1 system fully represented.
	Closed or open flight deck.
	Note: Choice of systems simulated is the responsibility of the organisation seeking Approval or re-approval of the course).
		Suitable for:
		- Selective system management credits (except for pilot manual control handling skills) as follows:
		  - part of an approved conversion/transition course.
		  - recurrent training/checking.

LEVEL 2	Type specific All applicable systems fully represented. Closed flight deck. Type specific or generic flight dynamics (but must be representative of aircraft performance). On board instructor's station. Significant sounds. Control of atmospheric conditions. Navigation Data Base (sufficient to support aeroplane systems). Adequate test capability. Primary flight controls which control the flight path and be broadly representative of aeroplane control characteristics.	Suitable for: - Systems management initial and recurrent training, checking and testing (except for pilot manual control handling skills). - CRM training as part of an approved course. - LOFT (Route and area familiarisation only where at least a Level A simulator visual system fitted).

Technical Requirement and Credits

Qualification Level	General Technical Requirements	Maximum Credits
FNPT 1	Generic Package Non Motion Non Visual Instructors Control (refer to AMC-FTD 007 for detailed requirements).	Suitable for instrument flight general handling and navigation procedures for training in the following courses: - Airline Transport pilot (A) integrated course. - Commercial Pilot/Instrument rating integrated course. - Commercial Pilot (A) integrated course. - Commercial Pilot modular course - Instrument rating single-engine course. - Instrument rating multi-engine course - PPL (A) course. - Ab-initio Instrument Flight experience (Approx. 20 hours). The maximum number of hours which may be credited for the issue of a license or rating is specified in JAR-FCL. No other Training or Checking Credits.
FNPT 11	Visual System Generic, Class or type Package	As for Type 1 plus: - Airlines Transport Pilot (A) integrated course, phase 5 (MCC-course) including checking. - Ab-initio Instrument Flight experience (Approx. 40 hours). - Crew co-operation Training. No other Training or Checking Credits.

FNPT 1

Device	Minimum Technical Requirements	Maximum Credits
FNPT Type 1	1 A cockpit/flight deck sufficiently enclosed to exclude distraction, which will replicate that of the aeroplane or class of aeroplane simulated and in which the switches and all the controls will operate as, and represent those in, that aeroplane or class of aeroplane. 2 Instruments, equipment, panels, systems, primary and secondary flight controls sufficient for the training events to be accomplished must be located in a spatially correct flight deck area. 3 Lighting environment for panels and instruments sufficient for the operation being conducted. 4 In addition to the flight crew members' stations, suitable viewing arrangements for the instructor must be provided. These must provide an adequate view of the crew members panels and station. 5 Effects of aerodynamic changes for various combinations of drag and thrust normally encountered in flight, including the effect of change in aeroplane attitude, sideslip, altitude temperature, gross mass, centre or gravity location and configuration. 6 Navigation equipment corresponding to that of the replicated aeroplane or class of aeroplanes, with operation within the tolerances prescribed for the actual airborne equipment. This shall include communication equipment (interphone and air/ground communications systems). 7 Control forces and control travel shall broadly correspond to that of the replicated aeroplane or class of aeroplane. 8 Complete navigational data for at least 5 different European airports with corresponding precision and non-precision approach procedures including current updating within a period of 3 months. All navigational aids should be usable, if within range, without restriction and without Instructor intervention. 9 Engine sounds shall be available. 10 The following shall be available: a) variable effects of wind and turbulence. b) hard copy of map and approach plot. c) provision for position freeze and flight freeze. d) Instructor controls necessary to perform the training task. 11 A Qualification Test Guide which shall be submitted by the Operator in a form and manner that is acceptable to the competent Authority and which conforms to AMC STD 3A.030 (para 1.6). 12 Stall recognition device corresponding to that of the replicated aeroplane or class of aeroplane. See also Note 2 below Table 3.	Credits in accordance with JAR-FCL. (In order to be used for aeroplane type or class-specific training, testing and checking, the device must also be qualified as a Flight Training Device (FTD) or Flight Simulator).

FNPT II

<table>
<tr><th>Device</th><th>Minimum Technical Requirements</th><th>Maximum Credits</th></tr>
<tr><td>FNPT Type 11</td><td>As for Type 1 with the following additions or amendments:

1 The flight deck, including the instructor' station, shall be enclosed.

2 Circuit breakers shall function accurately when involved in procedures or malfunctions requiring or involving flight crew response.

3 Crew members seats shall be provided with sufficient adjustment to allow the occupant to achieve the design eye reference position appropriate to the aeroplane or class of aeroplane and for the visual system to be installed to align with that eye position.

4 A generic ground handling model shall be provided to enable representative flare and touch down effects to be produced by the sound and visual systems.

5 Systems must be operative to the extent that it shall be possible to perform all normal, abnormal and emergency operations as may be appropriate to the aeroplane or class of aeroplanes being simulated and as required for the training. Once activated, proper systems operation must result from system management by the crew member and not require any further input from the instructor's controls.

6 The Instructor's station shall include the following controls:

a) representative crosswinds.
b) a facility to enable the dynamic plotting of the flight path on approaches, commencing at the final approach fix, including the vertical profile.

7 Control forces and control travels which respond in the same manner under the same flight conditions as in the aeroplane or class of aeroplane being simulated.

8 Aerodynamic modelling shall reflect:

a) the effects of airframe icing.
b) the rolling moment due to yawing.

9 Significant cockpit/flight deck sounds, responding to pilot actions, corresponding to the aeroplane or class of aeroplane being simulated.

10 A visual system (night/dusk or day) capable of providing a field-of-view of a minimum of 45 degrees horizontally and 30 degrees vertically, unless restricted by the type of aeroplane, simultaneously for each pilot, including adjustable cloud base and visibility. The responses of the visual system and the flight deck instruments to control inputs shall be closely coupled to provide the integration of the necessary cues.</td><td>Credits in accordance with JAR-FCL.

(In order to be used for aeroplane type or class-specific training, testing and checking, the device must also be qualified as a Flight Training Device (FTD) or Flight Simulator).</td></tr>
</table>

FNPT II MCC

Device	Minimum Technical Requirements	Maximum Credits
FNPT Type 11 MCC	For use in Multi-Crew Co-operation (MCC) training - as for Type II with the following additions or amendments: 1 Turbo-jet or turbo-prop engines. 2 Performance reserves, in case of an engine failure, to be in accordance with JAR-25. These may be simulated by a reduction in the aeroplane gross mass. 3 Retractable landing gear. 4 Pressurisation system. 5 De-icing systems. 6 Fire detection / suppression system. 7 Dual controls. 8 Autopilot with automatic approach mode. 9 2 VHF transceivers including oxygen masks intercom system. 10 2 VHF NAV receivers (VOR, ILS, DME). 11 1 ADF receiver. 12 1 Marker receiver. 13 1 Transponder. The following indicators shall be located in the same positions on the instrument panels of both pilots: 1 Airspeed. 2 Flight attitude with flight director. 3 Altimeter. 4 Flight director with ILS (HSI). 5 Vertical speed. 6 ADF. 7 VOR. 8 Marker indication (as appropriate). 9 Stop watch (as appropriate).	MCC credits in accordance with JAR-FCL.

C545/062/98

The CMC Leopard project

I CHICHESTER-MILES
Chichester-Miles Consultants Limited, Welwyn, UK

OUTLINE

The CMC Leopard project for a 4-seat 500 mph twinjet was launched by the British company Chichester-Miles Consultants Ltd., (CMC) in 1982.

The program had a clear commercial objective: the profitable marketing of a 'minimum executive jet' through the early decades of the 21st Century.

A very extended low-budget Research and Development program, including flight tests on two prototype/development aircraft from 1988 onwards, has been aimed at refinement of the airframe pending the availability of the definitive turbofan engine.

The second development aircraft currently flying in the UK (Figs 1 and 2) is installed with two Williams International FJX-1 engines of 700lb. nominal takeoff thrust, and will be retrofitted with FJX-2 turbofans when these are available for experimental flight.

THE DEVELOPMENT PROCESS

In preparing this paper, it has been suggested to me that, although the commercial and marketing context should be explained, the main interest is likely to be in the technical or more exactly technological aspects of the Leopard project.
This presents an opportunity to describe a development process which is quite familiar to our small R&D team associated with the work over at least a decade. The process (Fig. 3) clearly includes a number of activities extending from those which are very close to marketing at one end to the most detailed technicalities at the other.
In simple terms, it must start with the Concept: the choice of basic seating capacity and

accommodation standard, the performance targets, and the consequent target price. Secondly, what we might call the Architecture: the size, style, and configuration of the aircraft. Contemporary aerodynamic standards are highly influential here. Then we move into the broad Design Definition: the layout of structure and systems and a closer look at aerodynamic detail, piloting requirements, and ergonomics, and design for production and servicing, leading on to design and analysis and decision-making at the most detailed stage. Technology, on this model, is the fourth and central element, continually interacting with the others.
It is an important, but not necessarily the dominant element in a project which relies as much on the marketing, costing and traditional design judgements which have had to be arrived at by a process only slowly changing over the last twenty years.

THE CONCEPT

The initial Leopard concept emerged from the author's contacts with Dr. Sam Williams of engine manufacturers Williams International in Michigan, in the mid-70s, on acoustic research proposals from British Aerospace Hatfield (previously De Havilland Aircraft). The proposals, for a twin-turbofan adaptation of the single-engined Student jet trainer, (Fig.4), based on the Williams WRI9-3 turbofan of some 600–700lb static thrust, were not proceeded with. But, on its establishment in 1978, Chichester-Miles Consultants Ltd., (CMC) turned its attention to the commercial possibilities of small civil aircraft powered by the WR19-3 and quickly came to the conclusion that these engines, in a twin installation, could in principle make a 5:1 downscaling of the contemporary executive jets (as exemplified by the DH 125 and Learjet23 of the time) a real possibility. The research team and its advisers also foresaw the beginnings of a potential demand for such a 'minimum executive jet'.
At that time, twenty years after the decisive introduction of the jet transport onto the World's airlines, air travel had largely become jet travel.
In contrast the vast field of general aviation, based as it was largely upon the piston engine, was perceived as increasingly anachronistic.
Particularly in the 4–6 seat category a real market opportunity seemed to be emerging for a new, very small, personal/business jet.

The market thinking was in fact reminiscent of that which led – through a similar downscaling process – to the executive jets, such as the DH125, of twenty years earlier. But of course, the aircraft size was now falling towards a size bracket as small, or smaller, than that of an earlier generation of light aircraft pioneered by companies such as De Havillands in the 1930s. In recognition of that, the new project adopted, at least partly, the name of the Leopard Moth of 50 years before.

Around 1980, when the Leopard concept was arrived at, the 4-6 seat segment of the general aviation market was not only the most popular (accounting for some 55% of the world-wide GA fleet) but it included an element of high-performance single and twin-propeller aircraft (perhaps 10% of the fleet) which were sufficiently expensive and productive for their owners to consider eventual replacement by small jets in the \$ 1. 0m–\$ 1.5m bracket. (1998 dollars).
However, the marketing approach was not only to encourage a move upmarket by the propeller community, but also to offer a true 'minimum executive jet' for those who needed the performance of a Learjet but not its seating capacity or the associated costs. These considerations defined the basic concept:

- 500 mph cruise speed

• A selling price (1998) of around $ 1.35m

Realistically, the specific price of such an aircraft was unlikely to be less than $ 600/lb empty weight, so that the project target became 2200lb empty, with a design gross weight around 4000 lb.

These speed, price and weight targets together implied a carefully-packaged four seat aircraft of very compact dimensions.

Wing loading and hence stalling speed follow the conventional executive jet formula and this defines the other main characteristic of the concept: twin-engined propulsion. For civil aircraft, both propeller and turbofan, single-engined layouts must demonstrate adequate crashworthiness following engine failure, and this probably implies stall speeds (at maximum T/O weight) in the 60-75 knot range. True executive jet performance however implies T/O stall speeds in the 80-100 knot band. The virtual elimination of total powerplant failure therefore requires two engines (or more) and the Leopard concept follows normal practice in this respect.

MARKET PROSPECTS

The projection therefore is for a carefully-designed and very compact 4-seater 500 mph aircraft which, in production form could sell (at today's money values) at around the $1,300,000 - $1,400,000 mark. (See Section 10)

We are not aware of any direct competitor. Indeed, in the absence of any aircraft at all in this size/speed category, the market itself is a potential, not a visible one. This is therefore a new product: will it generate a new market?

We can have no real evidence on the point. But evidence has to be at least suggested in as logical a way as possible. We have tried to do this ourselves and we have tried to involve others. consultants, agents, distributors, prospective purchasers. .with a proven expertise in the present marketplace.

There is no guarantee about any of this, but some kind of general view has developed over the fifteen years or so since, as the first tentative step in the Leopard programme, we introduced people to an early mock-up of the design.

Generally the response from the General Aviation community has been positive. In addition there has been a not insignificant military interest.

The concept, the styling, the projected performance and price (See Fig.8) have been received with interest, in some cases with enthusiasm. On the other hand the response has not been uncritical. The accommodation and passenger access are frequently seen as very 'sporting', limiting the clientele to the more dashing and image-conscious archetypal Ferrari or Porsche owner.

And we would not disagreed markedly with that kind of assessment in that, although its unique position may attract a wide spectrum of purchasers both civil and military, the best sales prospects lie in the civil personal/executive category with successful 30-40 year olds for whom performance and style is both a business and a leisure asset.

Consequently the Leopard must be viewed, like all high-performance private aircraft, as something of a specialist product. Its penetration into the civil general aviation market is

likely, if measured in percentage terms, to be modest.
However, by the standards of larger commercial aircraft, the current general aviation fleet world-wide is very large, over 300,000 units in all and even a modest penetration into the replacement market for a limited category can result in quite healthy levels of production.

And all the evidence shows that technological advance can produce decisive market penetration, or market shift, in established sectors.
A good example lies at the higher end of the general aviation size spectrum, say 10 seats or more (including crew). where over the past 30 years or so the dominant proportion of the fleet has moved steadily to the turboprop and the turbofan. A similar trend seems now to be underway in the 6-10-seat category, with a mix of relatively inexpensive turboprop singles and twins (Meridian, TBM7OO, KingAir C9OSE) new Citationjet, Premier, and J30 turbofan twins and, on the horizon, the innovative single turbofan Vantage.

With the development of the 'small jet' technologies exemplified by the Leopard, there are real prospects of 'trading up to a jet' for current operators of light twins (e.g. Baron) and high-performance piston singles (Mooney, Malibu).
With well over 30,000 aircraft in this more upmarket sector of the 4-6 seat category, a progressive move from the current piston-propeller fleet to a new fleet of turbine-powered aircraft, including a sizeable proportion of turbofans, seems highly probable for the early decades of the next century.

The prospect is likely to be attractive to a number of manufacturers: the unique status of the Leopard seems unlikely to persist beyond the end of this decade. Nevertheless, the market advice given to us over 15 years ago, for a potential level of sales in the primary US market of some 100 units per year for the Leopard and any direct competitors does not seem unrealistic for the period 2000-2030.

AIRCRAFT CONFIGURATION: THE 'ARCHITECTURE'

The concept of a 500 mph, M=0.76 twin-turbofan 4-seat personal/executive jet weighing empty no more than about one tonne, is a very demanding one. The general layout or architecture is determined, as always, by aerodynamic and structural considerations, but these are particularly influenced by scale effects.

In the general aviation field, there has been much interest, over the last ten years or so, in somewhat innovative configurations, particularly for higher performance designs, including turbofans. A major aim of many of these has been the offsetting of the main wing centresection structure to pass aft of the cabin, preserving adequate depth in a space which is particularly limited on small aircraft.
One arrangement has featured the canard, or foreplane, layout: another, forward sweep. Both are effective in achieving a clear cabin space.
Another area of innovation has been in the powerplant empennage relationship.
Buried and semi-buried engines in association with differing tail heights.. and V-tails.. can produce attractive aesthetics and/or low wetted area.
In another configuration area.. landing gear positioning and retraction.. recent small-aircraft trends have followed the military in the adoption of fuselage-mounted (and in-fuselage retraction) main gears of modest track.

In contrast the Leopard configuration has followed a more-or-less conventional pattern. (See Fig.5). The wings (A) are of a sweepback, thickness and aspect ratio which takes full benefit from transport aircraft aerodynamic research over the last 30 years but which maximises cabin length by resort to a concentrated spar, rather than a continuous box beam, across the centre-section. Similarly the empennage surfaces (B) are of conventional sweptback form, the tailplanes (J) being very low set for strong longitudinal stability at and beyond the stall.

The engines (C) are installed, in conventional executive-jet fashion, in separate pods mounted off the rear fuselage, giving ease of access and, in the longer term, retrofit. Intake location is as far back as possible, (consistent with adequate separation from both wings and tailplanes), to reduce cabin noise. Engine height is dictated by vertical clearance of intake flow over the inner wing, and of the exhaust plume over the low tail.

The fuselage is tightly-packaged and encloses a volume only just sufficient for four occupants, baggage, fuel, avionics and other systems. This determines the location of the main landing gear(D) well out on the wing, with retraction into the inner wing (F), rather than encroaching into fuselage space.
On the ground the fuselage is relatively low-slung, subject to adequate tail clearance for rotation, to ease passenger access. However, main landing gear length, defined by energy absorption requirements, fixes wing height relative to the ground such that the configuration emerges as a mid-wing one. This does, of course, tend to minimise wing/fuselage interference effects, improving flap effectiveness and underfuselage drag, in comparison with the very low wing mounting position adopted on many contemporary cabin-class executive jets.

In contrast to those larger aircraft, a somewhat unusual feature of the basic Leopard configuration is the adoption of a non-circular fuselage cross section(F). It is accepted that this must entail a penalty in structural weight on an aircraft pressurised to a high differential. However, for comfortable accommodation the classic circular cross section gives rise to a larger fuselage frontal area and also a proportionately longer fuselage length, since a minimum length/diameter ratio is fixed by high-Mach No. flow conditions. As with many light aircraft, and most military types also, minimum fuselage drag appears to go with a closely-wrapped non-circular fuselage cross section.

The last, and perhaps most controversial feature of the fuselage configuration is also reminiscent of military practice: the hinged canopy(G).
Comfortable access to two seat rows in a very shallow fuselage is. as it is with the equivalent sports car very difficult to arrange. On the Leopard the structural implications of conventional or unconventional side doors, allowing for pressure as well as flight loads, would be formidable. With the hinged canopy the structural design is perhaps simplified, and with tight latching, in-flight strength and stiffness is good. But pressure loads are very high at the design pressure differential of 9.6 psi.
The complete canopy is large and heavy enough to require powered actuation, but opens to a wide angle to give relatively unimpeded access to occupants (including on occasion, stretcher cases) and other loads.
The whole assembly, with bonded-in load-carrying transparencies, is a structure of some complexity and has needed careful analysis and manufacture.

SCALE EFFECTS

These last are particular examples of features which fit the requirements of a compact, tightly packaged design. A small efficient high-performance aircraft is a considerable design challenge because of the problems of scale itself. The Leopard could never be a scale model – say a 40% linear scale of a BAe 125 – .since, although weight and drag could be about right, real people could not be accommodated in a fuselage diameter reduced to about 31 ins. Volume for fuel and for avionics similarly shrinks in relationship to surface area (which roughly determines airframe weight and drag) in line with the fundamental 'square-cube' law. This was perhaps the most pressing design problem which was only resolved, and that not completely, by the settling around 1980 upon a very tightly-packed configuration with accommodation and styling of a particularly sporting Ferrari-like, character. The design is so densely-filled that it would be difficult to see how one might achieve the basic specification to carry four people and baggage over 1500 nm at M=0.75 out of a 3,000 ft field, with anything smaller or lighter.

The purchaser is not of course directly interested in smallness and lightness, but he is interested in low purchase price and low operating cost. Broadly these tend to go together, and both reflect aircraft size and weight.

However, simplicity is equally important in the control of production cost (and hence purchase price) and in attaining reliability and ease of maintenance, all of which directly affect the cost of operation. And here, small scale offers potential advantages.
Small aircraft offer good scope for design simplification, since the fundamental square/cube effect which would make small scale structures and systems disproportionately light, can be 'traded in' for less complex and sophisticated designs at a weight which is still quite acceptable.
The Leopard is a typical example in its relatively simple near-monocoque structure, manual pushrod flight controls and very simple electric retraction, flap, and trim systems (there are no powered hydraulics).

These features are typical of the kinds of design features which are found in low-performance light aircraft generally. They result from the emphasis upon low cost and the opportunities of small size and are as directly applicable to a very small high-performance aeroplane as to the conventional low-performance one.

The search for simplicity on the Leopard has, however, generated some design features which are less conventional. There are three interesting examples.(Fig. 5)

Firstly, in flap design. The maintenance of a practical landing distance (at the aircraft's typical landing wing loading of 55 lb/sq.ft) of less than 3,000ft could imply quite complex flaps, lift spoilers and possibly thrust reversal. Instead, we have gone for simple full-span large-chord plain flaps(H) with a very rapid flaps-up lift-inversion/drag capability for maximum wheel downloading and enhanced braking. These flaps are also available for airbraking across the whole flight envelope.

Secondly, and following on from the full-span flap concept, both roll and pitch control are vested in high-aspect-ratio slab-type taileron surfaces(J). The all-moving formula (shared by the fin) is structurally simple and stiff, and exceptionally clean aerodynamically.

A third example is in the undercarriage retraction system. On such a small and 'shallow' aircraft, it requires little imagination to see the danger to all occupants of any wheels-up landing. Accordingly leg extension is always by gravity, drag and spring bias. Retraction only requires actuation, by a single electrical unit(K) and tension cables (a disconnect feature covering both power failure and actuator jamming).

Returning to technologies directly related to scale, during the development period of the aircraft a number of technical advances have been coming together which are highly-favourable (and in some cases essential) to very small civil jet aeroplanes.

An obvious case in point is avionics. Miniaturisation of electronics has been gathering pace at a rate that already allows a full all-weather fit (including EFIS) in the very limited space (and the correspondingly limited weight and cost budgets) of the most compact four-seater.

On the aerodynamic front, in 1980 the understanding of natural laminar flow (which develops most extensively on small aircraft) and of transonic flow computations, led to the first known example by ARA Bedford of a 3D wing for Leopard exploiting both types of flow. The drag reductions represent a very useful bonus from downscaling, which are further enhanced by the remarkable aerodynamic cleanliness which can be achieved using the thick-skin composite structure (GRP & CRP) adopted for the Leopard airframe.

The high quality finish of composite construction has been well-demonstrated; particularly on sailplane and 'homebuilt' light aircraft designs since the early '80s. The structure database filled out in the intervening decade with increasing use on major components both civil (Airbus) and military (Harrier).

Composite construction avoids the cost penalties of extensive compound curvatures, which tend to be a feature of small aircraft fuselages, and of the more sophisticated advanced wing designs. (See also Section 8)

In summary, the practicality of the concept for a very small jet aircraft such as the Leopard rests upon the solution of the fundamental problems of small scale, but can also benefit from simplified design which would be uneconomic in larger sizes and aerodynamic and structural technologies which are particularly advantageous at small scale.

But of all the small-scale/low cost technologies none has been as crucial to the evolution of the 'minimum executive jet' or will be as influential to the new small-jet category in the future, as the 'minimum turbofan engine' in the thrust class below 1000lb.

THE ENGINE STORY

Although small-engine technology was the catalyst for the Leopard concept in the first instance, in fact the Leopard story to date has been one of a rather hesitant progress from the very limited capabilities provided by experimental engines towards the fully-certificated engine/aircraft combination essential for civil sales.

The Leopard has certainly not been the first aircraft to be launched at the prototype stage without a 'proper' engine. In the military and research fields this is by no means uncommon.

However CMC's decision to launch a prototype/proof-of-concept aircraft in 1982 was in the face of the inability of engine producers generally to supply units in the required 600-1000lb category, allowing only two much smaller engines seriously to be considered. These were the French Microturbo TRS18 of 220-292lb SLST and the British Noel Penny Turbines NPT-301 of 318lb SLST.
Although the former had been cleared for both the Microjet 100 and Caproni-Vizzola C22J light jet trainers, the rather higher thrust capability of the NPT-301 and convenient location of the NPT plant in Coventry swung the balance to the British engine.

The NPT-301 flew in the Leopard 001 prototype in December 1988, which over the next 2½ years was able to explore low speed handling and performance of the basic Leopard configuration, the aircraft being limited to a gross weight of around $^2/_3$ the design weight viz: 2700lb.

By mid-1989 NPT had joined GM-Allison in a submission of a turbofan engine, the GMA/NPT 1204 for the multinational MSOW standoff weapon programme, an engine programme in which CMC was very closely involved, and from which a civil variant de-rated to rather less than 1000lb SLST was planned under the designation NPT754.

The dismantling of the MSOW consortium was a foretaste of the progressive cutback in military programmes which was well underway in 1990, and was a factor in the commercial failure of NPT and its entry into Receivership in late 1991.

This was a major setback for CMC, since NPT had been relied upon for the eventual NPT754 and for the monitoring and support of the current short-life NPT3O1 S. With the failure of the engine company, all flight tests on Leopard 001 ceased: the aircraft has remained mothballed at CMC ever since.

Fortuitously it had emerged in September 1991 that updated versions of the American Williams WRI9 turbofan (on which, as has been explained, the early Leopard project studies were based) could be made available (under the designation FJX-1) for experimental flight trials in Leopard development aircraft 002.
These engines, at a nominal take-off thrust rating of around 700lb, are now installed in the 002 airframe and are powerful enough to produce a performance at least approaching that projected for the full production Leopard, at gross weights up to the design takeoff figure of 4000 lb.

Accordingly, we anticipate that Leopard 002, essentially a pre-production standard airframe, albeit installed with interim engines, will, over the next year or two, demonstrate something of the commercial capabilities of the design, in addition to covering most of the performance and handling requirements of JAR23.

There are nevertheless still some years to go before launch of the Leopard into the marketplace.
As in the past development programme, the timescale of that launch will be dominated by that of a fully-certificated long-life high-reliability civil turbofan at a thrust rating (below 1000lb SLST) suitable not only for the Leopard but also for complementary aircraft of a larger capacity which we foresee for the first decades of the 21st Century.

The interim FJX-1 engine has always been seen by Williams as the forerunner of such an engine, designated FJX-2, which will exploit that company's technical experience in small turbines generally and the commercial qualities of the well-received 1900lb thrust FJ44 in particular.
We would assess the appearance of such an engine as a certificated production unit as very probable in or around the year 2003.
Certainly CMC's very long-term commitment to the Leopard concept, and a conviction that aircraft in this general category are set to transform the standards of an important segment of General Aviation, must imply a similar conviction that the powerplants, on which such a transformation must be based, will be certificated in response to the emerging commercial demand.
And that conviction has been strongly reinforced, over the past few years, by the AGATE and GAP initiatives of NASA aimed at the radical transformation and revival of the whole American light aircraft industry.
From CMC's perspective, drawn from a 20-year R & D effort in the UK, these NASA initiatives and in particular the NASA/GAP support of Williams on the new FJX-2 turbofan, are seen as the absolutely crucial technological advance towards the new breed of small turbofan aircraft.

Apart from low weight, SFC and noise targets, a basic NASA objective with this engine is the achievement, by the most modern production techniques and very high production volumes, of unit costs reduced to a fraction of those experienced to date. Insofar as these costs are bound to be volume-sensitive, the dissemination of evidence, in favour of the new small jet class of aircraft, could be as important as the development of the aircraft technology itself.

In the longer term, normal competition in the field should stimulate demand, leading to the associated economies of volume.
However, with flight demonstration of the prototype FJX-2 engines scheduled for 2000, the timing and specification of the production engine will require a degree of co-operation amongst airframe manufacturers world-wide to demonstrate a high level of demand.
The Leopard development therefore continues to have important promotional functions as well as technological ones

THE TECHNOLOGIES OF INTEREST

Even on a small and relatively simple aircraft, the technologies of interest in aerodynamics, materials, systems, electronics, in design and analysis methods, as well as in propulsion – all these have been developing rapidly over the past two decades and can produce a reassessment within any of the elements of development, including that of the basic Concept.

This is hardly surprising: the original Leopard concept of around 1980 was itself heavily influenced by the technological advances of the time. Indeed in many respects the characteristics of the Leopard design were fixed by the technologies of the early '80s.

The outstanding example is the structural layout. Around 1982 a firm decision was taken to adopt an all-composite structure. The judgement was made against the background of increasing acceptance of composites on light aircraft and sail planes (and cars and boats) on the score of lower-cost manufacture in complex curvatures, fatigue and corrosion resistance, and an impressive surface finish. There were no great expectations, on the early glass-reinforced hand-layups, of weight benefits, particularly when allowance had to be made for production, environmental, and damage strength margins. However, a modest incorporation of carbon reinforcing helped to keep weight down. Carbon fibre composites certainly exhibit very high strength/weight ratios and stiffness (E) in the ideal state. But their sensitivity to production defects and damage has been judged by CMC to reduce their advantage over glass composites such that their higher costs cannot always be justified in areas. .such as the fuselage. .where stiffness is of subsidiary importance. Consequently the carbon fibre usage is at present limited to around 40% of the structure by weight. Apart from a small element of aramid (Kevlar) where minimum density is the criterion, the remainder is in glass composites.

However, carbon is the predominant material in both wings and tailerons.
On the former the concentration of wing bending material in the inner wing adjacent to the landing gear cutout, culminating in overlapping centresection (rear) spars, is achieved at acceptably low weight by careful outboard wing-box torsional/flexural 'tailoring' and by good damage-protection (and hence increased allowable stress) in the relatively solid spars themselves.
This particular composite design meets the configuration requirement for an uninterrupted cabin space and wing-mounted landing gear in a way that would be more difficult with a light-alloy structure.

A second example of composite 'tailoring' is in the design of single-piece full-span flaps on which a very high torsional stiffness is combined with good bending compliance, permitting actuation from the flap root only whilst maintaining flutter-free operation over the whole flight envelope. A more extensive use of carbon (and aramid) on 002 and subsequent aircraft, and the general progress in composite technology.. preferably with reducing 'superfactors' as experience builds. should lead in to useful weight advantages as development proceeds. Wet lay-up by hand ('bucket and brush') has proved a sensible prototype technique. Its modest capital demands in comparison with the higher-temperature and pressure autoclave techniques (or the even more expensive automated layup methods) have made this a feasible option where production volume is, or may turn out to be. .low.

However, recent medium-temperature, non-autoclave methods using prepreg materials present a further 'middle-of-the-road' process which we are considering for at least the early buildup stages of production.

These options exemplify the progressive development possibilities of composites which, in contrast to the radical design and production changes required in moving from metallics to composites.. a discontinuity having now to be faced across a broad range of both civil and military production could lead to a very long-term stability in technique equivalent to the 70-odd years dominance to date of the light alloy semi-monocoque formula.

At what date that across-the-board change might be made. .if at all. will be decided on a case-by-case basis, but in the light aircraft field there are signs that composite construction is already a real alternative to metals and may within the decade be the preferred option.
If this assessment is correct it has an important technical implication. In comparison with light alloy construction, the statistical data base, both in terms of aircraft build and more particularly in hours flown and maintenance cycles performed, is still narrow and this is reflected in a justified conservatism as regards strength margins.

This is highlighted in Fig.6 where the rather fanciful tensile strength/weight concept of 1'g' breaking length (for a vertically-hung rod of material) is shown for metallic and composite materials, current 'de-rating' of composites is only too apparent.

With experience, production quality will improve, less vulnerable design features will emerge, environmental effects will be better understood. Many, if not all, the variabilities will diminish together with the design factors to accommodate them, so that, as in the past, the wider adoption of the technology itself leads to its improvement: the technology becomes self-fulfilling.

Neither at its early concept stage nor throughout its extensive development to date has the selection of all-composite construction been beyond all question. Illustrations have been given of particular features of the Leopard design which would be difficult in metal. Certainly a metal airframe would be different. But even if the advantage of composites were today marginal their potential in the future looks set to outpace anything which metallic developments could offer at similar production cost.
Structural tests have already shown reserve factors approaching 2 on important parts of the Leopard airframe (i.e. overall factor of nearly 3 on limit loads) so that there may well be scope for really useful weight savings as the design is refined.

In summary a reasonable standard of composite technology, particularly suited to the design of the Leopard and, not least, to the limited design and manufacturing resources to hand, is in place, and should show significant improvements over the period ahead.

As regards the other technologies of interest, some, such as miniaturised low-cost avionics emerge from continuing military and light aircraft programmes, particularly in the USA.
Others, on the other hand, though well-enough developed and available for aircraft of larger size or lower 'density' sit uneasily within the Leopard's dimensional, weight and cost restraints.
In three areas a new technology standard would be, if not absolutely necessary, then enormously beneficial.

These are:

(a) Electrics
Despite miniaturisation of avionics, the weight and bulk of batteries, cabling, relays, etc., are particularly significant on a small, but relatively sophisticated, aircraft. Current power systems are electric rather than hydraulic and it is expected that developments in actuator size, weight, cost and safety provisions will result in a much superior system to the interim one installed today.

(b) Airconditioning
Although not yet installed in Aircraft 002, all ECS ducting is in place and space provision made for a combination bleed-offtake/heat exchanger system associated with vapour-cycle refrigeration.
However, the concept of a high-flying and very small turbofan aircraft is some way away from the design points of light aircraft systems now available.
As the small-turbofan sector develops, innovative turbofan-based ECS systems could command an expanding market.

The automotive industry, with its widening systems requirements, very high production volumes and associated reliability and cost benefits is, from the standpoint of the Leopard design seen as the possible source of a number of technologies, including the two above.

(c) Brakes, wheels and tyres
Current mainwheels and tyres are of very small diameter, consistent with stowage within an acceptable cutout in a very small wing.
Tyres are by Dunlop, running at a 'large executive jet' pressure of 170 psi.
Wheels and brakes are, however, standard or near-standard items, relatively bulky when retracted, due to a standard light aircraft 'side disc' arrangement.
A more advanced and less bulky design, based possibly on military fighter technology, might improve both braking and aerodynamic performance.

THE RESEARCH AND DEVELOPMENT PROGRAMME

The Leopard R & D programme has been, and still is, determined by a number of factors which differ from those experienced on larger programmes mounted by larger aeronautical companies.

These include:

(a) Exclusively private financing
(b) A very extended timescale set by outside suppliers.. specifically engine suppliers.
(c) Anticipation of, rather response to, technological and market developments.

All aeronautical investment is long-term, and a 20-year (1982-2002) R & D timescale, consistent with a 1980's 'early start' and a 2000+ new engine availability, could only be financed at a relatively modest annual spend rate. For technical work, these conditions are not all unfavourable. A long timescale allows a less-hurried sequence of decision-making: a small development team eases communications, flexibility and learning. But small resources give little margin for failures and setbacks. Technological anticipation.. innovation. carries a relatively high risk. There is however a case that we can control or reduce this risk if we can integrate design, manufacture and particularly testing into one combined effort, involving every member of the development team in all three. This is practical with a small outfit but perhaps not with a large one.

Of the three functions, testing is the essence of an R & D programme and the Leopard programme has been characterised by structural, systems and flight testing on two prototypes (001 and 002) over 14 years to date.

Much of this test programme has been typical of prototype work generally: a mix of ad hoc 'design proving' with demonstrations of compliance with Airworthiness Requirements (Jar23 and FAR23).
Points of interest include:

Structural Testing
The major prototype structures. effectively the whole airframes of both 001 and 002 were 'proof tested' several times before flight at 1.25-1.50 times limit loads. These factors (metallic

aircraft require only 1.00) took account of variabilities in the composite elements (due to manufacture, defects, temperature and moisture) without driving the metallic elements, such as spigots, bolts, etc., beyond the yield point.

In addition, ultimate tests to destruction were carried out on a number of vital components (main spars, canopy latches, etc) achieving factors of 2.50-3.00 on limit load.

No high-cycle repeated-load (fatigue) or pressure tests were performed on either aircraft. Such tests.. and/or residual static strength monitoring. will be required for clearance of the production machine, and a procedure decided upon to accommodate the possible variabilities in composites without overstressing of the non-composite parts of the structure.

Systems Testing

The functional testing of fuel, electrical and control systems has followed conventional patterns

Aeroelastics

001 flutter clearance was based upon Ground Resonance tests at COA Cranfield, analysed by standard BAe programmes at BAe Hatfield. The clearance on 002, provided by Stirling Dynamics at Bristol, builds on the current Stirling programme. Present (June 1998) clearance is for the designated Vc=250 KIAS up to medium altitude, with progressive enlargement of the envelope as flight tests proceed.

Flight testing

The flight tests of 001 were carried out from the previous RAe airfield at Bedford, and at Cranfield. The latter has been the site for tests on 002. In addition flight displays were mounted at a number of sites, including SBAC Farnborough '90 on 001, and Oshkosh '98 on 002.

All flight tests to date have been confined to 250 KIAS and 25,000 ft altitude and have investigated low-Mach No. handling and performance.

A comprehensive flight instrumentation has been partly installed to date.. including a test nose boom. .and flow-visualisation tests made with still photography (to be augmented by video recording).

However, the main outputs of interest have been pilot's assessments of the handling qualities as likely to be perceived by private pilots of average ability.

In general stability and control at takeoff, in the cruise and in approach and landing are satisfactory. Braking and lift dump on landing are effective and general ground handling and manoeuvrability good.

However, at its present stage of development the aircraft exhibits a typical swept-wing wing-drop at stall, both clean and with flaps and this must be corrected. Gear extension effects need to be alleviated, and further development is required on low-speed roll control and on control harmonisation.

A number of the configuration objectives, however, appear to have been attained, including a good ride quality from the relatively highly-loaded swept wing, low noise levels from aft-mounted engines and strong stall recovery from low-mounted tailplanes, and forgiving touchdown stability from the wide-track landing gear.

Flight testing now moves on to higher Mach Nos. and altitudes. In view of the aerodynamic innovations incorporated into the Leopard design (wing geometry, all-moving manually-controlled empennage surfaces etc) this testing will proceed, over a still extending period, in gentle small steps.

All initial testing on 001 was carried out by Angus McVitie, soon joined by Chris Chadwick. The latter now operates as CMC Chief Test Pilot, supported by Chris Yeo.

COSTS AND THE MARKET DEMAND

The Leopard R & D programme has naturally been focussed upon technical issues. .the proving of the technology, the testing of the structure, the demonstration of flight performance and so forth.

But, as important in a competitive and highly price-dependent marketplace is the reliable assessment of costs and prices.
How much will a production Leopard cost?
What will be a reasonable selling price?
How sensitive will the market be to price variations?

In this paper a number of references have been made to technical choices which have been much influenced by considerations of cost. A target datum price of $ 1,350,000 has also been set.

This figure is highly sensitive to prices charged by the suppliers of items. engines, avionics, electrics, ECS etc. .which in total are estimated to account for some 70% of the basic production cost. Nevertheless the development programme to date has generated useful data on the possible variation of these 'bought in' costs, and in particular realistic statistics on early 'airframe' costs, the associated learning curve, return on investment and overhead costs generally.

The datum target price of $ 1,350,000 appears reasonable (at mid-1998 prices) on the basis of this evidence and on the assumption that bought-in costs do not depart too much from the current statistical norm.
Is this a sensible and marketable price?

Fig. 7 presents a variation of market price trends, based upon existing aircraft in the US fleet, varying with cruising speed and cabin spaciousness (as indicated by cabin depth H). The trend lines are normalised into price/seat (including crew) and suggest that the datum Leopard target price should be at least consistent with current market indications.
It is possible that the price could vary significantly from the datum figure, not only, as is all-too-common upwards, but also.. in response to the NASA/Williams aims of low-cost turbofan engines. .downwards.

A study by CMC takes account of the rapidly accumulating number of aircraft which might be candidates for replacement by aircraft such as the Leopard, as price is steadily reduced. Expressing this effect in terms of elasticity of demand-to-price, the index comes out as about -4 i.e. a price increase of 10% would reduce sales by 40% (and vice versa).

It must be admitted that some advisers would predict a much lower price sensitivity taking account of the 'exclusivity factor' which maintains demand as price is raised, in a number of 'luxury product' markets.

Such an effect might indeed enhance the profitability of a stylish product such as Leopard. It would not much reduce the pressures to minimise production costs both by design and by technology which is a theme of this paper.

FURTHER DEVELOPMENT

Current plans (Summer 1998) are for certification, production and sales launch of the definitive FJX-2-powered Leopard in 2003.

The programme over the intervening five-year period will include a range of technical and commercial activities directed to that objective as follows:

(1) Further research and demonstrations based on Aircraft 002.

We have already identified a wide range of flight test objectives which can be investigated on this aircraft using the existing FJX-1 interim powerplants. Many of these relate to handling, and to the requirements of JAR23/FAR23, but it is not intended to attempt Certification of this aircraft either with the current FJX-1s (which are in any case non-certificatable) nor with the new FJX-2s which may be retrofitted towards the latter part of the 5-year period.

Apart from Flight test, there are a number of useful advances and modifications which are planned to be made on this development aircraft, as indicated earlier.
These chiefly relate to systems, and to the upgrading of the aircraft to full pressurisation/ECS status for which structure and space provisions are already designed-in.

Functional and structural pressure testing will be part of the continuing 002 R & D programme.

(2) Design definition, construction, test and certification of production Leopard '003'

The present intention is to achieve an improved design by evolution from Aircraft 002 and by taking full advantage of the advances offered by the new FJX-2 turbofan engines. The process will be analogous to the development of 002 from the earlier 001, although the step is seen as a smaller one in view of 002's closer approach (e.g. in weight) to the likely production specification.
003 will, however, be aimed at more ambitious targets in airframe weight and cost, reinforcing the very large improvements in these areas expected from the new powerplants.

(3) Reconstruction of the Leopard company.

CMC in mid-1998 remains a privately-funded Research and Development company with resources sufficient to pursue a useful part of the 002-based programme outlined in (1) above. Progression towards the production aeroplane and its certification, and, in particular the setting up of production and service operation will require a major reconstruction or transformation of the company.
Investigation and negotiation on the options available will clearly form a major part of our activities in the near future.
The technical implications of these commercial activities are of course very significant: we would hope to maintain the compactness and economy of the continuing research but further design refinements aimed at ever-improving weight, performance and cost are likely to encounter a steepening 'cost/performance gradient'. .a law of diminishing returns.
The R & D programme since the early '80s has been within the capability of a very small company, but its full exploitation and the achievement of the highest technical standards for the 21st Century will need something of a redesign of the company itself.

Future Possibilities
It may not be unreasonable to look, in conclusion, at longer-term possibilities which might open up following.. or possibly even coincidentally..with the successful conclusion of the Leopard R & D programme and the production and sales of the aircraft described.

The range of possibilities might emerge from the following:

(a) The establishment of the small 4-6 seat commercial jet as a new category in General Aviation, in which the 4-seat Leopard will have made a useful contribution..

(b) The achievement, as a result, of high-volume production of the Williams FJX-2 turbofan in the sub-1000lb thrust class, and at relatively low price levels.

(c) The maturing or certificated composite structures.. including Leopard fully exploiting the high strength/weight potential of these materials to attain a new low in structure weight percentages and airframe costs.

(d) The availability of a new generation of aircraft systems. electrics, avionics, ECS, power systems. suitable for the whole range of small civil jets.

Under these conditions CMC can foresee and has to some extent studied two aircraft possibilities closely related to the basic Leopard 4-seater civil aeroplane (Leopard Jet 4) in its production form.

The first is a direct 120% scale rearranged to seat 6 persons in a Malibu-like arrangement. There is considerable market interest in an aircraft of this class and adaptation of systems and powerplant from the smaller 4-seat Leopard could lead to considerable economies in weight and cost. Although some uprating of the basic FJX-2 turbofans would be beneficial, a 500 mph maximum cruise speed might be possible at the standard rating at medium altitude.

The second possibility is in the civil training role: a straightforward adaptation of the civil 4-seater with military-style canopy, side-by-side seating (with an ejector seat option) and

strengthened high-g structure.
Still an aircraft of great compactness, this concept, with the twin-FJX-2 powerplant unchanged from the Leopard Jet 4, could demonstrate considerable advantages over larger single-turboprop trainers in terms of performance, engine-out safety, and most remarkably, fuel consumption.

Both these derivatives of the basic Leopard theme have attracted interest, and exemplify options which might emerge from the basic Research and Development programme which still continues.

In a competitive market-driven world such 'variations on a theme' may be necessary fully to cash-in on the inevitable rise in R & D costs and the difficulty in maintaining a lead in any particular technology. For the benefits of technological development depend on its fairly widespread adoption. The point has been made with respect to the composite data base and its bearing on achievable strength/weight ratios: also on the need to 'share' a given engine technology in order to reap the economies of volume.

This paper has tried to give a picture of the technologies lying at the centre of a new small aeroplane. In the event its market prospects over the next 20 or 30 years will probably depend more on the attractiveness of the overall concept and the efficiency and soundness of engineering design.

Fig. 1 002 Development aircraft in flight

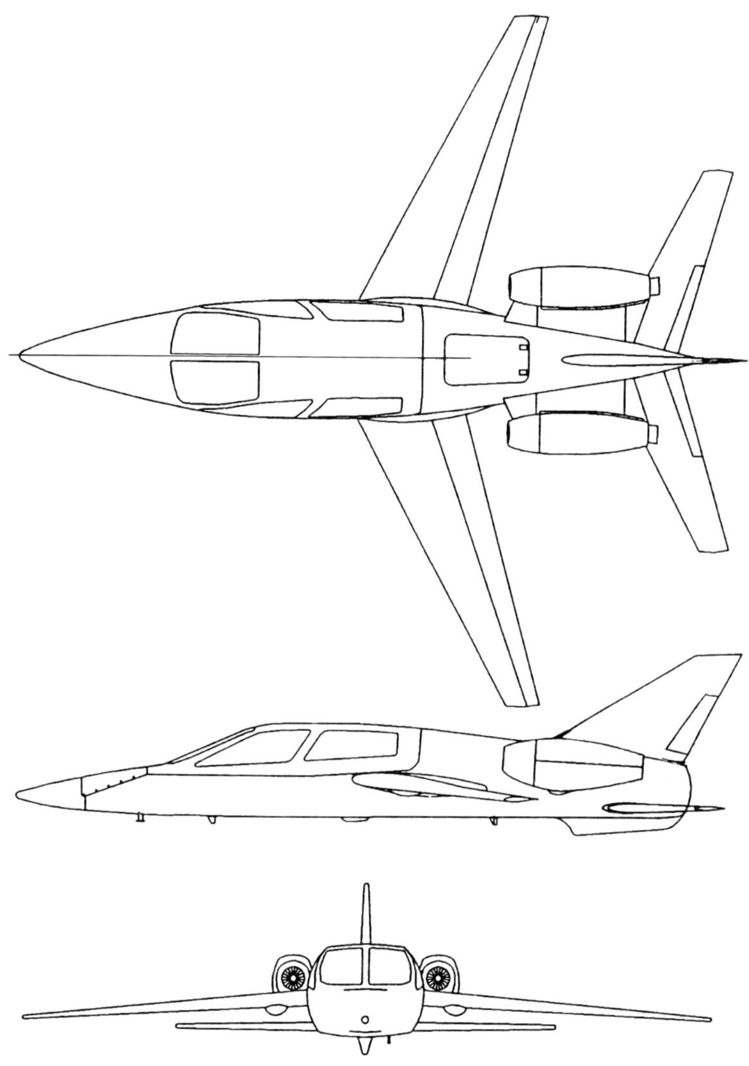

Fig.2 General arrangement

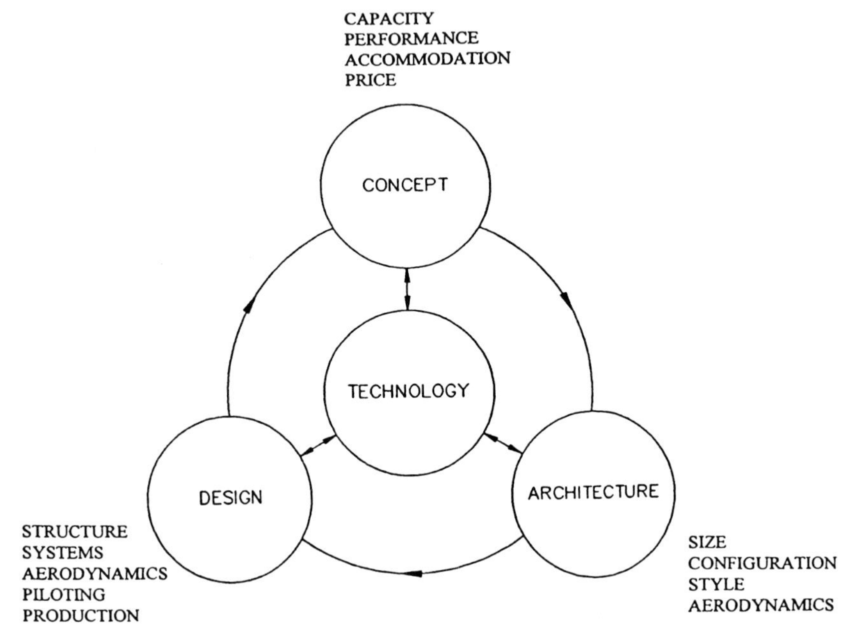

Fig.3 The development process

Fig.4 The 'student' jet trainer

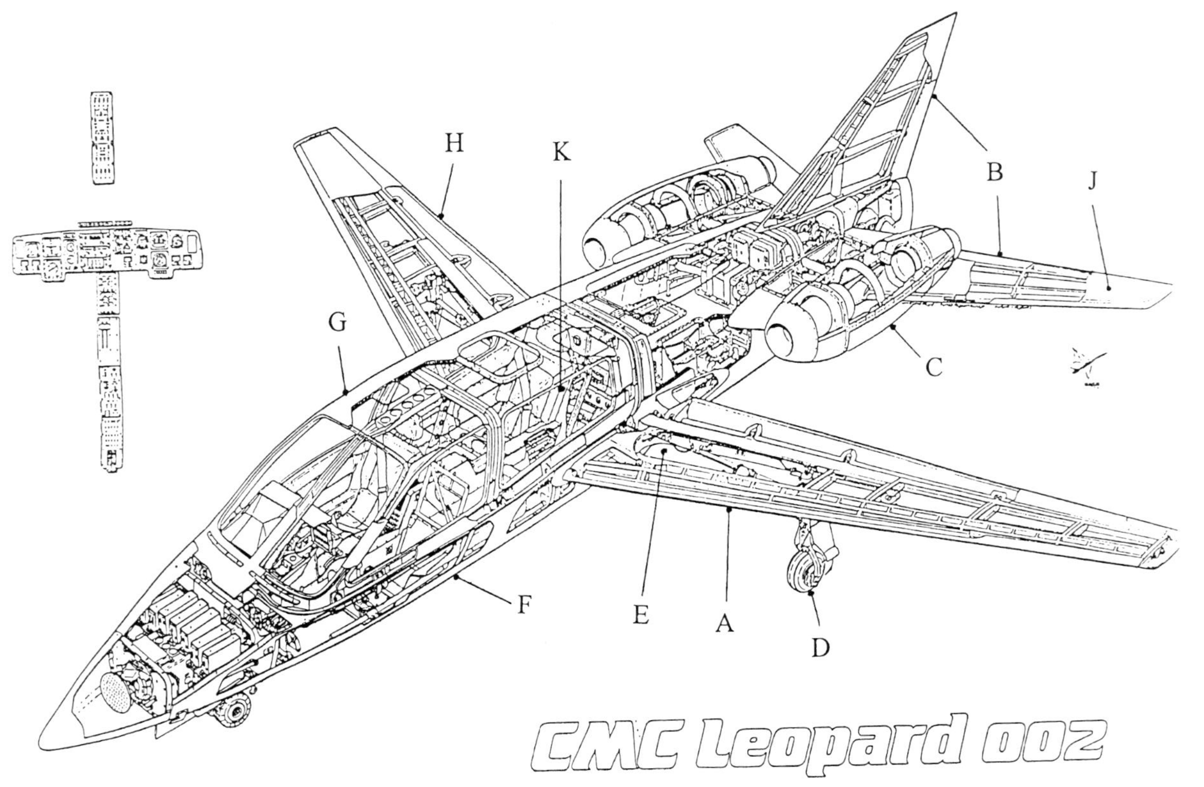

Fig.5 002 Cutaway drawing

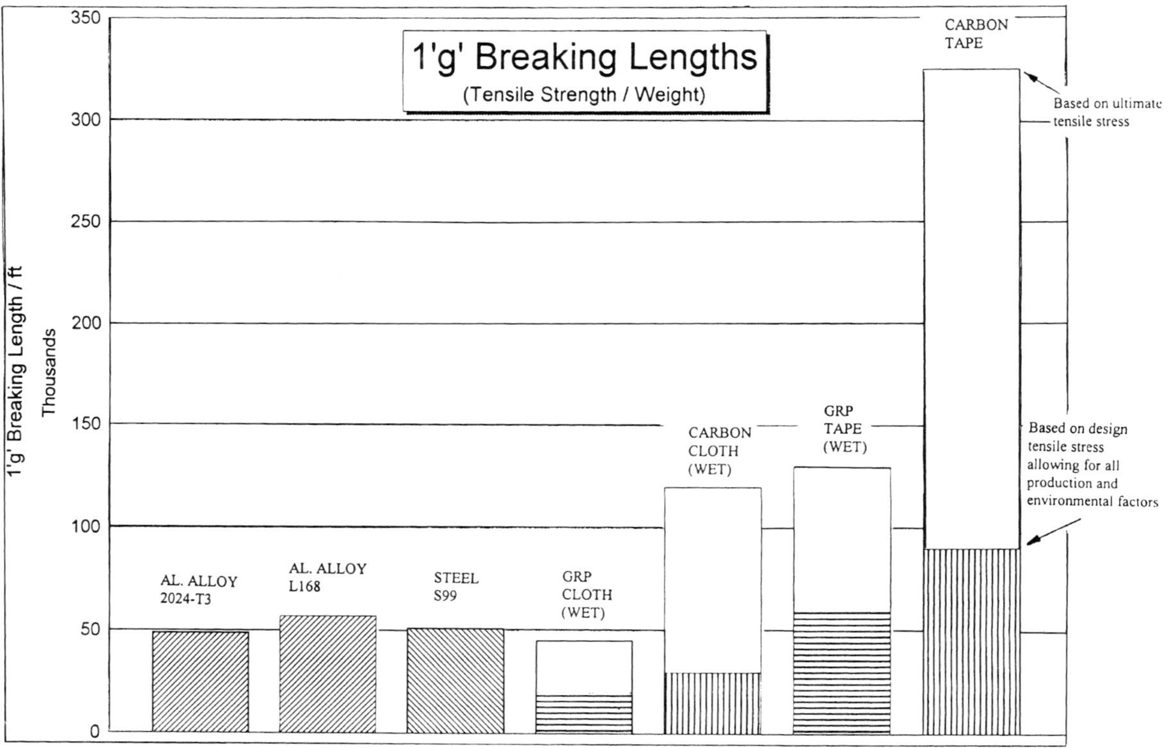

Fig.6 Strength/weight of various aerospace materials

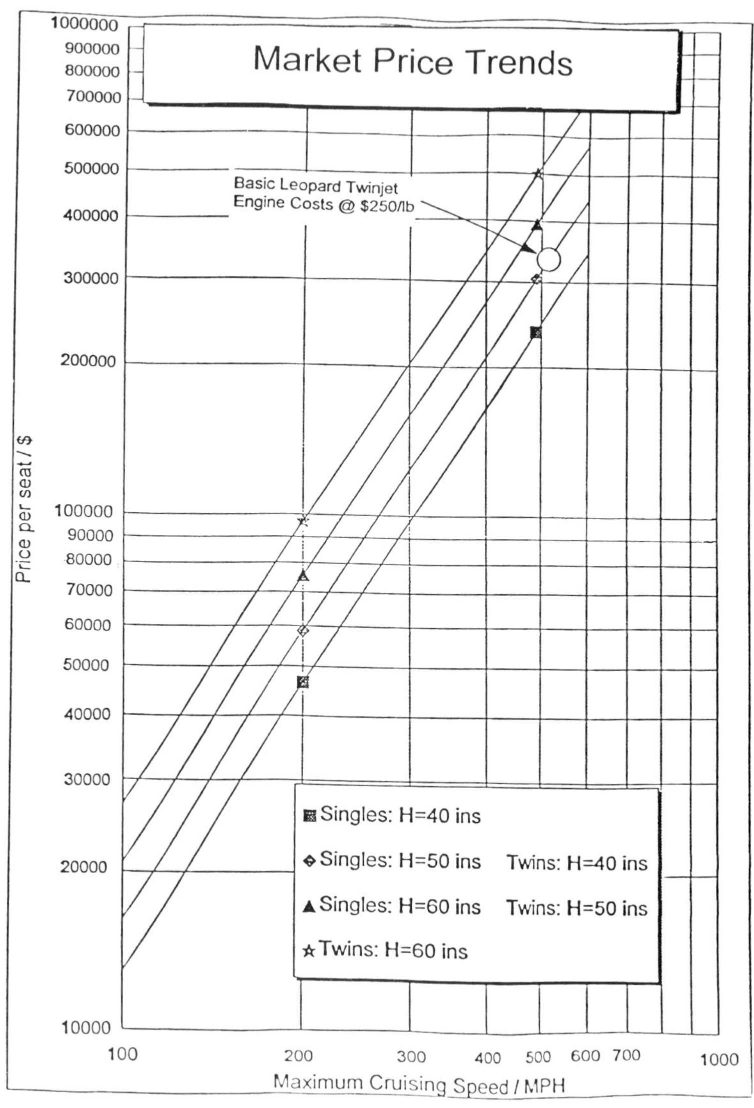

Fig.7 Market price trends

CMC Leopard

Development Aircraft 002

DESCRIPTION

- The CMC Leopard is a new high-performance 4-seat twinjet aircraft. It offers the personal and business user all the benefits of travel-on-demand in an efficient and economical size, but with all the smoothness and comfort of the high flying 500 mph executive jet. And for the discerning customer, aerodynamic refinement is expressed in a sleek and sporting form.
- Production aircraft will be certificated to FAR Part23/JAR23

Projected technical details are as follows:

- Performance: Cruise speed M 0.76 at 45,000 ft.
  Field length (SL,ISA) Takeoff 838m (2750 ft).
  Landing 747m (2450 ft).
  Range (full payload) 2775km (1500nm/1725sm).
- Dimensions: Wing Span 7.62m (25 ft 0 ins).
  Length 7.88m (25 ft 10 ins).
  Wing Area 6.00m² (64.5 sq ft).
- Weights: Empty equipped 998Kg (2200lb).
  Max T/O 1814 kg (4000lb).
- Structure: All-composite, using glass/aramid/carbon reinforcement. Landing gear: Main: oleo. Nose: rubber in compression.
- Systems:
  - Vapour-cycle airconditioning. Pressurisation from engine HP bleed. Cabin max diff. 0.65 bar (9.6 psi).
  - Electrics: 24V for U/C retraction, flaps, trim and canopy operation.
  - Avionics: Bendix-King EFIS with conventional standby.
  - Anti-icing: AS&T liquid system for wings and tailplanes
- Aerodynamics: Laminar/supercritical wing and empennage design by ARA.
  Pitch and roll control by all-moving low-mounted tailerons.
  Yaw control by all-moving fin.
  Full-span 27% chord plain flaps/lift dumpers.
- Propulsion: 2 Williams International FJX turbofans, aft mounted for low cabin noise. Aluminium alloy nacelles.

DEVELOPMENT STATUS

- The prototype aircraft (001) carried out flight trials from December 1988 to September 1991. Noel Penny Turbines (NPT) adapted 2 NPT301 turbojets for this prototype.
- Williams International supplied the interim 30kN (700lb) thrust class FJX-1 turbofan for the first production-configured aircraft (002) which commenced flight tests in April 1997
- Production aircraft installed with the definitive Williams FJX-2 turbofan are planned for marketing through the period 2003-2030.
- The overall concept and design of the Leopard project, and the programme management and financing are by CMC Ltd, England.

AIRCRAFT PRICE

- Maximum selling price for the equipped production aircraft, excluding taxes at September 1998 price levels is $1,350,000

Fig.8 002 Specifications and features

Civil Avionics
and Systems

C545/060/98

Opportunities and challenges in the next century of flight

C M JONES AIAA, GAMA
Rockwell Collins, Indianapolis, USA

SYNOPSIS

The next century of flight presents significant opportunities and challenges for improving flight operations, efficiency and safety through developments in aviation electronics. These challenges require new levels of thinking that build on insight into the man-machine interface dynamics inherent to flying. They also involve making the global transition from a system of airspace control grounded in procedure to a system of airspace management grounded in the technological capability of electronically gathering, analyzing and acting upon real-time information.

OPPORTUNITIES AND CHALLENGES IN THE NEXT CENTURY OF FLIGHT

Over the next few years the focus of the world's aviation community will gradually sharpen to reflect on the magic of a 12-second flight across a 120-foot stretch of a remote, North Carolina sand dune. On December 17, 1903, a power-driven, heavier-than-air machine lifted man into free, controlled and sustained flight. The single surviving photograph of that historic event doesn't do justice to the meaning of those few moments, much less the momentum they generated as civilization entered a new century of unprecedented change. Clearly, the world has never been the same.

I would encourage each of you to spend a few hours reading and reflecting on Orville and Wilbur Wright's own personal accounts of the many years of time, effort and frustration they invested in their undertakings as what Wilbur termed "students of the flying problem." As you will see, the Wrights regarded the man within the machine not as an aviator, but as an engineer. I was particularly struck by the fact that, in those memoirs, the Wright Brothers dwell at some length on the importance of developing methods and means of collecting and assessing the most accurate, real-time information available on their aircraft and its performance.

That is exactly the function of the advanced avionics systems found in today's high-performance aircraft – connecting pilots with their aircraft and the airspace that surrounds them in flight. Modern avionics systems rely on space-based data links, sophisticated sensors, computer analysis and advanced display technologies in performing these flight-critical tasks, all capabilities unknown in the era of the Wright Brothers, an era when electricity was little more than a curiosity. Nonetheless, the Wright Brothers understood the critical need to monitor aircraft performance, and they devised their own methods of doing so. With rain and snow along the wind-swept North Carolina coastline delaying their first efforts at flight, Orville and Wilbur Wright passed the time by developing what could be considered the world's first avionics system. Let me quote Orville Wright directly from an article published in Flying, the Aero Club of America Bulletin 10 years after that flight:

> "It had been disagreeably cold for several weeks, so cold that we could scarcely work on the machine some days. But now we began to have rain and snow, and a wind of 25 to 30 miles blew for several days from the north. While we were being delayed by the weather we arranged a mechanism to measure automatically the durations of a flight from the time the machine started to move forward to the time it stopped, the distance traveled through the air in that time, and the number of revolutions made by the motor and propeller. A stop watch took the time; an anemometer measured the air traveled through; and a counter took the number of revolutions made by the propellers. The watch, anemometer and revolution counter were all automatically started and stopped simultaneously. From data thus obtained we expected to prove or disprove the accuracy of our propeller calculations."

From first flight to the present, gathering information has been essential to flight management. Nearly 100 years later, data collection and analysis remain the challenge of modern avionics systems, as we continue our efforts to make flight safe, reliable and predictable. As were the Wright Brothers, those of us involved in avionics development very much remain "students of the flying problem." To open this session on Civil Avionics and Systems I want to quickly review where avionics have been, where they are today and where they seem to be headed in a world that continues to change as quickly as it was changing in the era of Orville and Wilbur Wright.

The forerunner of Rockwell Collins – the Collins Radio Company – pioneered radio navigation through development of airborne transceivers and the methods of quick and accurate radio frequency tuning in flight. During the Second World War, more than 26,000 100-watt airborne transmitters built by Collins Radio were used by U.S. and British pilots to establish and maintain the air superiority that won that war. Advances in electronics technology developed during the war carried over to civil aviation in the late 1940s and early 1950s, including air traffic control radar, instrument landing systems and VHF radio communications, all aiding in the control and separation of civil traffic in increasingly crowded skies. A new system of navigation called VOR – VHF Omnidirectonal Range – was installed through ground stations to replace limited coverage, low-frequency beam navigation systems. Simple, reliable and easy-to-operate aviation electronics – avionics – provided reliable position, steering and weather information to pilots. To keep the process as simple as possible for pilots, new flight director systems were developed. These new systems replaced five cockpit instruments with two, giving pilots for the first time a clear visual presentation of the information they needed for en route navigation and for precise instrument approaches.

This pioneering effort in systems integration led to development of more advanced flight management systems, which allowed more precise mission management. As large air carriers began introducing jet service, Distance Measuring Equipment was introduced. At the same time, the military embraced TACAN -- tactical air navigation – as its favored approach to precision navigation. Ultimately, a common VORTAC system was established in a divergence and eventual convergence of military and civil avionics capabilities that would be an omen of things to come. Ultimately, the space race brought satellite technology that continues to revolutionize navigation and communication capabilities. A continuously orbiting constellation of artificial stars – the Global Position System – continues to provide precision navigation and location information, while satellite communications provide air-to-air and air-to-ground linkages through space-based voice and data capabilities that are still being developed. And situational awareness has benefited greatly from advancements such as predictive wind-shear weather radar, enhanced ground proximity warning systems, traffic control and collision avoidance systems and the ability to link all of these systems together so that they function seamlessly as one capability. I think it's fair to say that, over the past 50 years, avionics developments have made incredible progress in connecting pilots with their aircraft and the airspace that surrounds them, making flight today as safe and reliable as its ever been.

So, with that historic summary, what opportunities await us in the future? In my view there are a number of major challenges that will require new levels of thinking. One involves improving safety through development of aviation electronics that build on insights into human factors – the man-machine interface dynamics inherent in flying. Another – which is perhaps as immersed in politics and economics as it is engineering – involves making the global transition from a system of airspace control grounded in procedure to a system of airspace management grounded in the technological capability of electronically gathering, analyzing and acting upon real-time information. Beyond these two major, systemic challenges are two additional challenges. One involves bringing together in new avionics design and development the lessons learned within both military and civilian aviation. The other involves extending what we've learned in development of advanced flight deck avionics capabilities beyond the flight deck to the rest of the aircraft.

Human factors engineering speaks directly to safety. With the evolution of air space management away from rigid, procedural ground control toward more free flight autonomy, more and more responsibility is being transferred to the flight deck. Pilots are gathering, processing and analyzing more information than ever before as digital data links deliver flight planning, routing, weather and messaging information. The avionics systems that gather this information must also display it in formats and symbologies that answer questions instead of posing them, making the increasingly complex task of flying increasingly intuitive, predictable and simple for those at the controls. In engineering solutions, those designing and developing next-generation avionics systems must understand that the most complex operating system to be found on the flight deck of any high-performance aircraft is the pilot.

Their design efforts must put the human at the front end of the process. Unless avionics engineers get the human factors issues right, they won't get the safety issues right.

Making the transition to an information-based global airspace management system is a challenge that is less about developing new technologies than it is about agreement on the best ways to phase into use capabilities that already exist. In the era of free flight, real-time information freely exchanged from air-to-air and air-to-ground through space-based data links will replace speculation. Today aircraft flying oceanic tracks are separated by 100 nautical miles not because they need to be, but because of uncertainty over their exact positions in an antiquated Air Traffic Control system that relies on periodic voice reporting of position rather than continuous data-link surveillance and monitoring. The technology required to replace guesswork with hard data exists. What's not yet in place is the operational consensus on how best to put that technological capability to work in a way that enhances efficiency and cuts costs by allowing safe passage of more aircraft through any given airspace.

Addressing that issue will also present the challenge of blurring the lines between civil and military avionics capabilities. Much has been made in recent years, at least in the United States, of ongoing acquisition reform efforts that have facilitated quick and affordable technology transfer of commercial avionics capabilities to military aircraft. Rockwell Collins has helped to pioneer this effort and we are now heavily involved in bringing advanced avionics to the flight decks of the world's military aircraft fleets. Rather than re-engineering the wheel, we are installing into these military aircraft the same flight management, GPS navigation, weather radar, terrain awareness and other capabilities than have proven reliable in thousands of civilian aircraft. And while much has been made of this technology transfer from civil aviation to military aviation, it's clear that in the emerging era of data-intensive Air Traffic Management, there is also military-to-civilian benefit being transferred in areas such as data-link capability. As a former fighter pilot, I can assure you that the level of data-intensive avionics developed and proven reliable in the electronic battlefield environment is critical not only to mission success, but to mission survival. Is it not just as important for a civilian airline to know the real-time location and status of each of its aircraft worldwide as it is for those engaged in aerial combat to have complete situational awareness of every asset dispatched into the fray?

Another significant challenge will be keeping the lid on costs. The most advanced capabilities are useless if no one can afford to implement them, which is among the realities driving the acquisition reform movement. Smaller, lighter, better have always been the collective mantra of avionics development. More recently, we've added value benefits and we have begun designing avionics that can meet both near- and long-term needs through open architecture that achieves capabilities growth through software upgrades instead of expensive and operationally disruptive hardware replacements. At Rockwell Collins, we've evolved beyond providing avionics components to offering avionics systems – totally integrated capabilities that offer both civil and military customers total solutions. We've also begun developing capabilities for the entire aircraft, not only the flight deck. In this information age, data-intensive operations don't begin and end in the cockpit. We're now integrating the front of the aircraft with the back of the aircraft through in-flight entertainment and information systems for both airline and business aircraft. We're also participating in field trials of integrated information system technologies that provide microwave links between flight deck file servers and terminal area databases. We see significant operational efficiencies and savings in the wireless exchange of flight planning data and aircraft system analysis and maintenance data.

These are some of the opportunities ahead -- challenges worthy of our time and attention as "students of the flying problem." Let's overlay some reality on this discussion by exploring some of the barriers that must be confronted in meeting these challenges.

One very important issue revolves around the lack of consensus on international standards for Communication, Navigation and Surveillance in the emerging Air Traffic Management environment. Clearly, in the process of bringing the international aviation community to agreement on how best to proceed, getting there will not be even half the fun. What's needed?, When is it needed?, and Who's going to pay for it? are all questions begging answers. The U.S. has one way of thinking, Europe another. Africa and the Asian Pacific regions – where the existence of very little air traffic control infrastructure could in fact be a blessing in disguise – present their own issues, not the least of which is the question of costs. Agreement on even interim steps toward the era of Free Flight, such as channel spacing requirements for voice communications, have proven difficult as U.S. and European regulators each opt for different approaches to solving the same problems. International certification bureaucracies present their own challenges as new avionics technologies are evolving at such a fast pace than those tasked with assuring their reliance and safety are having difficulty keeping up. The explosion of capability within the micro-electronics industry has created a parts obsolescence dilemma. Avionics providers draw on a tradition of designing aircraft systems engineered for decades of reliable use. Today, we find ourselves doing business in an era when technology cycles are measured in months, not years, as we conceptualize avionics systems being designed and sold for long life-cycles.

The obstacles we confront also include an aviation industry paradigm that equates new capability with new hardware, despite the clear advantages of designing systems that can be quickly and affordably upgraded without flight deck overhauls. The rigorous certification environment in which new avionics software development is undertaken often induces delay and increases cost. And our customers, sometimes unaware or unconvinced of its value, can be hesitant to pay for functional upgrades that involve software replacement only.

Finally, I would suggest that one of the most formidable obstacles we confront in meeting the challenges I have outlined this morning is our own entrenched way of doing business. In my years of involvement in the aviation industry, I have never seen a greater demand from both the civil and military avionics markets to keep costs down. Airlines are establishing long-term, preferred supplier agreements with avionics providers, under fixed-price contracts which necessitate cost-reduction efforts. Aircraft manufacturers are insisting on first-time flight test success that requires continuous improvements in quality and reliability. Our military customers are now demanding not only mission-critical performance, but best value and low-cost maintainability. In delivering these business benefits to our customers, we must learn how to improve quality, shorten cycle times and introduce cost-sensitive production processes that eliminate waste and add value.

As you can see, the challenges are many, as are the obstacles. But, as dedicated "students of the flying problem," I have faith that our industry will persist and succeed, as did those two bicycle mechanics from Dayton, Ohio.

Let me close with another bit of insight from Orville Wright's retrospective look at what he and his brother accomplished. Written 10 years after the historic flight that changed the world, it speaks to the fruits of relentless application of vision, experience and perseverance – three traits we must have to confront head-on the challenges of a new era in aviation:

> "With all the knowledge and skill acquired in thousands of flights in the last 10 years, I would hardly think today of making my first flight on a strange machine in a twenty-seven mile wind, even if I knew that the machine had already been flown and was safe. After these years of experience I look with amazement upon our audacity in attempting flights with a new and untried machine under such circumstances. Yet faith in our calculations and the design of the first machine, based upon our tables of air pressures, secured by months of careful laboratory work, and confidence in our system of control developed by three years of actual experiences in balancing gliders in the air had convinced us that the machine was capable of lifting and maintaining itself in the air, and that, with a little practice, it could be safely flown."

C545/032/98

Integrated utilities control for civil aircraft

J P HALLER BSc, **D V WEALE** BSc, and **R G LOVEDAY**
Smith Industries, Cheltenham, UK

Synopsis

The Boeing 777 has introduced new technology in a number of areas, one being in the aircraft electrical distribution system. The 777 Electrical Load Management System (ELMS) introduces electronic control and monitoring into the distribution function. Advantage has been taken of this to provide smart load shedding and to incorporate control of various utilities functions, while providing comprehensive status reporting to the crew and fault logging for maintenance purposes. The ELMS electronics is packaged in Line Replaceable Modules such that functional elements may be easily replaced following failure. In this regard it is an example of an Integrated Modular Avionics rack, replacing Line Replaceable Unit packaging methods. The lessons learned and benefits achieved from this technique are reviewed.

Building on this very favourable experience, the paper examines new philosophies for the integration of aircraft power and utilities control functions. Additionally, improving requirements capture and design methods, the adoption of 'open architectures' – such as will support multi-vendor environment– and the emergence of new technology will all significantly contribute to better systems in the future.

The paper examines these issues and looks forward to the benefits which future aircraft such as A3XX will gain from these development trends.

1 BACKGROUND

The Boeing 777 airliner set new but carefully measured standards for system integration. It embodied many of the benefits of system integration, but these were implemented in a way that acknowledged the short time scales and the practical constraints of developing and certifying new systems. The Honeywell Aircraft Information Management System (AIMS), GEC Flight Control, and Smiths Industries Electrical Load Management system (ELMS) were notable examples. Now the industry is ready to take the next step.

2 INTRODUCTION

The Smiths Industries ELMS integrates much of the functionality associated with primary and secondary electrical distribution as well as some specific utility control functions.

The electrical distribution is separated into three lanes, for electrical and system segregation. Each lane has a Primary Power Panel, which takes in power from the main a.c. power sources, and acts as the main a.c. bus distribution centre for aircraft loads >10kVA. Apart from "smart" Electrical Load Control Units (ELCUs), the Power Panels are non-intelligent. The sources used to power the primary a.c. buses are selected by the Generator Control Units (GCUs) and the Bus Power Control Unit (BPCU), while load switching is performed in response to external commands. The power panels set new standards for physical integration through the use of laminar power backplanes and plug-in contactors.

Each lane also has a Management Panel, which acts as the distribution centre for lower power a.c. and d.c. aircraft loads. Each Management Panel contains an Electronics Unit (EU), which contains the ELMS Integrated Modular Avionics (IMA), which provides the intelligent processing for system control and monitoring both for its own panel and for its associated Power Panel.

In addition to the three Power panels and three Management Panels, the system includes a Ground Service/Handling Panel. The complete system is illustrated in Figure 1.

Figure 1 - B777 ELMS System Integration Rig with the seven panels

The Management Panels interface to many of the utility systems (see figure 2).

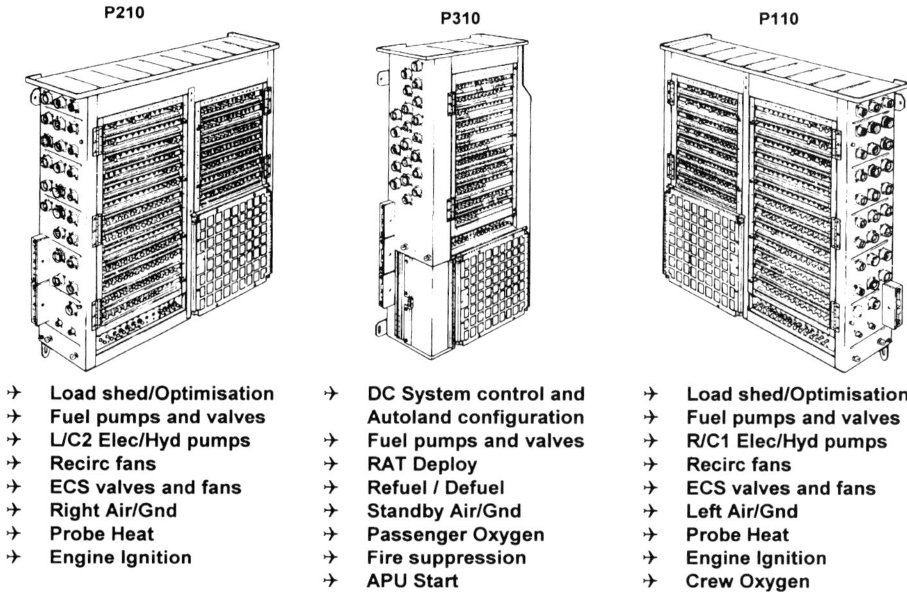

Figure 2 – Summary of Utility system interfaces associated with each Management Panel

The EU in each of the Management Panels is a classical IMA solution: identical racks, orientated appropriately for the distribution panel they fit, with common modules for processing, signal input and output (I/O), and the gateway on to the aircraft ARINC 629 System Busses.

An innovative part of the Management Panel packaging is the IMA rack being integrated into the distribution system. The rack is a physically separate entity, but fits within the distribution panel. This allows a control and monitoring assembly, and secondary electrical distribution system to be installed as a pre-tested single entity on the aircraft, helping reduce the electrical system on-line build time.

3 THE ELMS IMA RACK

The racks support functions up to critical level, by selective system implementation in hardware. This has allowed the software development costs to be contained to those associated with Essential Systems. The rack itself (see figure 3) is fitted with a passive back plane that carries the busses, and I/O signals which pass through passive filter Line Removable Items (LRIs) contained to one end of the rack.

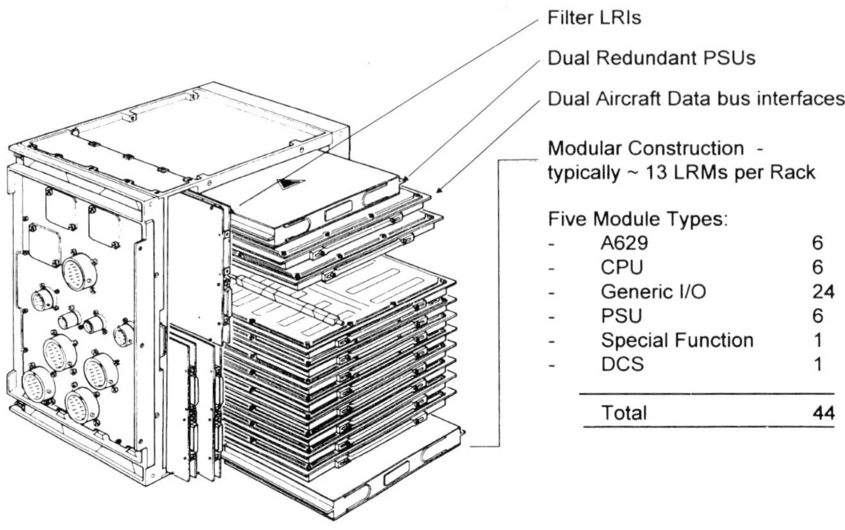

Figure 3 - ELMS IMA rack with LRMs

Each rack contains: two Power Supply LRMs; two Gateway LRMs to provide access to the left and right aircraft system busses; two Core Processor ('CPU') LRMs; six Common I/O LRMs and in the centre lane (P310) a dedicated module for the d.c. System (DCS) and Special Function module (an adapted I/O module for non-standard signal inputs). All modules are line replaceable and physically protected against handling damage by LRM shrouding.

The racks have common software Operational Flight Program that is downloaded over the ARINC 629 System Bus into all three racks (and into both processors within a rack). This is configured for the application functions, within a given rack, by common loadable Configuration Tables. The application software is abstracted from the physical hardware by the use of an operating system, which is again common to all computing lanes.

The experience with a 44-module system (6 module types) per aircraft, is one that builds reliability experience at a rapid rate compared to conventional dedicated Line Replaceable Units (LRUs) solutions. Any weaknesses are flagged up very rapidly in service. The Smiths Industries system was optimised, from day one, to work in a purely convection cooled environment, and has exceeded all reliability expectations. Reaping the benefits of a low dissipation design, with good thermal management, benign power supplies, and the careful handling of the 500+ I/O signals typically has LRMs achieving in excess of 200,000 operating hours MTBF.

4 ELMS IMA ARCHITECTURE

The key to ease of validation of the system was the architecture. A dual system, configured in a hot standby arrangement was selected for active components and back plane in each rack (see figure 4).

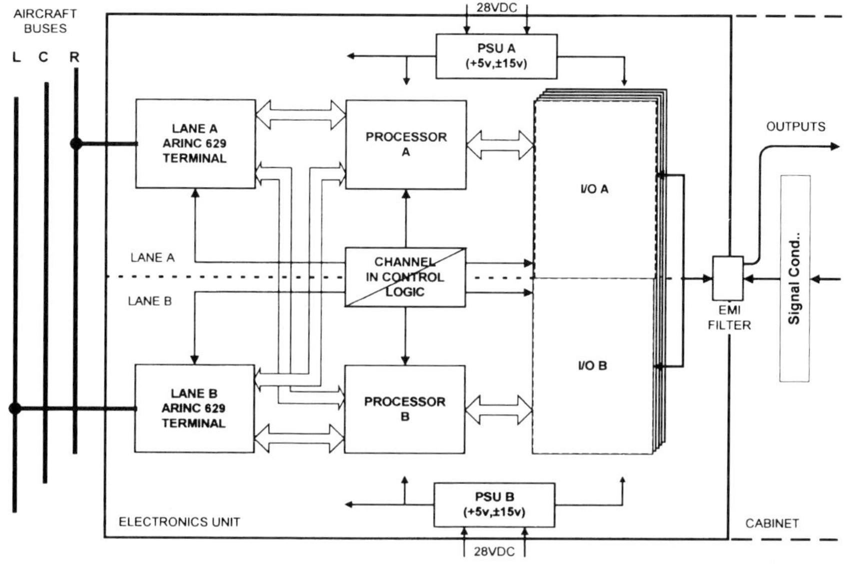

Figure 4 - Rack Architecture

This arrangement gives very high availability, moving into the region of planned only maintenance. Currently ELMS has a 99% probability of functional availability for ~200 hours after first failure. The system response to multiple failures; the first being an external system level failure (e.g. power busses or ARINC System Bus) is failure⇒operational, failure⇒operational, finally failure⇒passive. There is no loss of functionality after first rack failure (and in fact ELMS has addition reversionary modes for graceful degradation after second and further failures).

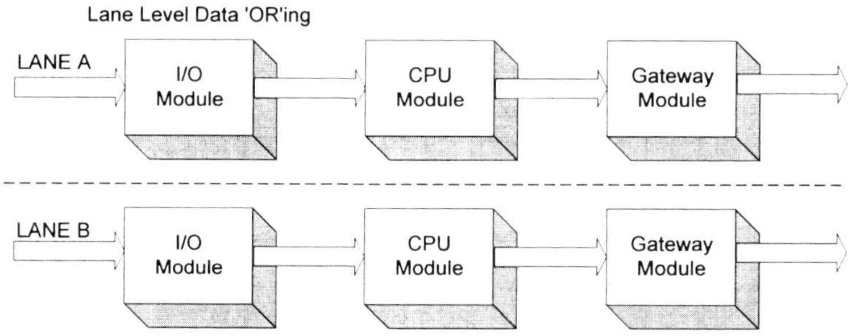

Figure 5 – Lane level data 'OR'ing

As reliabilities increase, architectures can simplify, in turn increasing reliability - a case of having your cake and eating it! The ELMS Lane A **OR** Lane B method of operation (see figure 5) is a simple, segregated, verifiable solution and suited to parallel backplane. Seemingly, more *flexible* solutions that utilise Lane A **OR** Lane B data selection at each module (see figure 6) increase the system complexity with minuscule operating benefits in the utilities environment and are only suited to serial backplane busses. The ELMS solution provides signal availability of better than 1×10^{-10} per hour against better than 1×10^{-11} per hour for the module level source selection. What these figures disregard, although both impressively high, are the single path elements of the system and the common mode failures cases taking these into account there is no effective improvement in availability. Also looking to the future; as hardware continues to reduce in size and cost and the reliabilities increase further the ELMS solution most readily integrates into fewer modules and ultimately one lane per rack.

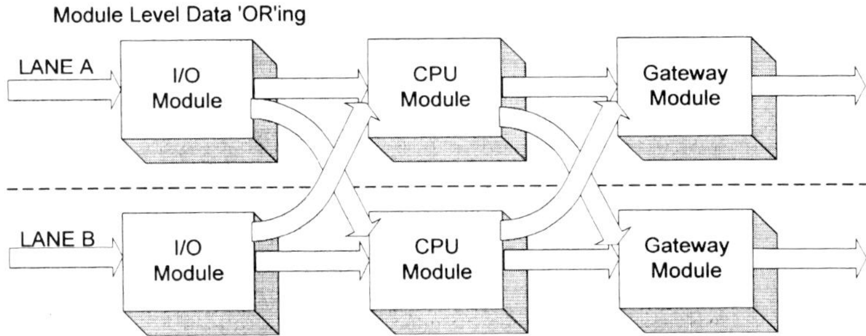

Figure 6 – Module level data 'OR'ing

Key to substantiating this viewpoint is the fact that external world interfaces for power and utilities primarily use discrete simplex signals for input and output. Many Utility controllers, and even many purely electrical systems are simplex, and, standalone, cannot cost effectively justify any monitoring electronics or interface onto the System Bus. The ELMS integrated solution provides a cost-effective method of enhancing the availability, functionality, visibility, and maintainability of these systems while reducing discrete wiring by bringing more systems onto the aircraft data busses. This distinguishes them from Flight Control and Management where the system interfaces are data busses which are dual, if not triple redundant. These fundamental differences allow *domains* to be identified within the aircraft systems that recognise the different needs.

One experience on ELMS is the need for Built In Test Equipment (BITE) extending out into the aircraft systems. As operation experience builds Mean Time Between Replacements (MTBRs) are improving and the operators' experiences of the LRU being the least reliable component is reversing. It is the sub-system components and plant that may need to be suspected first, no matter how unattractive the thought. Integration helps by gathering system data with little, if any additional aircraft installation. Considerable first line diagnostics can be obtained through status and current measurements. However, even here nothing is free and systems analysis of reliabilities should be done to apply the BITE appropriately to systems that need and are going to use it, while avoiding overloading the system with unnecessary data, which may serve to confuse rather than inform.

The penalty some would observe is that this architecture does not readily support all Utility Functions within the integrated environment, perhaps through lack of processing power, signal types or inherent Control and Monitor capability. Technology progress addresses some of these but the use of the *embedded* system solves the rest.

The ELMS d.c. standby system, or DCS, module is an example of an embedded function taking advantage of residing in an integrated rack. The function is implemented in a simple state machine this has no standalone system overheads and allows rapid development of the system. However, it now has dual redundant power supplies and a dual communications interface with the System Busses through the use of the rack resources. Its circuitry is further simplified by running the BITE functions on the Core Processors and enabling self-test through interlocks with the hardware. This allows the DCS system to have manually, and automatically, initiated BITE.

5 LESSONS LEARNT

When you look back to recall what you have learned, it often seems obvious, but that makes it no less important. So in no particular order and by no means comprehensively:

- Keep the solutions simple – deal with problems one at a time, and do not try to solve everything at once.
- Understand the differing needs of the domains – utilities, flight control, flight management do have fundamental differences that need to be considered first
- Plan the integration process very carefully - so that everything is checked once, with no duplication but without missing anything.
- Design to keep reliability up by providing a benign environment for the electronics - keep power consumption down to minimize heat, dissipate unavoidable heat, and protect from transients.
- Standardize as much as possible and make it fool proof to use and operate - in the pressure of the moment no one has time for quirky features
- Make BITE a system consideration – do not specify extensive BITE on the electronics, which is the most reliable element, and ignore the peripheral items such as sensors and actuators, otherwise the MTBURs will never come down
- Ensure status and maintenance messages give clear and useful, and do not clutter up the system with irrelevant information.
- Make best use of generic building blocks – it is the key to making an integrated solution cost effective

6 PRAGMATIC CONSIDERATIONS FOR THE FUTURE

Integration of Utility Systems has taken substantial steps forwards in recent years - the use of common modules and a core operating system has been demonstrated within the ELMS 777.

The optimum level of integration for the future depends on a number of factors including:

6.1 PHYSICAL

Technically it may be possible to integrate systems into fewer racks. However, the real word considerations such as tolerance to particular risks such as explosion, fluid contamination, engine and tyre burst damage will mean that multiple racks are always desirable. Also too much centralisation will increase aircraft wiring and connectors, particularly for discrete I/O intensive Utility Systems and Power systems.

6.2 RACK ARCHITECTURE COMPLEXITY

In trying to create an architecture suitable for the most complex or critical functions we are in danger of driving the architecture complexity upwards unnecessarily. The use of embedded functionality (benefiting from the rack environment, but not fully integrated with the central processing) within an IMA rack can bring the best of both worlds.

There are differences between different domains; the Avionics functions (e.g. Flight Management) are processor and memory hungry, and rely heavily on databus I/O, whereas the Utility system (e.g. Environmental Control Systems) tend to have a much greater proportion of discrete digital and analogue I/O and require relatively small amounts of processing power.

In reaching the optimum architectures for the future we need to recognise the differences between domains so that simple cost-effective solutions are not excluded. The design of an integrated architecture to be effective for the majority of functions whilst allowing some functions to exist as embedded LRM's within the rack, rather than drive the core architecture complexity, even accepting LRUs as the best solution in some instances.

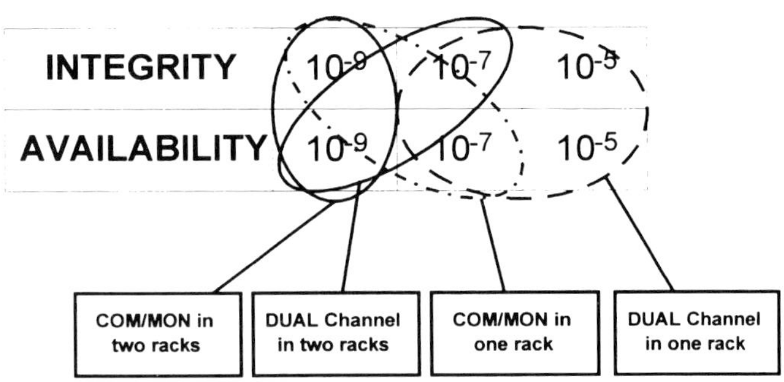

Figure 7 - Solution mapping

What the system architect needs is a solution that allows all sub systems needs to be mapped on the racks (see example figure 7). As shown in that diagram, a dual system in a single rack will provide an availability of better than 10^{-5} and with DO-178B level "A" software, an integrity of 10^{-7}. This simple architecture can therefore support many systems. However, if it is necessary to provide higher integrity, a separate command and monitor lane must be used, while availability can be increased by the use of two racks.

6.3 OPENNESS

Driven by the desire of air travellers for convenient low cost travel, aircraft manufacturers need to contain new development costs. Ownership costs also have to be reduced to make the introduction of new aircraft viable. Unless any IMA solution is to be the prerogative of only one company, an open solution is a key issue to further integration being achievable with low risk and at affordable cost.

6.3.1 3RD PARTY APPLICATION S/W

Within the host rack there are resources that can be used by others system suppliers. However, those resources need to be easily available, and implementation has to be as transparent as possible to the software supplier, otherwise each new host he places his system in will require virtually a new start.

6.3.2 EMBEDDED FUNCTIONS

For technical, integration, and commercial reasons embedded systems need to be included. The aim must be to achieve this through a simple and non-proprietary physical and logical interface.

6.4 COMMON MODULES

From a technical standpoint it should be possible to increase the use of common modules across the total Utility and Avionics domains. Greater standardisation would bring benefits to the airlines from reduced spares holdings, while allowing airframers the possibility of multisourcing standard modules. The key issues to confront are business related, as traditionally the hardware recurring costs have been loaded with development costs of both the hardware and the software it contains; also it will mean greater centralisation of interface/module specifications at either the airframes or a small number of Tier 1 suppliers.

6.5 COMMERCIAL CONSIDERATIONS

Integration within the Utility domain certainly brings substantial benefits as multiple LRUs are swept up into a few Utility IMA racks. For the future, the cost benefit of increasing the complexity of the Utility system architecture in order to create one standard architecture across the aircraft would need to be carefully weighed. In addition, we must not lose sight of the trends of development and production costs which are increasingly dominated by the systems and software elements, while the hardware recurring costs continue to fall. Thus in global terms, focusing exclusively on common hardware/architectures will ultimately drive up the total aircraft costs if this is only achieved at the expense of software complexity.

6.6 SAFETY/CERTIFICATION

As we move forwards to greater levels of integration, openness and common modules, we must not lose sight of the inherent dissimilarity/dependence created by separate LRUs. The integration process must very carefully consider the problem of combined criticality, as loss/inadvertent operation of the integrated IMA rack can now affect multiple functions. For this reason, combined with the arguments surrounding the architecture complexity, we favour some embedded (not fully integrated) functions within the rack.

6.7 MODULE LEVEL INTEGRATION

As we have stated large amounts of discrete I/O and relatively low amounts of processing power dominate Utility systems. This picture leads to increased module level integration where SMART I/O modules provide both I/O interfacing, processing and databus access. This trend also provides very simple partitioning between software functions when required.

7 THE WAY FORWARD (AT LEAST FOR THE NEXT GENERATION)

The horizontal integration process has risks. It can be driven by the most complex system needs and common sense will delay the day that all aircraft functions reside in a single unit. It is not a truism that simpler systems are included, at little or no extra cost; rather they acquire the overheads of the most complex system, if not in the implementation, in the maintenance, support, tools, and configuration control. In the real world benefits are more directly achieved through attention to the placement of control and distribution elements. This approach simplifies the aircraft wiring by moving systems out of the Avionics bay, where appropriate. Compact modular racks can thus be placed in accessible locations, providing zonal distribution, and dissimilarity from other systems.

7.1 VERTICAL INTEGRATION

Smiths Industries believes that vertical integration is the natural extension of the ground breaking ELMS solution, where the high power distribution was integrated with the power management, control and monitoring centre. On the ELMS, conventional power protections and distribution (circuit breakers and relays) were used. The next step is to bring that functionality entirely into the rack, not a popular technology mixture for electromagnetic compatibility integrity. However, Smiths has already established these techniques in other product areas as well as the ELMS. The next step is the blending of the power bussing technology and backplane bus solution into a fully integrated power and utilities rack (the two domains that interface most directly with the power distribution system).

From the experience with ELMS, Smiths believe the use of embedded functions within Utility racks is a critical success factor, as opposed to concentrating on full integration. Embedded systems address special needs at a primary function level, but can benefit from the rack environment for secondary (BITE, monitoring, aircraft system bus i/f) and support functions (power, environmental protection). This accommodates high integrity functions and also high speed closed loop systems that could drive the performance requirements and possibly combined criticality of the integrated rack up.

This simplifies the verification/validation task for system suppliers. Complex systems that require specialist test environments or large plant can be developed out of rack in parallel with the generation of the integrated environment resources.

7.2 OPEN INTERFACES

This approach will only offer the required flexibility in development and service if the open interfaces presented are robust and easy to manage.

These interfaces are:

- The backplane
- The core software i/f for applications

A goal should be to set a minimum level of standardisation points on the basis of the maxim *standardisation kills innovation*. This will allow a system architecture that can evolve and adopt the latest technologies within the modules.

The mechanical and environmental requirements are secondary in as much as they are interface challenges to which aerospace companies have always had to conform.

The backplane i/f needs to be technically capable, but more importantly technology independent, with minimum constraint on the module operation; also based on a well-defined, understood, acknowledged, and available standard. Although commercial standards appear to offer attractive performance, they are less controllable, rapidly obsolescent, often have additional processing overheads, and are not conceived to provide the integrity levels desired. This leads the search back to aircraft data busses or developments of them; ARINC 429, ARINC 629 being the obvious candidates in the civil arena, as both these have a wide experience base and available tools to be implemented easily by many independent suppliers.

The aerospace application software interface standards are all proprietary today, but there is a strong move now to standardise a powerful subset of APEX interface services based on ARINC 653. This will enable a wider use by application suppliers and easier code transportability between platforms.

8 CONCLUSIONS

For the next generation of Civil avionics there is tremendous potential to make significant advances over the current avionics systems in cost, reliability, simplicity and ease of use. An important step towards achieving this goal on future aircraft will be the replacement of the plethora of different hardware implementations found on current systems with harmonious sets of hardware designed to serve as common platforms for diverse aircraft applications. Careful System Engineering will be required, utilising the best of new technology, established techniques and tools, to map the functionality onto these common platforms, and to optimise the level of integration, while devising solutions that can grow and evolve for a considerable period of time.

9 ACKNOWLEDGEMENTS

The author would like to thank the directors of Smiths Industries PLC for the opportunity to present this paper. Also the contributions and advice from the co-authors and others not mentioned. The views expressed herein are those of the authors and do not represent the official view of Smiths Industries PLC.

C545/033/98

AIRBUS new avionics generation

H SUBRA DE SALAFA and **Y SAINT-UPERY**
Aerospatiale, Toulouse, France

SUMMARY

For AEROSPATIALE, the control of the avionics architectures design on the AIRBUS families is the result of thirty years of experience. Major steps were taken with each new program, from Concorde to the AIRBUS A340 : change from analog to digital electronics, introduction of digital data exchange between interconnected systems, electronics and systems integration.

The main criteria which govern the choice of avionics architectures and their design: dependability requirements, development and certification plan, cost of ownership targets, ... have been considered all along these continuous improvements.

AIRBUS Industrie is actively working to prepare evolutions of his aircraft avionics architectures. The objectives of these evolutions are of two sorts: improvement of the cost of ownership of the aircraft and addition of new capabilities such as those related to the changes in the ATM environment.

For both new systems as CNS ATM, and the enhancement of existing systems such as the Display System, AIRBUS Industrie offer is twofold. Although AIRBUS Industrie has of course a certain freedom to design optimum avionics architectures for the future aircraft families such as the A3XX, AIRBUS Industrie also proposes retrofit solutions for all the in-service AIRBUS fleets to continue to operate and take benefit of these new capabilities.

It is within this context that AEROSPATIALE, within AIRBUS INDUSTRIE, has defined and initiated the AIM-FANS (AIRBUS INTEROPERABLE MODULAR FUTURE AIR

NAVIGATION SYSTEM) which provides a modular and upgradeable solution. AEROSPATIALE is also developing a New Display System.
These both systems will be installed on the existing fleet and on new aircraft families. They are based on new equipments developed by extensively taking into account modern avionics concepts such as the portability and the robust partitioning of software and the downloading of this software onboard the aircraft.

For new aircraft, the capacities offered by electronic technologies enable us to go further. The future of avionics is based on integrated modular avionics whose basic principle is extended standardization of avionics components and sharing of these components between several aircraft systems.

For any new AIRBUS, and especially for the A3XX, AEROSPATIALE and the AIRBUS partners are already actively working on a global onboard electronics optimization approach based on these principles.

INTRODUCTION

For AEROSPATIALE, the control of avionics architecture design is the result of 30 years of experience. Major steps were taken with each new program from Concorde to the Airbus A330-A340 family through new added capabilities on one hand and through an enhanced control of the avionics architecture design on the other hand.

AEROSPATIALE is now preparing the future through the development of enhanced and new systems for the existing families. AEROSPATIALE and AIRBUS INDUSTRIE are ready to take benefit of this experience to develop a new avionics for a brand new aircraft such as the A3XX.

After a rapid historical roundup of AEROSPATIALE past avionics developments, the presentation details the main criteria for avionics architecture design, underlines the main drivers of the evolution of avionics architectures, considers 2 examples which are the AIM-FANS solution and a New Display System for the Airbus families, and which already rely on new avionics principles. The last part discusses new avionics and its application on the A3XX.

MAJOR STEPS

CONCORDE: 1st flight in 1969, entry into service in 1976: A modern system approach.

The development of CONCORDE at the beginning of the 60's was marked by the incorporation of many technological innovations in the avionics field: fly-by-wire controls, digitally-controlled air intakes, auto-pilot with, on the basis of CARAVELLE experience, all-weather landing capability, inertial navigation, ...
The avionics onboard CONCORDE uses analog electronics and, in very specific cases, digital electronics (e.g.: inertial system with a program of 4 Kbytes).
Extensive experience has been acquired in the formalization of the design, the development and the tuning of complex systems with, especially, for the first time, a modern system approach demonstrating the upholding of targets derived from safety analyses.

Thus, the dual-dual control/monitoring architecture concept was incorporated on CONCORDE. This concept is still used today for all AIRBUS auto-pilots.

AIRBUS A310: Entry into service in 1983: Towards the all-digital concept.

AEROSPATIALE has demonstrated initial control of interconnected digital systems and development of high-level safety software with the introduction of a completely new cockpit with CRTs (Cathode Ray Tubes) for a two-man crew, the flight management computer, fly-by-wire controls on the secondary control surfaces and other avionics systems (4 million bytes of software and an installed processing capacity of 60 MIPS).

AIRBUS A320: Entry into service in 1988: The modern aircraft standard.

The A320 represents an important technological step with, especially, in the avionics domain: fly-by-wire controls, full authority digital engine control, integration of the auto-pilot with flight management, ACARS, centralized maintenance, etc. (10 million bytes of software and an installed processing power of 160 MIPS).

The A320 very short development program has been allowed by the use of a System Workshop based, for the first time in aeronautics, on formalized specifications enabling automatic generation of software and demonstrating the conformity of the software developments with DO 178 quality requirements. The flight tests used new real-time facilities such as telemetry.

AIRBUS A340: Entry into service in 1993: The first four-engine AIRBUS.

The A340 is the first four-engine AIRBUS. All system attainments of prior programs are reused. Its avionics calls on highly integrated digital electronics, sometimes remote (20 million bytes of software and an installed processing capacity of 250 MIPS). On account of its long-haul mission, priority is given for the first time in the AIRBUS families to functions such as GPS navigation, satellite communications and soon FANS.

With this program and its sister program, the A330, a new step is taken in terms of design and development control maturity with new design support means such as validation by multiple-equipment and functional simulation upstream of the integration benches used for each program since CONCORDE, and extended automatic software generation,...

MAIN DESIGN CRITERIA

Experience acquired over these thirty years enables us to underscore and recall the main criteria governing the choice of avionics architectures and their design.

Dependability requirements associated with the criticality of the functions:

- regulatory or operational availability targets enabling the number of occurrences of the functional chains to be defined from the estimated equipment reliabilities,

- integrity targets which will lead to different implementation choices such as segregation between functions (e.g.: navigation and surveillance, etc.), command/monitor architecture (e.g.: auto-pilot, etc.), hardware and software dissimilarity (e.g.: fly-by-wire controls, etc.) and completely deterministic operation (e.g.: all critical functions, etc.).

Development and certification plan:

- definition and justification dossier including: description notes, safety analysis (PHA, SSA), simulation, etc.,
- realisation, verification and validation activities: qualification of hardware (DO160) and software (DO178), partial tests, integration tests: IRON BIRD, flight tests and tests on simulator, route proving, etc..

Life cycle:

- very short development with numerous changes in definition requiring a large number of software versions. Only an organized process and advanced methods and tools enable this to be achieved (formalized specifications and automatic coding, validation by upstream simulation, ground and flight test means, etc.),
- a very long life (30 years) with, therefore, new functions to be incorporated and obsolescence problems to be anticipated.

Cost of ownership targets:

- hardware and software development costs and recurring cost targets which lead to search for common elements, to the software portability requirement, to the family policy, etc. and, of course, to the implementation of efficient methods and means,
- maintainability targets: efficient detection of failures to obtain a low rate of unjustified removals, reliability, repair costs, etc.,
- upgradeability, that is the capacity to safely incorporate modifications (right at first time) and at reduced costs.

Good evaluation of these criteria and their relative importance enables us to get the most out of the new technologies available whilst controlling the risks in three main fields: quality, cost and lead-times.

CHANGE FACTORS

AEROSPATIALE and AIRBUS Industrie are willing to continuously improve their aircraft avionics. The motivations are of two sorts. First of all, there is the improvement of the cost of ownership of the aircraft. To contribute to this objective, the development costs especially for software must be tightly controlled. Then, the second motivation is the ability to offer to the customer the possibility to operate AIRBUS aircraft taking fully benefit of the planned changes of the ATM systems in the various regions of the globe.

To prepare these evolutions, we have assets at our disposal: the new technologies available and our development methods and means. We have a target: to implement robust, reliable and upgradeable architectures and principles based on experience acquired: redundancy, dissimilarity, functional segregation.

Concerning the new functional CNS ATM requirements, our reply is twofold as, although we have of course a certain freedom to design optimum avionics architectures for the future aircraft families such as the A3XX, we propose also solutions to retrofit the new capabilities required to enable all the in-service AIRBUS fleets (more than 2000 aircraft in service by the year 2000 and much more by 2005) to continue to operate in an ATM 2005 environment.

For other systems such as the display system, AIRBUS Industrie is also developing and proposing new capabilities to Airbus customers whilst improving the cost of ownership of this system.

CNS ATM REQUIREMENTS: FROM ATC TO ATM:

New functional requirements... a difficult transition.

The requirements which are to be satisfied:
- RVSM over the Atlantic
- BRNAV and VHF 8.33 in EUROPE
- FANS 1/A in the South Pacific
- TCAS / S Mode transponder in Europe

The requirements which are theoretically defined:
- ICAO CNS ATM 1
-

The requirements which will come:
- ADS B/CDTI
- GNSS2
- FREE FLIGHT
- etc.

within the framework of announced strategies:
- in EUROPE: ATM 2000+
- in North America: NAS 3.0

or still yet to come...

THE AIRBUS ONBOARD FANS SOLUTION: AIM-FANS

AIM = AIRBUS INTEROPERABLE MODULAR

AIM-FANS is an AIRBUS solution as it defines a change in the avionics for all AIRBUS aircraft, both aircraft under production and in-service aircraft (retrofit).
It is Interoperable, that is, capable of all environments throughout the world.
It is Modular in the sense that it is adaptable to the needs of each customer and capable of future development steps.
AIM-FANS meets two main criteria: the man-machine interface ergonomics and maintaining attainments at best cost.

In that context, and as for other major avionics AIRBUS systems, AEROSPATIALE has the responsibility to design the whole system : definition of the functions and of the architecture of the system, design of the operational concepts and of the man-machine interface, coordination of the equipments development, integration and validation of the system.

AIM-FANS introduces the following new equipments:

For Navigation:
- two multi-mode receivers (MMR) with integrated GPS function and choice of approach means: ILS, MLS and/or DGPS,
- two second generation FANS compatible flight management computers (FMS).

For Communications and Surveillance:
- three VDR receivers (mode A) capable of evolving towards the 2/3 modes by software,
- a "hosting platform" ATSU housing the ATC, AOC and ADS functions in separate software partitions capable of ARINC 622 / ATN dual stack and a "dual" configuration.

For the MMI:
- two DCDUs dedicated to written communications.

The new AIM-FANS computers: the FM and ATSU widely integrate modern avionics concepts.
They use the robust partition concept enabling several independently certified applications (functions) to be hosted on the same processor.
They incorporate a standard interface between the operating system and the application software (ARINC 653 standard for FM and POSIX standard for ATSU) facilitating software portability.
They have a high processing power thanks to their 32-bit processor with floating point computation (AMD 29040 or MOTOROLA 68040 for FM, Intel 486 for ATSU).
The software is completely downloadable onboard with provision in each of the new units for implementation of the new high-rate ARINC 615A standard (ETHERNET) now being elaborated.
The AIM-FANS solution is an ARINC 600 LRU modular and upgradeable solution for the existing Airbus families.
The developments made for this solution will be widely reusable in the new architectures of the future programs implemented into Integrated Modular Avionics.

THE AIRBUS NEW DISPLAY SYSTEM

AIRBUS has recently decided to prepare a New Display System to improve the cost of ownership of this system and to offer enhanced and new features.

The A320 and A340 families are the targets for the installation of this New Display System which will be, in a second step, widely installed on the other new Airbus families.
The New Display System consists in the replacement of the Display Management Computers and the Display Units by new units with a minimum impact on the aircraft. It will be installed both in forward fit and retrofit conditions.

The new Display Units are no more CRTs (Cathode Ray Tubes), as on the current AIRBUS aircraft families, but LCDs (Liquid Crystal Displays) flat panels. The LCD is now a mature technology which presents very good optical performances, even better than the CRTs in sunshine conditions, and larger display area for the same outside equipment size.

The New Display System will greatly improve the maintenance costs and will provide benefits in terms of necessary volume for their installation, power supply requirement and weight.

The New Display System will, in addition to the conventional flight, navigation, engines, aircraft systems and warnings displays, also introduce new functions :

- SMGCS (Surface Movement Guidance and Control System) : display of airport maps and aircraft movement on ground,
- CDTI (Cockpit Display of Traffic Information) : in-flight traffic surveillance display
- CFIT (Controlled Flight into Terrain) : terrain information display
- Weather Forecast data display

Additional features will be introduced by the New Display System : video recording of the displayed images, hard copy of the display, "windows" like capability : windowing, icons, scroll bars, pointing device, bitmap and 3D displays, ...

The new Display Management Computers and Displays Units will be based on up to date and powerful technologies featuring new avionics solutions : high throughput processor, large size memory, partitioned software, fast downloading of all the software.

These computers will allow an easy stepped introduction of the here-above identified new functions and even other new functions as soon as they are defined.

The technologies selected in this development will be reused in new avionics future architectures as well as the application software which will be fully portable.

FUTURE AVIONICS: INTEGRATED MODULAR AVIONICS

The capacities offered by electronic technologies now enable AIRBUS Industrie to go well beyond the ARINC 600 LRU solutions of the current aircraft families.

The future of avionics is based on integrated modular avionics.

The basic principle is advanced standardization and sharing of resources between several aircraft systems.

In conventional avionics, the main functions are ensured by independent computers which each include a power supply, data acquisition and exchange capacities, processing resources (microprocessor and memory) and resident software supporting the functional software.
Each computer acquires the data it requires. The various computers are developed independently of each other.

In integrated modular avionics, a rack consisting of modules connected by a backplane bus will accommodate several functions. The power supply will be common to these functions. Also, acquisitions will be made once and shared by the various functions that also share common processing resources.

Modular avionics leads to reductions in weight, volume and electrical power consumption.
Apart from the reduction in costs obtained by the sharing of the resources (reduction in onboard electronics), these new concepts also enable a reduction in:
- development costs by reducing the number of different items of equipment,
- recurring costs by the increased number of equipment of same type,
- operating costs by reducing stocks (less onboard equipment), reduction in unjustified removal rate (improvement in reliability, capability to detect and identify failures, fault tolerance enabling deferred maintenance in some cases),
- development costs by the independence of the software and hardware, standardization and downloading.

Optimization of data exchanges with the use of multiplexed "high throughput" communication networks as ARINC 629 (2 Mbits/s), ETHERNET (10 Mbits/s) etc. are associated with these concepts.

The expected advantages are significant. However, we must not conceal that the sharing of resources between avionics systems comprises a rupture and that new problems concerning the control of the complexity of the interdependencies created between these systems arise:
- at the design and certification phase level: the platform common to several systems must be covered by its own specification but must also be considered in the analysis of the various systems (safety and failure mode analysis in particular),
- at integration and validation phase level: several functions must be integrated onto a common platform. Specific means must be developed to enable the development, integration and validation of the various functions to be conducted in parallel,
- and on the industrial scale, concerning the sharing of the development between several suppliers: some of them will no longer develop computers but will be limited to supplying application software and peripherals and sensors dedicated to their functions.

FUTURE AIRBUS AVIONICS

Beyond AIM-FANS and the New Display System, new steps of integrated modular avionics concepts implementation will be taken in the future, on the AIRBUS aircraft.

Already today, for all new programs, all avionics suppliers propose highly integrated modular avionics platforms.

For the A3XX on which AEROSPATIALE and the AIRBUS partners are already actively working, several integration domains are considered:
- cockpit avionics: perimeter close to the B777 AIMS (which includes the following functions : displays, the flight management, auto-thrust, communication management, data recorders interface)
- cabin systems: air conditioning, pressurization, temperature control, etc.,
- aircraft utility systems: landing gear, fuel, electrical power, etc.,
in an overall optimization approach to onboard electronics.

The architectures and technologies will allow :
- an efficient implementation of the functions
- a reduction of the cost of ownership of avionics
- flexibility and growth potential
- easy airline customization
- long life duration of the avionics architectures.

For this future brand new development, AEROSPATIALE will take benefit of his 30 years experience of avionics systems overall design gained from Concorde to the A340 and even through on-going avionics enhancement programs as the AIM-FANS and the New Display Systems.

C545/051/98

Using technology for business advantage – automatic test equipment

K RANDALL
British Midland Engineering, Derby, UK

I was asked to present a paper looking at the use of Automatic Test Equipment and its place in an Airline environment. With an investigation into how automatic test could reduce costs.

Automatic test for commercial airline operators has always been a costly investment, and not to be entered into lightly without a lot of investigation into the real need. Historically Automatic test equipment has been provided only where a need has been identified in the repair workshop, and only then if the through put of equipment has deemed it cost effective, or an automatic test is the only way to test the equipment. In this case the cost has to be absorbed.

In most cases the time taken to carry out a test and the complexity of the test, have been the main reasons to create the need for Automatic test methods. In the main automatic test methods have been used in the repair workshop to define the problem and support the repair, and then carry out the final certification test. This has always been accepted, this paper will look at another way automatic testing may be identified as a useful cost effective method of reducing the overall costs associated with Commercial Airlines.

Automatic test, (a name meaning any method that uses computer technology to control an automatic test method, that can replace the manual testing methods that have been traditionally used), has been used for some time as an on-aircraft test, and as a test method in the workshop to enable a components serviceability to be established, prior to return to service.

Let us first of all look at the on-board use of automatic test (known as Built-in Test Equipment BITE), this has developed along with the commercial development of computers and not always with the aircraft technician (Licensed engineer) in mind, as the commercial market is always the main market in component terms. Some of the design idea's are coming from this area, with the computer games market also having its impact.

Aircraft systems have become mush more reliable as technology and construction methods have advanced, and therefore design methods used for on-board test must take the technician simply and straightforwardly to the right conclusion to enable the problem to be identified and a fix established. If the route he has to follow is complicated he will just change something and probably add to increasing burden of component 'No fault found' after test, statistics.

Design engineers have to get as close as possible to providing a simple statement of the required action. The defect information can be kept in memory for retrieval later, as this is not always needed by the technician at the time.

We have just mentioned 'No fault found' this must be one of the greatest cost generators in the airline maintenance business, and we, the technicians in the maintenance business call on the designers to find ways of limiting this phenomena, by designing BITE systems that really help the technician make the right decision.

Over a number of years the 'no fault found' issue has been debated on many occasions, and many possibilities have been given as to the reason, with the recognition that most of these reasons were a probable contributory cause. The industry have tried to find many ways to reduce the 'no fault found' costs, a lot of these costs are generated by our inability to come to terms with, and reduce the reasons, causing 'no fault found'. This area of component maintenance continues to be a major financial burden. A number of programs have had reasonable success in reducing this phenomena, but the industry has recognized that 'unable to confirm the defect' continues to be the reason a large percentage of components are returned from the repair shop. This situation remains what ever action we take and therefore has to be managed one way or another.

On aircraft test can lead the engineer to conclusions that necessitate removal of a piece of equipment that may in fact be serviceable, or due to the lack of training and understanding, the technician decides to remove a component even if the BITE messages indicate that the system is serviceable.

The cost of returning the 'no fault found' units to a serviceable condition can be very high, an average cost being £800 - £1000 per component. But this is not the end, the actual cost after taking into account the need to keep a number of extra components on the shelf, to facilitate the rotations through the component repair facility, shipping costs, AOGs etc. could amount to a much greater sum.

So how can we manage this condition. One way is to be in a position to declare the component serviceable at the time it is removed. This of cause can be done by understanding the technology behind the BITE messages and not removing a component unnecessarily. This needs focused training and retraining, and a recognition by management of the need by providing time and understanding. But even so a substantial number of components will be removed unnecessarily.

This is where the use of Automatic test equipment, could prove to be a way of reducing cost by using technology to good advantage.
If an installation of automatic test equipment was able to be provided at a location close to the airport, and therefore accessible, the component could be quickly tested and returned to service. The speed at which the test could be made can be seen as clearly reducing costs, especially if the alternative is returning the component to a repair facility at another location.

In trying to provide an installation capable of providing the service required, technical difficulties can be overcome but it is the need to be able to provide a slick and effective receiving and dispatch system, that is the key to maximizing the process. The speed with which the automatic test equipment can be used to accomplish the test, compared with the manual equivalent is also very important, but so is the need to recognize the importance of keeping the test equipment as a filter station, and not used as a repair station. This is a major issue if turn round times are to be kept to a minimum, and this is the key to providing an acceptable budget plan, that satisfies the money providers.
The return on investment needed to support the budget plan for an ATE installation, is satisfied by reducing the need for spares, to be kept on the shelf, this will be accomplished by screening out the NFF (no-fault-found) parts and returning them to the aircraft or quickly back on the shelf. .
The use of automatic test equipment in this way would mean dedicating expensive equipment, staff and facilities to this function, but when demonstrated (after a number of tests had been carried out), that these actions significantly reduced the number of components having to be sent for repair, and therefore reduce costs, especially where an airline out sources the repair work. It could be shown that the investment in the equipment is more than justified.

To further enhance the process, the ability to network the test equipment with the equipment manufacturer could be considered, although this level of software integration may not be acceptable as airlines would probably need to retain control. But it would be worth developing the concept, as this could reduce the time scale when a problem with software has been identified.
At the present time, the time taken to introduce a new LRU to be tested on the equipment is unnecessarily long, due to the need to 'shop verify' the test on a line by line basis in accordance with the manufactures component manual. The Atlas computer test language identified in the component maintenance manual, in most cases, is not used by the manufacturer for his workshop testing and therefore has to be verified by each user. This has to be done by the ATE manufacturer to prove the test program and again by the receiving user because of the certifying process. A cleaner method that can be used by both the manufacturer and the ATE user is much needed by the Airline community.

The Atlas language is dated and many contend it is a test specification language not a computer language. A more robust language like 'C' could be used which would lower the test costs to airlines because of its more general use, and would solve many problems. Atlas test program writers are few and far between. How can you have equipment test specifications developed from the component manufacturers own internal test documentation and used to release the component from his facility, and a different specification (ATLAS) used in the component maintenance manual, developed only to satisfy some other requirement. These extra costs have to be passed on.

What is needed here is a fresh look at test languages (specifications) this is being done by the industry, and development of the next generation ATLAS 2000 may have the answer to reducing costs within the ATE user community.

C545/046/98

Augmentation of head-up guidance by fusion of data from independent sources

G R SLEIGHT BSc, CEng, FRAeS
GEC Marconi, Rochester, UK

SYNOPSIS

The advent of Head-Up Display (HUD) systems to civil aircraft has been long in gestation. The technology factors which limited the introduction of such systems are discussed. Such HUD systems allow reductions in take off and landing minimums not otherwise possible with the existing avionic fit on the relevant aircraft. The data presented to the pilot is, necessarily, considerable to handle this complex and safety critical task.

A unique feature of a HUD system is it's ability to show spatially related information in the correct and real world position. The impact, and status, of work to fuse data from imaging millimetric wave radar, infra red sources and terrain data bases for HUD display is discussed. Such supplementary systems can further enhance safety in the critical phases of let down, approach and landing.

INTRODUCTION

The advent of Head-Up Display (HUD) systems to commercial aircraft has been a long time in gestation. The genesis of such systems lies in work done in the early 1960s (some 35 years ago!) at the Blind Landing Experimental Unit (BLEU) - the establishment which is now DERA, Bedford. This work was primarily focused at the establishment of hardware and control law criteria - (it predated software!) for fully automatic landing under the weather criteria of Category IIIb. From such work came the original UK aircraft certificated to such levels - the Trident and VC10. The scientific work done had a major influence on other developments for all weather landing at the time as applied to aircraft such as the DC10 and Tristar.

In parallel with the automatic first operable flight control systems, BLEU also carried out extensive trials of Head-Up Displays fitted to their Comet and other experimentally configured aircraft. This work led to the Belfast freighter aircraft (at that time on order for the Royal Air Force) to be fitted with Head-Up Displays to be used in conjunction with the automatic flight control system which gave a Cat III capability to the aircraft.[1]

Although popular with the aircrew and, technically adequate against its sub-system requirements, this early work, which included Long Beach based trails on various models of the Douglas DC9 typified by the installation shown in figure 1 did not lead to fruition. Although some airline technical reservations were expressed in the late 60's[2], they did not give the whole story which, with the wisdom of hindsight, it is now easy to see. These were primarily:

- Inadequate Integrity - single lane, non fault tolerant system, incompatible with Cat III operations.

- Inadequate Field of View - inability to maintain the actual velocity vector (especially in azimuth) in the HUD field of view, given a crosswind.

- Inability to overlay data from other independent sensors - such as FLIR, Radar and so on.

- Finally with these technical fragilities it became impossible to create a viable economic case due to the HUD system giving no increase in the operational capability of the aircraft.

In the early 90's (some 30 years later!) technical solutions to the integrity and field of view issues had been established. In addition military use of HUD systems had now evolved to the HUD being classed as primary flight instrumentation and very considerable flying experience gained from flight operations with imaging Head-Up Displays. The principle imaging source for such raster capable military HUDs being infra-red since their primary purpose was to aid low level night operations. These developments increased interest in the commercial sector not simply for Head-Up Displays but for the ability of such displays to show in a real world overlay mode the images from weather independent sensors. Serious trials were conducted of such systems the first involving a University of Maryland operated Cessna 402 and the second was a fully instrumented (fog characteristic measurement) FAA operated Gulfstream II. Sensors used were both imaging millimetric wave radar and forward looking infrared. Radar wave lengths which have been used in such trials are 35 GHz and 94 GHz. FLIRs have operated in either the 3-5 micron or 8-12 micron bands.

OPERATION

A unique feature of a HUD system is the ability to show information in the spatially correct position. This allow cues such as the aircraft velocity vector and runway position symbols to be generated and displayed in their correct spatial position however, to make this practical, the HUD itself must typically provide a field of view to the pilot of at least 30° in azimuth by 20° in elevation. Achievement of these large fields of view has become possible using novel optical systems especially those relying on selective wavelength imaging combiners. Nonetheless the installation issues of a practical system within the limited space available in a commercial flight deck are considerable, especially so for the jet more limited space in commuter aircraft/business jets. The ability to solve these installation issues is a key factor to a potential satisfactory installation and certificatable HUD system and the ultra compact design achieved by GEC Marconi for such equipment was a major factor behind their selection to supply such equipment to confer Cat III a capability to the Gulfstream IV and V aircraft and the new generation B737 aircraft. The compact installation achieved is shown in figure 2.

Coupled with real world cues, is the need to show considerable additional information such as pitch, roll, speed, height and so on plus a guidance related information much of which involves aircraft dependent algorithmic computation for the late stages of the approach, flare and touch down. A very considerable volume of work has been done over the years in simulators[3] to evolve the basis for acceptable symbolic guidance which is largely manufacturer or user independent. Typical approach symbology is shown in figure 3.

Clearly the integrity of all such HUD data is a major factor affecting use and certification. Electronics are now used to monitor performance by forward and reverse path computation such that the integrity, defined in terms of the probability of displaying hazardly misleading information to the pilot, equals that of the multi-function display viz. a figure of 1 x 10-9. Safety is further enhanced by a very high time between any failures plus the ability to show collision avoidance (TCAS) symbology and guidance. The incorporation of TCAS data is the first production implemented step of fusion into the HUD of independent sensor data.

Data from such sensors can be used in different ways - the most straightforward is by the generation of icons drawn cursively on the HUD to provide further guidance and cues to the pilot. The alternative is to provide a scanned video overlay of the real world scene (as derived from suitable sensor/s) to the basic cursively drawn display. These two approaches are often referred to as synthetic vision and enhanced vision. Although synthetic vision is a less desirable sounding implementation it has been with us in some form for many years in the shape of runway icons drawn on the HUD from ILS data plus other known aircraft parameters to establish the correct shape, size and orientation of such an icon.

Much additional useful cues for data and guidance can be presented in Head-Up from once the basic installation has been implemented. Use of TCAS data has already been mentioned. The biggest single cause of civil aircraft accidents remains that of controlled flight into terrain (CFIT). As the earth does not suddenly change shape and the technology currently exists to determine where an object is in space in three dimensions to an accuracy of a few metres at all times CFIT accidents are wholly avoidable from a technology viewpoint.

GROUND PROXIMITY WARNING

The requirement for a robust Ground Proximity Warning System (GPWS)to provide serious assistance in the CFIT problem is typically for a nuisance alarm rate of one in one million iterations. This is unlikely to be achieved by the existing GPWS mechanisms which combine barometric height and radar altimeter measured ground clearance. This is particularly true during low level flight over hilly terrain.

This is because such systems have no knowledge of the terrain being overflown, or the terrain about to be overflown. Consequently they generate warnings whenever steeply rising terrain is encountered, regardless of whether or not the aircraft can clear the rising terrain (see fig 4). If the aircraft can clear the rising terrain without changing its flight path then this is classified as a nuisance warning. Additionally such a system cannot look forward over the terrain to see what is in front of the aircraft. For example if the aircraft is flying towards steeply rising terrain a warning is not generated until it is actually over the steeply rising terrain, by which time it may be too late to avoid the ground as shown in figure 5. Consequently any system which is to avoid nuisance alarms, but still give a warning if the aircraft is in danger of flying into terrain must be referenced to the terrain about to be flown over.

The avoidance of nuisance warnings can be achieved by using a terrain database. A database has a low recurring cost and a system using a database can be almost totally passive.

In conjunction with its Terrain Reference Navigation (TRN) work, GEC-Marconi Avionics developed a Ground and Obstacle Collision Avoidance Technique (GOCAT).[4]This utilises the same terrain data employed by the TRN system, with the addition of obstacles. Based upon the precise three dimensional position information provided by TRN, GOCAT extrapolates the current flight path forward, assuming the pilot initiates a hard pull up. It then generates a search area covering a swathe of terrain either side of the predicted flight path. The extent of the search area is bounded by the navigation uncertainty and the allowed aircraft climb rate and energy status. This search area is compared with the terrain and obstacle elevation database and the terrain clearance calculated. The results from this process determine whether or not a warning is generated.

When performing these calculations GOCAT uses a terrain database so that it has a knowledge of the actual terrain being overflown, and the terrain about to be overflown. In addition to enabling GOCAT to be referenced to the real world, the use of a terrain database means that GOCTAT is passive and can take into account obstacles simply by adding them to the database. GOCAT also takes into account the accuracy of the navigation data, by enlarging the area of the database which is searched if the accuracy of the navigation data is poor. Consequently whatever the accuracy of the navigation system a check is made that the actual terrain to be overflown can be cleared.

By using the TRN position the search area used by GOCAT is minimised as it is checking for a safe clearance height over the terrain which may be potentially overflown. If a non TRN system is used then the search area must be expanded to take into account any difference in reference frames between the navigation system and the database. This enlargement of the search area in undesirable because it increases the likelihood of nuisance alarms.

GOCAT employs a terrain and obstacle elevation database and is not required to estimate the shape of the ground ahead. Therefore, nuisance and missed alarms due to unpredicted changes in terrain slope are completely eliminated.

The system operates on the assumption that the pilot knows what he is doing until it become obvious that he does not. Thus the system does not generate a warning to avoid the terrain until the pilot has to take an avoiding action, taking into account the climb performance of the aircraft, the energy state of the aircraft and the terrain safety margin. Warnings are generated in sufficient time to enable the pilot to act by accounting for pilot reaction and aircraft response times as shown in figure 6. The system also takes into account turning flight, and checks that both the turning track and a straight ahead escape route are both safe. This ensures that a wings level straight ahead pull-up is possible to allow for an initial reflex escape manoeuvre when a warning is generated - see figure 7.

An additional feature of GOCAT is that it will generate 'no turn' warnings to indicate the presence of high terrain and obstacles to the side of the aircraft's predicted flight path which cannot be cleared. In addition to these operating modes GOCAT includes a 'no turn' warning. If, for example, the aircraft is flying along a valley its current course may be quite safe but the sides of the valley may be too steep to clear safely. In this instance GOCAT will provide an advisory warning that a turn to either port or starboard would be unsafe.

The UK Ministry of Defence conducted an airborne evaluation on a twin engine turboprop aircraft. The results [5] showed good GOCAT performance over rolling hills and towards steeply rising terrain.

The database associated issues to be addressed in ground proximity warning include the likely minimum clearances to be achieved and the amount of processing power to be allocated to the task. In its current military implementation GOCAT employs a 200 metre grid of terrain elevation data, produced by data thinning DLMS DTED Level 1. Obstacle data is added to this grid. This allows GOCAT to iterate at 4 Hz with a lookahead range of 3 kilometres. Updates look beyond 3 kilometres out to 10 kilometres to ensure that no dramatic terrain changes create problems.

The availability of the terrain database for commercial operations is no longer perceived to be a problem as Jeppesen Sanderson have generated a low resolution (30 arc second grid) for virtually the whole world. Databases of higher resolutions for many areas can readily (if expensively) be generated from Earth Observation Satellite images (SPOT and ERS). Further terrain and obstacle data are expected to be provided by the proposed Space Shuttle Topography Mapper mission.

Procedures for verifying the integrity of database management processes from initial production through to installation in the aircraft are expected to follow the requirements being defined jointly by EUROCAE WG13 and RTCA SC-181 in DO-200A, and DO-201A.

Work is continuing to define the minimum information required to satisfy the database validation criteria which will in practice be defined by equipment manufacturers in agreement with the database providers and the certifying authorities.

GOCAT outputs (generated at 4 Hz) comprise GOCAT status and warnings. The warnings generated by GOCAT can provide the pilot with a simple and unambiguous 'pull up' warning with a back up 'no turn' advisory warning. These messages can be conveyed to the pilot in a number of ways:

- As an audible warning.

- On the Head-Up Display, with a 'pull up' cross, and 'no turn' warning arrows.

ADDITIONAL DATA FOR APPROACH AIDS

Another interesting recent concept, albeit one which failed to attract significant user support was the "APALS" system proposed by Lockheed Martin. This used standard weather radar to pick out significant known and distinctive cues as relevant to specific airports. From the computed orientation of such cues relative to the aircraft and known height it was suggested that precision approaches could be flown in weather to a pre-determined runway. This in practice has not proved adequately robust to operate in stand alone mode but could provide an additional integrity cross check to add further to differential GPS based approaches. Ongoing work on DPGS is intended to ensure approach guidance looks the same to the pilot as ILS and MLS data as portrayed on the HUD. As such for certification purposes DGPS will need to establish integrity matching that of ILS and thus any additional confirmation such as that potential available from an APALS type system comes only in the nice to have category which in general does not impress airline commercial management.

For many, especially US based, airlines, an analysis of their route structures shows that typically less than 10% of their movements are into Category III capable runways. To illustrate the scale of this issue these (non Cat III) runways include those at Washington, National and New York, La Guardia. DGPS holds the promise of allowing Cat III level approaches to all such airfields. However even if the aircraft system becomes adequately capable to allow Cat III landings (with DGPS) at such airports issues are about roll out, turn-off and taxi operations.

As such at least the lighting standard of runways will need to be that capable of supporting Cat IIIa to allow operation in this weather level with a HUD involved approach and landing. This is because it is a requirement for the pilot to see an adequate portion of the runway environment to complete the flare and roll out. Enhanced Vision systems offer the potential to compensate for lower runway lighting standards.

The late phase of roll out, turn off, taxi and gate parking requirements in low visibility (Cat IIIa) are also extremely challenging. NASA have recently conducted work [6]on how to utilise a HUD system to assist with such activity under their "Terminal Area Productivity" (TAP) programme.

In this programme airport related data was fused from three sensors: the airport surface detection radar, the aircraft tagging and identification multilateration system and the aircraft's automatic dependent surveillance broadcast system. This information coupled with position data blended from DGPS and aircraft inertial sources was used to update both detailed moving map information as shown on the multifunction displays and steering/taxi cue guidance data giving in its correct spatial position on the HUD. Icons representing runway/taxi way side lights were displayed along with cues to give tends, speed errors and alpha numeric information as relevant to the required taxi route. The taxi route instructions were sent to the aircraft by the ground controller by data link. Such information was given in parallel to the voice channel.

ENHANCED VISION

To turn to enhanced vision systems. Current EVS will not see through all weather conditions, all of the time however they perform well most of the time. The EVS work carried out some five years ago by Maryland University has already been mentioned. Following extensive demonstrations the FAA gave approval for hand flown approaches on their aircraft to Cat IIIa minimums (50' DH, 700 RUR) on Cat I ILS beams.

To operate with an EVS sensor, the HUD system needs to incorporate a raster mode in which a raster image from the sensor is displayed and overlaid with the normal HUD symbology and guidance icons drawn cursively in the flyback time of the video. This requires very careful design of the HUD system to achieve this functionality without inducing a cooling or reliability problem with the resultant power densities in the equipment.

Two forms of imaging sensors have been extensively investigated to date, Infra Red and Millimetre Wave Radar. Each has it's advantages and disadvantages. Infra red sensors offer good resolution at increasingly reducing cost but do not penetrate the weather as effectively as Radar under many conditions. The FAA/DOD/Industry Enhanced vision program investigated the performance of a 35 G Hz Radar and a 3 to 5 um Kodak FLIR in measured weather conditions. The program demonstrated [7]that pilots were routinely able to fly at Cat III a minima at Cat I facilities.

Recently, interest has increased in the use of IR to detect runway lights both during the approach and through landing and taxiing. Many fold improvements of visual range have been demonstrated on ground tests to date which hold the promise of cost effective terminal guidance where the use of Radar is not possible. This work offers particular promise when coupled with the growth in performance being achieved by uncooled IR systems. Such uncooled IR initially provided very poor resolution but recent gains in performance are such that such IR will clearly offer adequate performance for roll out and taxi with a sensor available at very modest cost.

Further work using a 94 G Hz imaging radar has been carried out by the GEC-Marconi company, Lear Astronics on a variety of platforms including the Maryland Cessna 402, a B707 and this activity is now transferring to a C130 fit. An image from the C-402 flights is shown as figure 8. The penetration achieved by the radar in the weather conditions tested is much better than FLIR and additional the 94 G Hz radar gives higher resolution than a 35 G Hz system due to the shorter wavelengths involved.

Such FLIR and/or imaging radar sensors offer a real capability of imaging the runway (and all taxiways) in a wholly robust manner and, in conjunction with an imaging HUD, overlaying the runway image in the spatially correct position. The detailed methodology of selecting the optimum sensor or data fusing the sensor image in a satisfactory way remains to be fully addressed as does the issue of when and how to transpose from the sensor scene to the weather attenuated "real" scene.

Nonetheless such enhanced vision systems offer the airlines a system which will be operationally useable in most conditions most of the time. This will allow throughput operations approaching those achieved in visual conditions and help eliminate the airport capacity bottlenecks which currently occur when visibility reduces below the level required for visual approaches. The underlying concept is that the vision enhancement provided to the pilot allows him to see the necessary runway cues and continue his approach "visually" in Cat IIIa conditions.

Enchanced vision thus offers:

- Reduced take off RVR minima
- Improved taxi at extremely low RVR values
- Improved flight path guidance on approach with velocity vector and other cues
- Improved energy management on approach
- Windshear guidance
- Improved vision capacity to help prevent runway incursions even in fog.

CONCLUSIONS - LANDING SYSTEMS FOR THE MILLENNIUM

The capabilities of this HUD system, made possible by recent advances in electronic, optical and sensor technology have moved us significantly closer to our goal of a fully integrated HUD equipped cockpit which will provide the pilot with the ability to operate his aircraft safely in adverse conditions with minimum workload.

We see this goal as an essential step towards the realisation of true Free Flight - providing the pilot with the means of safely operate autonomously in conditions of heavy traffic and poor weather conditions.

REFERENCES

1. "When will Head-Up go Civil" -B S Wolfe and G R Sleight Flight 26 December 1968

2. "Technical Reservations in HUDs" - Capt F L Wallace, Panam - Chairman of ATA HUD working group, ALPH forum - Seattle 1968

3. "Human Factor Issues in Head Up Display Design" -D Weinstraub and M Ensing, University of Michigan 1992

4. "GCAS - The CFIT Warning Concept for the Next Millenium" -A J Henley, C Hewitt and S Broatch (GEC-Marconi), ERA Technology Conference, London 21 November 1996

5. Air Warfare Centre SAOEU Trials Report - AWC/WAD/72/326/TRIALS Trial PULP Stage 1 5 January 1996

6. "Low Visibility Landings and Surface Operations" S Young and Denise Jones (NASA Langley), ICAO Journal Jan/Feb 98

7. "Synthetic Vision - Aircrew in the Fog" - M Burgess and Dr R Hayes, Proceedings of 11th IEEE/NASA DASC Seattle, Wa, October 1992.

1. DC-9 Installation - Elegant but very limited field of view.

2. Installation of HUD - Needs to be mechanically rigid, retain boresight integrity but with large head clearance.

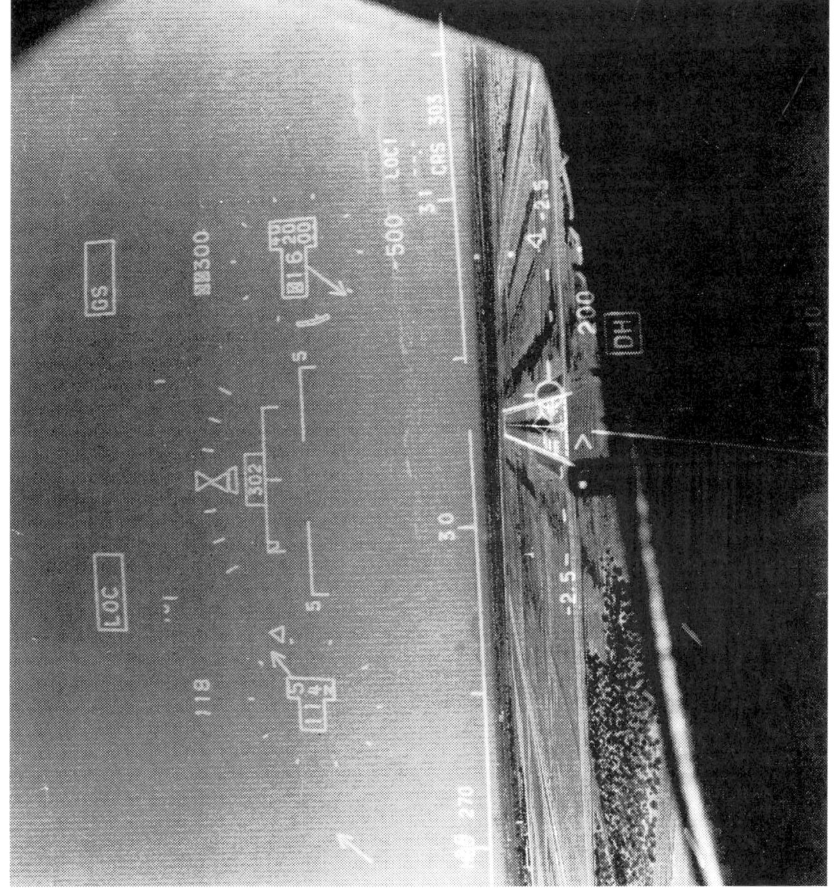

3. Approach symbology at 200 ft. Cat II decision height.

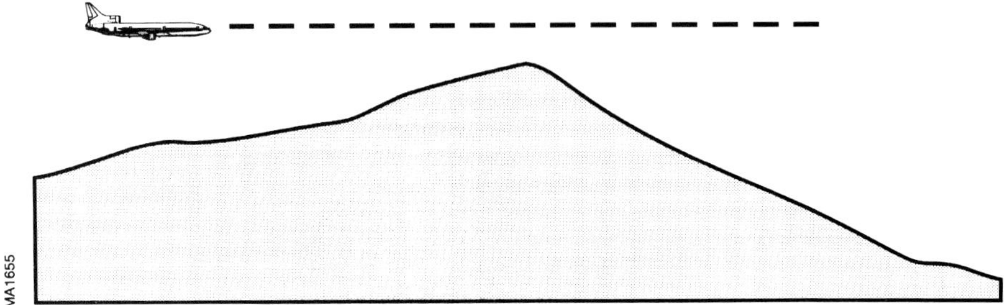

4. Flight with rising ground - no warning necessary

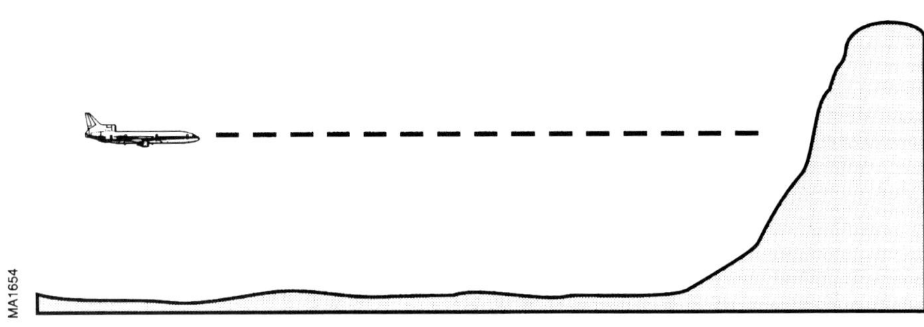

5. Flight towards a cliff - a warning is vital.

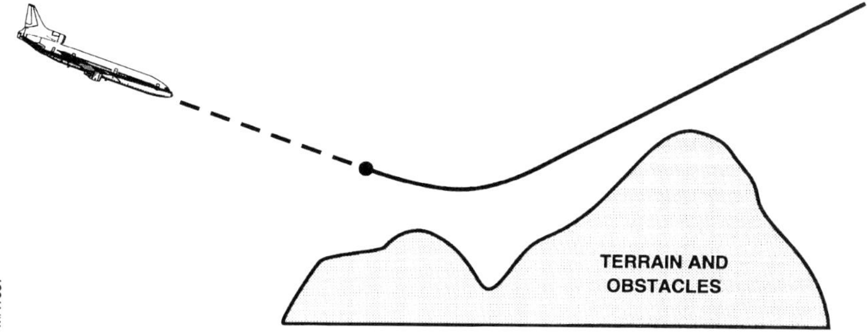

6. Warning for vertical plane recovery.

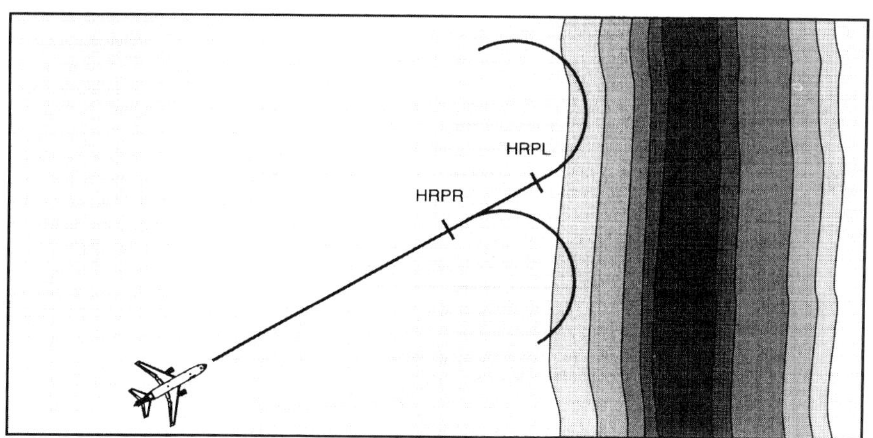

7. Terrain warnings for left or right manoeuvres.

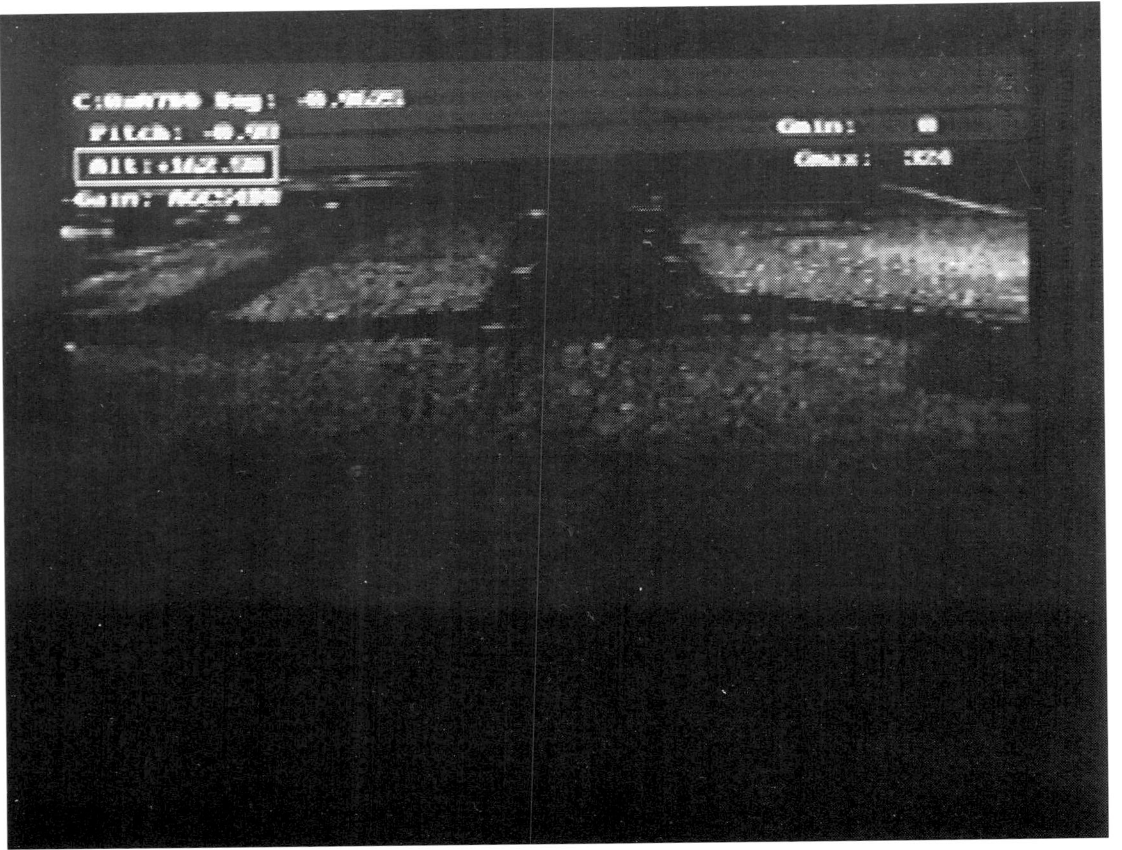

8. Actual 94 GHz imaging radar image for an approach into Point Magu, California.

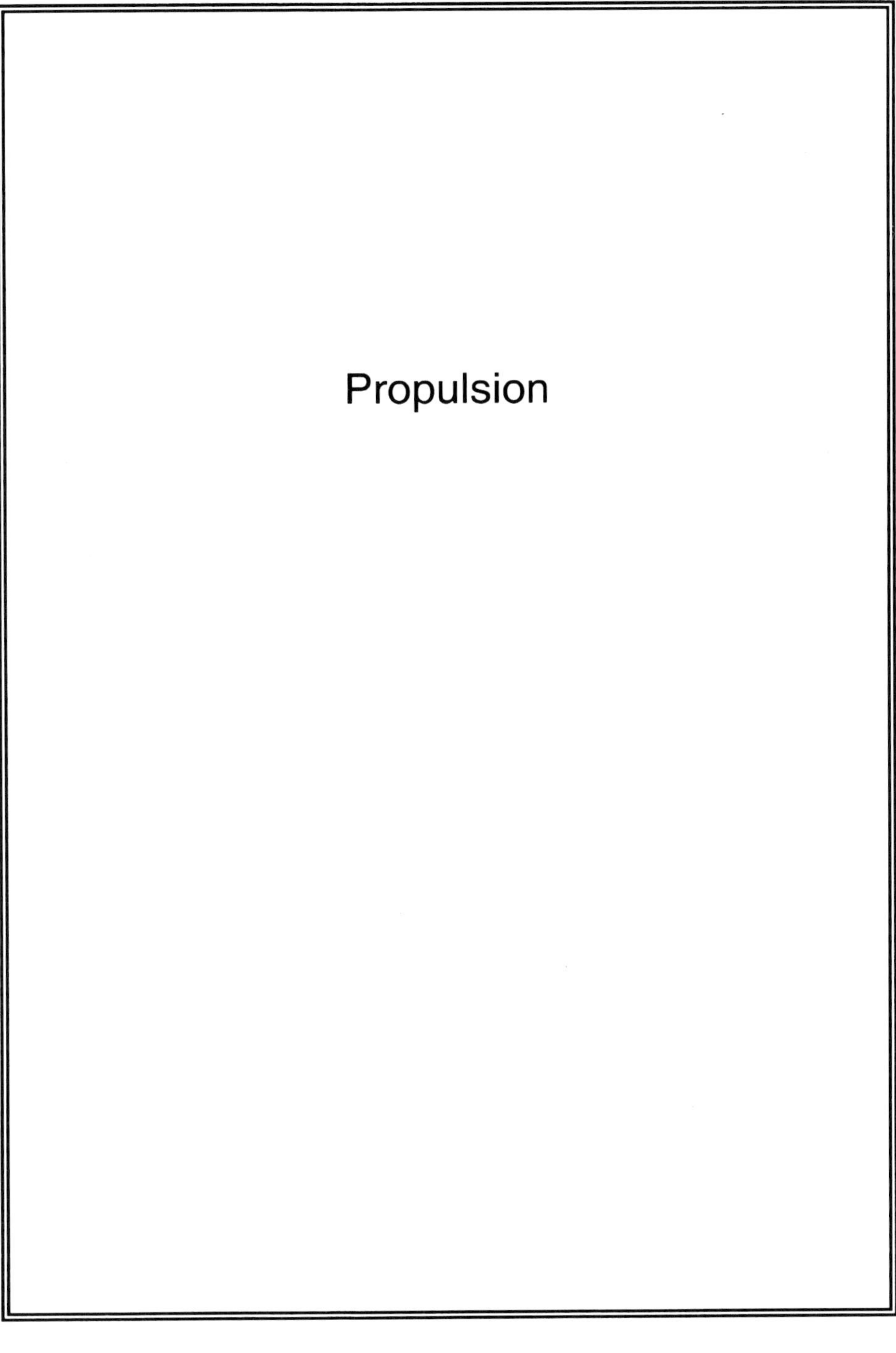

Propulsion

C545/023/98

The diffusion-bonded/superplastically formed wide chord fan blade and fan key system

A D LLOYD BSc and **G A FITZPATRICK** BSc, PhD, FIM
Rolls-Royce, Derby, UK

Synopsis
Faced with growing competition, and pressures on lead time, weight and cost, Rolls-Royce embarked upon a new method of manufacture for their wide chord, hollow fan blade, improving upon the successful first generation design.

The new technology employed solid state diffusion bonding, and involved a fundamental change to the design and analysis systems, and to the manufacturing processes.

A manufacturing and engineering team, through concurrent engineering, developed the necessary titanium fabrication technologies and a totally integrated, computer based design to manufacture system.

The team success has made a major contribution to the Rolls-Royce share of a very competitive market.

1. THE FAN BLADE DUTY

On modern large jet engines, the first row of blades, the fan, generates over 75% of the total engine thrust. The horsepower transmitted by a single blade on the latest Trent engine is equivalent to 50 family saloon cars. The fan tip travels at over 1600 km/hr, 40% faster than the speed of sound. At this speed the centrifugal load on each blade root is around 90 tonnes, equivalent to 10 double-decker buses.

A kilogram saved on the fan has a considerable gearing effect and can lead to many times that saving on the total engine weight. A 1% improvement in fan efficiency reduces the fuel burn of an engine by 0.75%. For a long haul operation this is equivalent to over 1 ton less fuel load, or 50 miles extra range, or 8 extra fare paying passengers.

But in doing all this, the fan blades must maintain their integrity under a range of mechanical disturbances. It must be capable of ingesting a flock of birds, with local blade loads of up to 20 tonnes, and continue to generate power to meet safety requirements. It must withstand sustained flight through hailstorms or large flocks of smaller birds, and in the unlikely event of a blade failure, its energy, equivalent to family saloon travelling at 288 km/hr, must be safely contained within the engine casing.

Under all these conditions, the strength and fatigue properties of the titanium alloys are used to their limit to exploit weight opportunities. It is not difficult to see, therefore, that the mechanical and aerodynamic performance of the fan plays a key role in the overall competitiveness of the powerplant and its operation in the service environment.

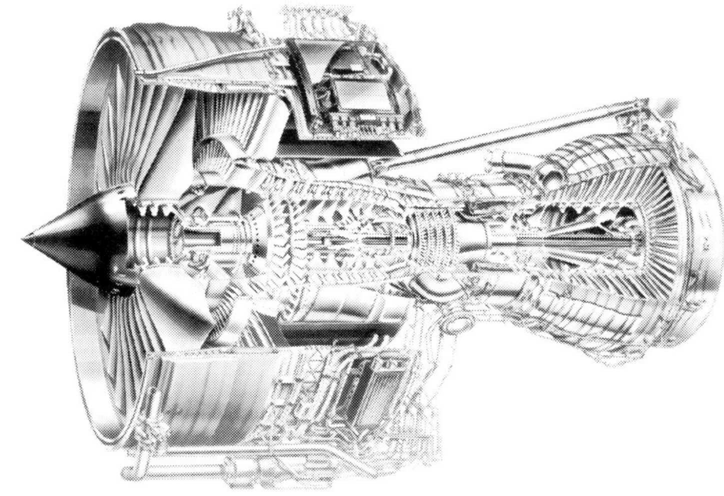

1. The Rolls-Royce Trent 800

2. FAN BLADE DEVELOPMENT

2.1 Clappered and Wide Chord Fan Blades

The first RB211 engines featured 33 fan blades with mid span snubbers to control vibration, but these interfered with airflow and efficiency. The first revolutionary step in improving performance came with the deletion of these snubbers, requiring an increased chord to restore vibration control. Such a blade in solid titanium would have been unacceptably heavy, hence the development of the hollow fan technology.

Rolls-Royce led the field in pioneering the design and manufacture of hollow titanium fans. The first generation design features a honeycomb core, and was introduced into airline service on the RB211-535E4 engine in 1984, followed by the V2500 and RB211-524 G&H engines, in the thrust range up to 276kN.

In the late 1980's and early 1990's, major commercial airlines identified the need to extend their service operations to longer flights, covering greater distances, carrying extra passengers even more economically. The large twin aircraft was, therefore, born with the consequent requirement for engines with more thrust, in the range 289kN to over 445kN. The consequence was that our American competitors, Pratt and Whitney and General Electric, developed engines with wide chord fans for the first time. This, in turn, led Rolls-Royce to develop an even lighter, high performance fan - the second generation design

2.2 1st and 2nd Generation Fan Designs

The first generation fan design is a fabricated titanium construction of an internal honeycomb core sandwiched between pre-formed external panels. Its key technology is low pressure diffusion bonding, which is a process developed specifically by Rolls-Royce for this application. Whilst this was innovative at the time, and has an excellent service record, it is both complex and labour intensive in terms of engineering and manufacturing activities. The parasitic nature of the honeycomb core, since it cannot support its own weight, leads to a relatively heavy blade.

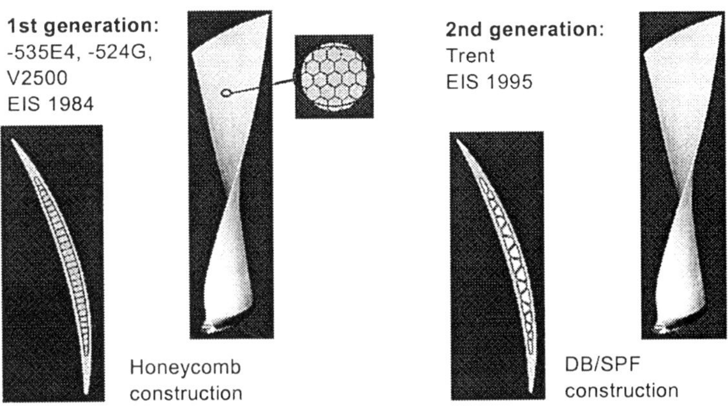

2. Wide Chord Fan Technology 1st and 2nd Generation Designs

The basic concept for the second generation fan blade was, therefore, to replace this core with an even lighter, self supporting internal structure, and to develop a simplified, more economical, production method, saving weight and cost, and reducing lead time. This involved a fundamental change to the established design and analysis systems, and to the manufacturing processes.

In particular it allowed the design of a continuous, load carrying, membrane core, and also allowed the key joining operations to be carried out by higher pressure solid state diffusion bonding.

The challenge was to apply these technologies to a complex three dimensional geometry and to deliver a product, capable of working in a severe service environment.

A totally integrated electronic system was required which allowed automatic transfer of data between subsystems in the design-to-make process. The objective was to provide a reliable environment in which to exploit design opportunities, and enable a lightweight, cost effective and high integrity product to be manufactured in the minimum time to market.

2.3 Technology Acquisition Process

Concept work began on alternative hollow designs in 1980. In parallel, laboratory programmes were developing joining and forming processes to produce high quality metallurgical bonds, a variety of internal core structures capable of fatigue and impact resistance whilst achieving dimensional accuracy. This demanded a thorough understanding of the diffusion bonding and superplastic forming processes, supported by the most up to date design and analysis systems.

Demonstrator programmes were launched to design, manufacture and test prototype components and provided the confidence to proceed with a Diffusion Bonded, Superplastically formed, or DB/SPF fan blade, on the Trent engine programme.

In 1992, the Company launched a totally integrated, computer based, design to manufacture system, the Fan Key System (FKS). Whilst the DB/SPF technology was applied to the Trent 700 and 800 fans using existing system architecture, the FKS system improvements were developed in parallel. This enabled the Trent to realise the immediate benefits of weight, cost and lead time.

Cultivation of the new technologies was achieved by the formation of an integrated team of engineers representing the key elements of the design to manufacture process. Central this operation is the Design Make Plan, which lays down the agreed cascade of deliverables necessary to give the final product on the required date. All new fan designs are driven through the team and manufacturing information is released at stages during the design to ensure full visibility of engineering intent and allow staged launch of material and tooling.

3. MANUFACTURE METHOD

3.1 Manufacture Method

The essential features for the production of the DB/SPF fan blade are

a) A simplified and economical method of manufacture. This has evolved from the first generation technologies and has lent itself to the automation of the key processes and their verification.
b) High pressure diffusion bonding. This is a less complex but more sensitive joining process, and demands component assembly in an environmentally controlled clean room.
c) A lightweight internal core produced by superplastic forming. This process exploits core material properties to produce a uniform structure with an enhanced capability.

The DB/SPF fan blade is manufactured as a fabrication of 3 titanium alloy sheets. The chemically cleaned panels have their mating surfaces coated with an inhibiting compound in a pattern derived from the intended core design.

This three-piece flat pack is then assembled and welded ready for high pressure diffusion bonding. A clean room provides the necessary environmental condition for these sensitive operations. The diffusion bonding heat-treatment cycles produces a single component with high quality joints at all uncoated surfaces.

This pack is converted to an aerofoil shape by a sequence of hot-forming operations. It is then inflated in customised rigs between contoured dies at elevated temperatures using an inert gas. This stretches the central sheet superplastically under strain-rate control, and uniformly produces the internal corrugated core. The external aerofoil shape is formed at the same time.

NC machining, followed by surface processing, completes the manufacturing sequence. Extensive process control and monitoring, and novel product assessment methods have been developed, as part of a progressive product assurance regime, to verify component quality.

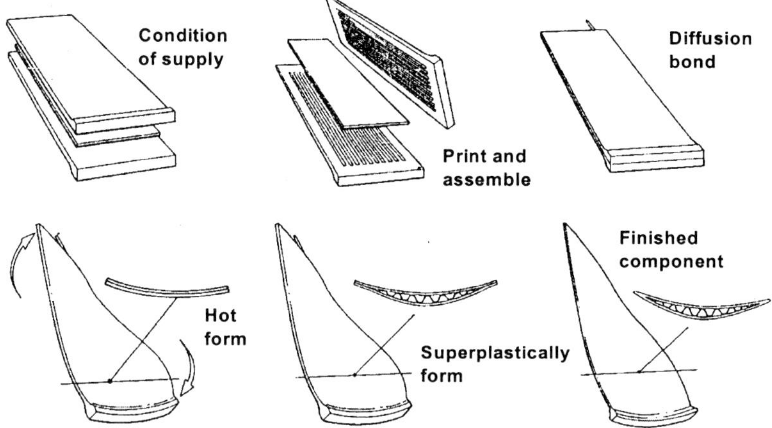

3. The DB/SPF Fan Blade Manufacture Sequence

A cross section of a DB/SPF fan blade shows a uniform core structure produced by superplastic forming and illustrates the positions of the diffusion bonds on the internal surfaces. No bond lines are apparent, as a result of atomic migration across the interfaces.

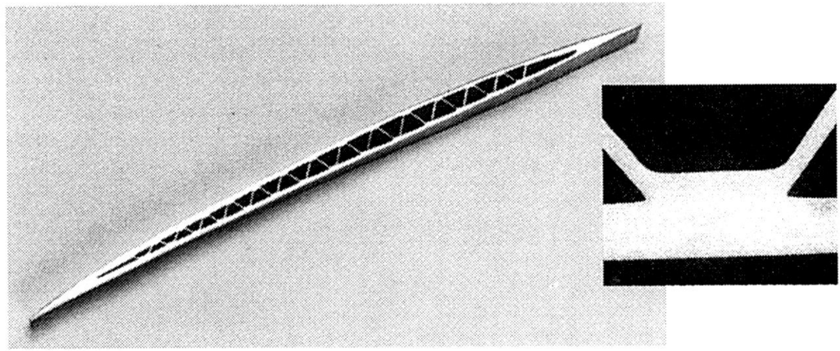

4. DB/SPF Fan Blade Core Geometry

3.2 Manufacturing Facility

A major investment in these latest technologies has taken place since 1990 at Rolls-Royce's Barnoldswick factory to provide an exclusive production facility for DB/SPF fan blades. All the processing plant is novel and has been designed in partnership with specialist British contractors. It has been installed in the factory to systems engineering principles for the most effective use of these unique resources and for optimised work-flow.

A special feature is the dedicated and isolated diffusion bonding cell. This incorporates a state-of-the-art clean room complex married to Rolls-Royce process technologies. Environmental conditions in here are continuously automatically monitored and controlled, and the staff operate to strict working disciplines. All the key operations, whether processes, machining or NDT, are NC programmable and digitally controlled. Importantly they are linked via a local area network to provide data for both component tracability and statistical process control. Visits by potential airline customers to this unique manufacturing facility are now a key part of sales campaigns.

4. THE FAN KEY SYSTEM

The successful development of the DB/SPF technology on the Trent engine, and the continuing demands for further improvements resulted in the launch of the Fan Key System initiative in 1992. It would link all the activities involved in the conception, design definition, evaluation and manufacture of the DB/SPF fan.

4.1 Process Outline

The Fan Key System is a fully integrated suite of specialist software tools, written primarily within Rolls-Royce. It straddles both Engineering & Manufacturing functions and comprises a number of subsystems :-

- An Aerodynamic system, utilising modern computational fluid dynamic tools for the design and evaluation of aerofoil geometry, and provides aerodynamic data;
- A Mechanical Design Geometry system, with highly automated, interactive CADDS based software, providing full electronic definitions;
- A Mechanical Analysis system, which uses advanced finite element software to analyse displacements, stress, life, vibration, and impact integrity

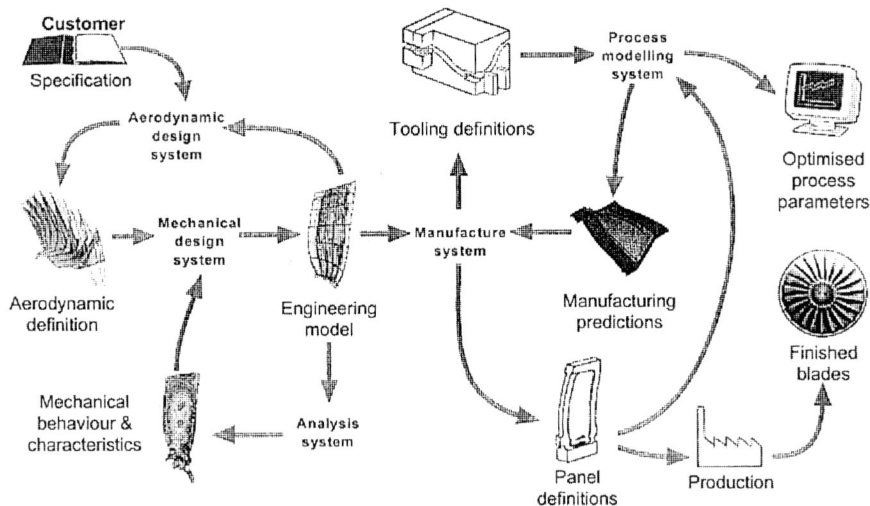

5. Fan Key System Process Outline

- A Manufacturing system, containing CADDS based software to 'reverse engineer' the manufacturing process, and generate tooling and component geometries, as well as NC and inspection data;
- and finally, Process Modelling, comprising finite element software to 'forward engineer' and predict the manufacturing process.

The total system is modular and highly automated, providing consistency, accuracy and quality. Its flexibility also allows it to accommodate evolving fan design styles. Integration of the design and manufacturing systems into one environment (CADDS) has resulted in a common currency of data, seamless operation and an increased level of quality assurance. It has reduced the number of interfaces, removed non-value added activities, and reduced the total product cycle time, resulting in significant cost savings.

4.2 Design and Mechanical Analysis System

The mechanical system includes advanced software which automatically meshes the blade ready for the various stress, vibration, and impact analyses. Finite element analysis and post processing packages are linked into the aerodynamic, performance, thermodynamic and materials systems, allowing rapid and accurate prediction of mechanical behaviour.

Steady state and vibrational stresses are mapped for the blade and disc assembly, and matched with material allowables to meet the specification objectives for component life and acceptable vibration characteristics.

The impact of medium and large birds are simulated using state of the art computer codes, in this case, Dyna 3D. This a quick and relatively inexpensive way of simulating a full fan assembly ingestion test. The Dyna 3D prediction is used as a design tool to reduce risk in the programme.

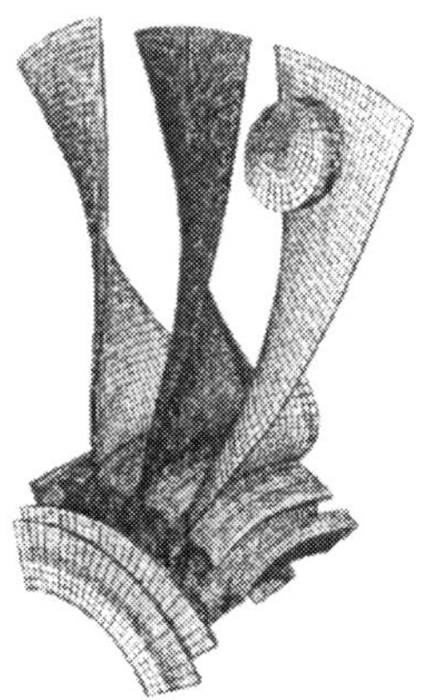

6. Dyna 3D Bird Impact Analysis **Dyna 3D Fan Blade off Analysis**

Another example of an advanced Dyna 3D simulation is a fan blade off analysis. A blade released during engine running impacts its following blade. Under this rare circumstance, the out of balance load, of some 90 tonnes, is reacted through the disc, shaft and engine structure. The release of another blade would place extreme loads on the system and, if designed for, would result in a very heavy structure. The objective, therefore, is to avoid failure of the following blade.

These state-of-the-art finite element analysis, calculates localised plastic strains and compares them with failure criteria. Avoiding failure, by thorough analysis, at this stage, can have a major effect on non -recurring costs in an engine certification programme.

Once the design solution converges to meet the specification, the final Definitive Engineering Model, or DEM, is released, providing a precise, unambiguous definition of the static fan blade. This is the starting point for the manufacturing operation.

4.3 Manufacturing System

The manufacturing part of the Fan Key System basically predicts the fabrication process, but in reverse. Starting with the finished hollow blade, it ends with a collapsed flat pack assembly.

The CADDS definition is received from the Engineering team and all the geometries required by the process, such as run outs, location features and processing allowances are added. From the progressive series of models, NC data is automatically generated for machining and verification of all the major tooling items such as the hot forming dies and for the machining of the panels. This represents a significant saving over the previous system.

In addition, simulation of the die machining processes also allows optimisation of tool cutting paths, radically reducing machine cycle times compared with traditional methods. The final flat developed state is then used to produce NC data and inspection drawings, tooling plattens to hold the material during the assembly operations, and the core pattern.

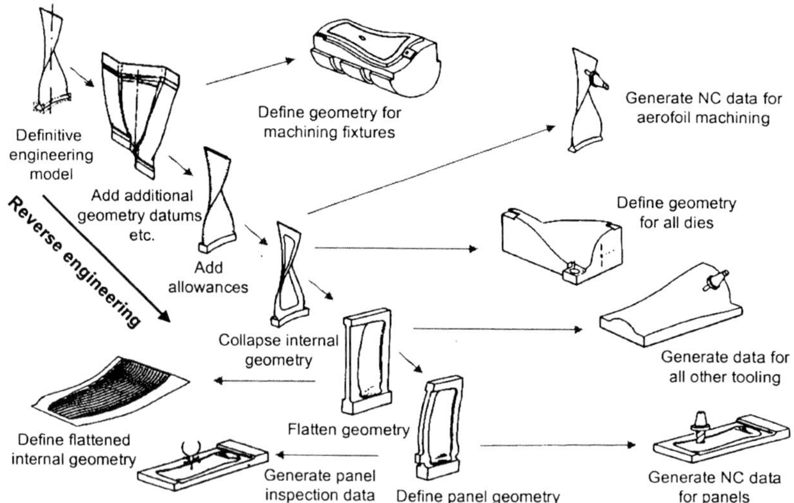

7. Manufacturing Part of the Fan Key System

The success of the manufacturing system depends on the accuracy of the empirical corrections used in the reverse engineering process. This is where process modelling can be of value.

4.4 Process Modelling

A process model is a computer based numerical simulation of a manufacturing process based on its fundamental physics, rather than an empirical approach. It "forward engineers" the process of metal forming, applying the same finite element analysis technology used to design the component. Firstly, from the fan key system geometry, finite element meshes of the flat pack and forming die are automatically created. The analysis then simulates the twisting and cambering displacements at temperature. Every stage in the process can be observed. By iterating around different twist angles and stop locations, the hot forming processes are optimised. The superplastic deformation of the internal titanium core is then simulated. This enables processing parameters to be optimised for the production of a uniform corrugated structure without necking or fracture.

In summary, the process modelling system ratifies the correct manufacturing geometries and parameters. Virtual iterations take place within the computer, before committing to hardware, saving time and cost as well as releasing shop floor capacity.

5. THE PAY BACK

Both the DB/SPF and the FKS technologies have been key contributors to the Trent engine competitive position. The benefits are a direct result of improvements to lead time, weight, and cost, as well as maintaining the Rolls-Royce reputation for technical excellence.

5.1 Lead Time Reduction

Lead times for the development and introduction of new fan designs have been reduced by 20%. For established production components, the improvement has been 40%.

The value of reduced lead time, for new applications, is that it provides a faster response to customer requirements, and an increased capability for the multiple launch of new designs. The benefit to manufacturing is to reduce inventory levels, with associated cost savings.

5.2 Weight Advantage

Compared with the first generation honeycomb design, the DB/SPF blade has achieved a weight saving of about 15% . When the effect of this weight on other components is taken into account, the fan makes a significant contribution to the overall weight advantage of the Trent engine. Translated into customer aircraft economics, this weight advantage contributes to overall fuel efficiency, increased aircraft range, and improved passenger payload.

5.3 Cost Savings

The introduction of DB/SPF technology has contributed by :-

a) Halving the number of operations needed to fabricate the component. This has directly led to a 30% reduction in operational costs.
b) Shortening the manufacturing lead-time by around 30%, yielding inventory savings.
c) Reducing fabrication tooling costs by 45%.
d) Avoiding further capital investment in the first generation honeycomb blade. Expansion of the honeycomb facility, to provide a capability at Trent size, would have required a further investment over and above that made for the DB/SPF fan.
e) In addition, the DB/SPF process has matured rapidly to achieve high production yields.

Implementation of the Fan Key System has also yielded cost benefits by:-

a) Further reducing lead time and inventory savings.
b) The faster introduction of new designs
c) Better utilisation of technical resources
d) Optimising manufacturing process cycle times, assisted by process modelling

At the end of 1997 the FKS investment had been covered and substantial ongoing savings are now being realised.

5.4 Trent Campaign Success

The fan has played a key role in the overall competitiveness of the Trent engine series, and customer confidence has been expressed by the success of sales campaigns.

Orders totalling over £8.6 Billion have now been placed by a variety of airlines preferring the Trent 700 on the Airbus Industries A330, the Trent 800 on the Boeing 777, and the Trent 500 on the Airbus Industries A340. The Rolls-Royce share of the world market on the Trent classes has risen to over 40%. The predicted world market in the Trent class of engine is estimated to be worth £80 Billion plus £40 Billion in spares.

Military applications have already been identified for the Joint Strike Fighter. Prototype trials have been carried out on static engine vanes. The technologies have also been diversified into industrial power applications and for a heat exchanger which now has entered the production phase within the Rolls-Royce plc group.

5.6 The Third Generation...The Swept Fan

Design and manufacture is now well under way on an advanced hollow fan concept for the Trent 8104 engine. This incorporate a more complex core structure and aerofoil shape, taking the technology one step further into the third generation.

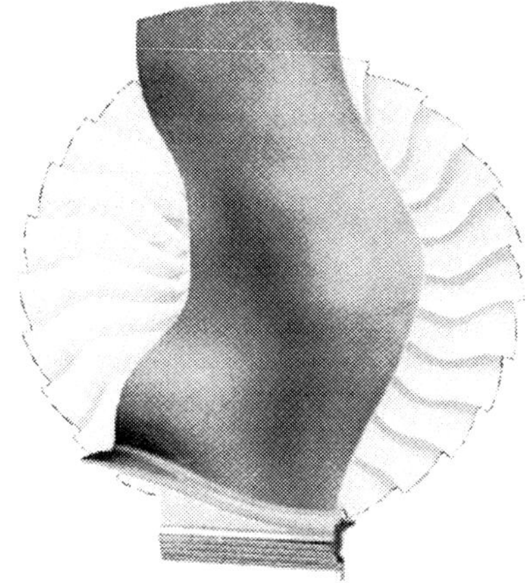

8. The Third Generation - The Swept Fan

6. ACKNOWLEDGEMENT

The authors would like to thank the IMechE, SBAC and RAeS for the opportunity to present this paper. They also wish to acknowledge the sustained and dedicated effort of many colleagues in Rolls-Royce.

C545/063/98

Development of emissions methodology to account for the global atmospheric impact of aviation

P MADDEN BSc
Rolls-Royce plc, Derby, UK

SYNOPSIS

Throughout the last decades, improvements in engine s.f.c. have been mainly achieved by the introduction of jet engines of higher bypass ratios and higher core thermal efficiencies. The introduction of these engines together with improvements in combustion technology have contributed to increased fuel efficiency and to reduced smoke, UHC, CO as well as limiting CO_2 and H_2O emissions. NO_x, however, has proved far more intractable both near airports and at high altitude due to the ever increasing cycle pressure ratios and temperatures. The current emissions regulatory regime is based on emissions production around the airport.. This regulatory regime is widely recognised as being unrealistic for modern high bypass ratio engines, and for engine emissions at altitude. This paper discusses the historical background to emissions legislation, the likely advances in combustor and airframe design, and the development of an emissions methodology to account for the global atmospheric impact of aviation.

1. INTRODUCTION

Smoke, CO and UHC emissions have been substantially cut since the first regulatory standards were issued in the 1970s, thanks to improved combustor design and the use of airblast fuel atomisation technology. The concern which in the past was focused on air quality in the vicinity of airports has now shifted to the global impact of aviation. Today, the aircraft engine emissions impacts of considerable concern are the effects of NO_x and CO_2 at high altitude.

This paper describes the historical background to engine emissions legislation, the likely future developments in low-NO_x combustor technologies, and how the traditional optimisation of the aero engine cycle affects near-airport and high-altitude emissions. The paper then examines whether the current regulatory regime developed to control emissions in the vicinity of airports is well adapted to reduce emissions, and considers the development of an emissions methodology to account for the global atmospheric impact of aviation.

2. THE ISSUE OF AVIATION'S EMISSIONS

2.1 Development of aircraft engine emissions regulations

Public awareness of air traffic as a nuisance has its roots in the rapid expansion of jet powered air services from the 1950s. Initially, the main concern regarding aircraft engine emissions was the impact on air quality in the vicinity of airports. As a result of this concern about ground level pollution, the International Civil Aviation Organisation (ICAO), following the US Environmental Protection Agency (EPA), developed standards [1] for the control of smoke and the gaseous emissions of oxides of nitrogen (NO_x), carbon monoxide (CO) and unburnt hydrocarbons (UHC). The standards were introduced through an engine certification scheme based on the Landing and Take-off (LTO) cycle (Figure 1).

Engine exhaust gaseous emissions are summed over the landing and take-off cycles, and divided by the engine rated thrust. The measurements are compared statistically against the regulations. The peak smoke measured at any condition is compared against the smoke rule.

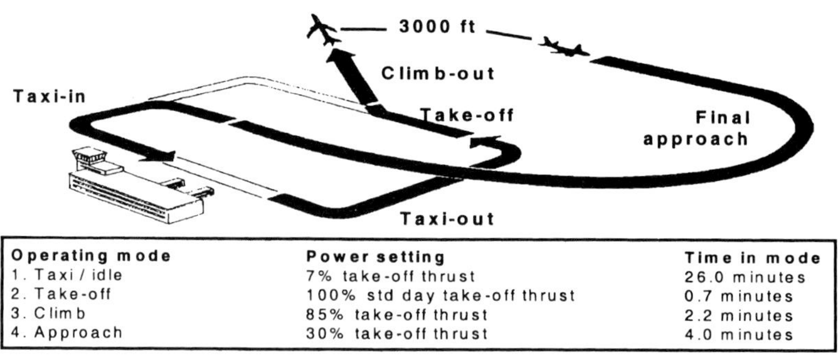

Operating mode	Power setting	Time in mode
1. Taxi / idle	7% take-off thrust	26.0 minutes
2. Take-off	100% std day take-off thrust	0.7 minutes
3. Climb	85% take-off thrust	2.2 minutes
4. Approach	30% take-off thrust	4.0 minutes

Figure 1. Engine Exhaust Emissions Landing and Take-Off Cycle

CO, UHC and smoke standards have been left unchanged since they were first issued. However, the stringency of the NO_x standard has been further tightened on two occasions. The second Committee on Aviation Environmental Protection (CAEP) meeting in 1991 led to a 20% reduction of the NO_x standard (referenced against a pressure ratio of 30). More recently, on the basis that reducing NO_x should not be undertaken at the expense of increasing carbon dioxide emissions (CO_2) [2], the fourth meeting held in 1998 has recommended increased stringency for pressure ratios below 30 with a progressively reduced tightening of the limit as the engine pressure ratio increases above 30 (see Table 1 and 2). An alleviation is also offered for low-thrust engines (below 89 kN), which, for reasons of physical size, have difficulty in incorporating the best low-NO_x technology [3]. The proposed limits are applicable to new engine types first certificated after 31 December 2003.

The adoption of such standards by the major aero engine manufacturing nations has reduced pollution over the LTO cycle, and has resulted in significant reductions in smoke, UHC (by approximately 90%), CO (70%) and more modest improvements in NO_x (10%) in the vicinity of airports. Such improvements have been made possible by the use of advanced single annular combustor technology and in a few recent cases staged combustion.

CAEE/CAEP Meeting	**Annex 16 Vol 2 Amendments**	**Standard (D_p/F_{oo})**	**Applicability**
1981 (CAEE)	Issue 1	$40+\pi_{oo}$	
1991 (CAEP/2)	Issue 2	$32+1.6\pi_{oo}$	
1998 (CAEP/4) Proposed	Issue 3	$19+1.6\pi_{oo}$ $7+2\pi_{oo}$ $32+1.6\pi_{oo}$	$0<1.6\pi_{oo}<30$ $30<\pi_{oo}<62.5$ $62.5<\pi_{oo}$

1. CAEE is the Committee on Aircraft Engine Emissions (set up by ICAO and superseded by CAEP).
2. π_{oo} is the overall pressure ratio at sea level ISA rated take-off conditions.
3. CAEP/4 presented standards are for sea level rated output >89kN.
4. The CAEP/4 standards for thrust below 89 kN are not shown. These standards vary linearly between the CAEP/2 standard at 26.7kN thrust and CAEP/4 at 89kN thrust.

Table 1: ICAO NOx, recommendations

Meeting	$20\pi_{oo}$	$30\pi_{oo}$	$40\pi_{oo}$	$50\pi_{oo}$	$62.5\pi_{oo}$	$70\pi_{oo}$
CAEE	datum	datum	datum	datum	datum	datum
CAEP/2	-20%	-20%	-20%	-20%	-20%	-20%
CAEP/4	-20%	-16%	-9%	-4%	0%	0%

NB: CAEP/4 discrepancies are shown relative to CAEP/2 levels.

Table 2: Evolution of ICAO NOx recommendations

The 1995 CAEP/3 meeting (which failed to reach international agreement) and the 1998 CAEP/4 meeting emphasised the need to reduce NOx emissions over the entire flight cycle. The LTO certification regime was the only immediate way to achieve further NOx stringency, and the major discussion was on the potential NOx/CO_2 trade-offs new regulations would introduce. Figure 2 shows the development of the NOx rules.

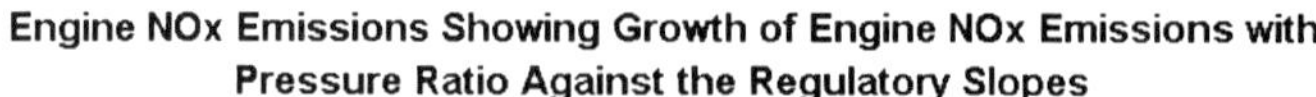

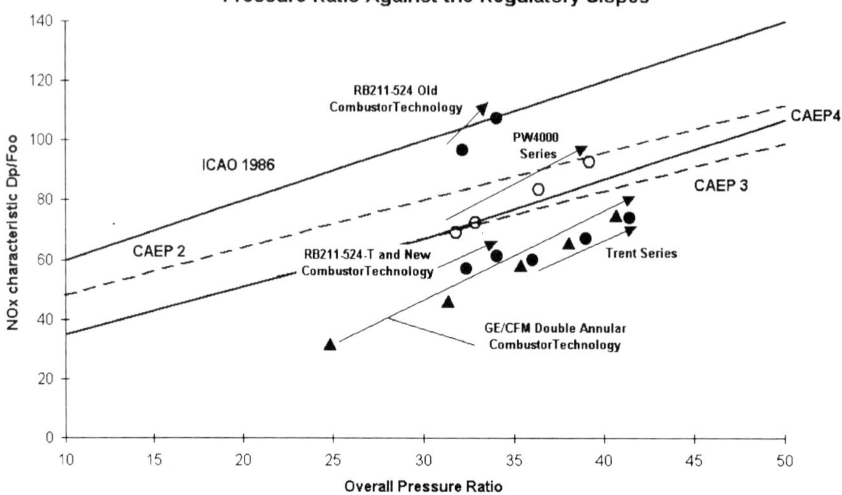

Figure 2. Development of ICAO NOx Emissions Rules

Unlike rules for the other gaseous emissions, CO and UHC, the NOx rule has alleviation for higher engine pressure ratios which is a clear admission that it is harder to control NOx at higher engine operating conditions. The original slope of 2 was superseded by CAEP2 which introduced a slope of 1.6 with pressure ratio (to maintain a 20% reduction throughout). The CAEP3 meeting tried to maintain the slope of 1.6 but this was not internationally agreed on the grounds that it would not allow engine manufacturers to grow the engine pressure ratio to minimise CO_2 emissions. Figure 2 also shows how engine manufacturers' NOx emissions grow for a given technology, and this clearly shows a slope of greater than 2! The ICAO standards have always claimed to be achievable with current technology, so at CAEP4 an international compromise was agreed. The compromise was to revert to a slope of 2 above a pressure ratio of 30, whereas below 30 pressure ratio CAEP4 was identical to CAEP3. Assuming the CAEP4 rule is ratified by the ICAO council this year it will come into place for new engine types at the beginning of 2004 (ie. certified engines are not affected.

CAEP4 also identified that the future engine emissions work program [4] should be:-

(i) to promote operational improvements including better air traffic management and a reduction in engine running around airports (eg. running 4 engined aircraft on 2 engines whilst on the ground), and

(ii) to develop a new LTO emissions methodology to make a parameter some form of aircraft productivity, and

(iii) to develop a new parameter for regulating flight cycle emissions including the cruise phase of the operation.

This paper concentrates on the latter two aspects of CAEP work, and the effect this work has on the choice of combustor and engine design.

2.2 Aircraft-generated high altitude emissions

As a result of the early concern for poor airport air quality, the standards enacted by EPA and ICAO since the beginning of the 1970's have been designed to control aircraft emissions below 3000 ft (915 m). At the end of the 1980's, however, there was increasing concern that aircraft engine emissions may be contributing to global atmospheric problems.

The majority of man-made emissions are produced on the earth's surface and emitted in the atmospheric boundary layer (the part of the atmosphere up to 1 km). Some of these emissions will penetrate into the free troposphere. A significant amount of these pollutants will eventually return to the surface through wet or dry deposition. Emissions from aircraft engines constitute only a small proportion - in the region of 2 to 3% - of man-made and natural emissions into the atmosphere. However, such emissions could have a great impact because they are emitted into the highly sensitive upper troposphere and lower stratosphere where the gases are long-lived and where aircraft are the sole source of directly injected anthropogenic pollution.

Most of the predicted increase in upper atmospheric pollution will be generated by subsonic, high-flying, long distance traffic [5]. This is because, for a short haul flight of 900 km, total cruise NO_x emissions typically represent 20% of the total, whereas this rises to over 75% for a long haul flight of 8500 km. Additionally, engines aimed for long haul operation are typically bigger and have higher engine pressure ratios which increases the amount of NOx emissions at cruise conditions. A future High Speed Civil Transport (HSCT) supersonic aircraft fleet is unlikely to enter commercial service before 2015. At any rate, the total number of HSCTs will still be limited by the market which they can generate. Subsonic high-flying aircraft will therefore keep on dominating the air transportation system in the next millennium.

Provided that global economic development is not halted by world-wide crises, global air traffic as of 1997 is expected to at least double early in the next millennium at around 2015 ([6] & [7]). Technological developments will ensure to some extent that the emission growth rate will be lower than the fleet growth rate. At current and expected combustor and engine technology levels, the net result will still be an absolute increase in global emissions.

Looking further into the future, the UK department of Trade and Industry (DTI) have carried out a recent study as part of the process to predict the growth of global NO_x emissions assuming that low-NO_x technology is aggressively introduced. The DTI study assumed a growth in air traffic of 5.4% per annum in the 1990's declining to 1.7% by 2050, and used the ANCAT 1992 inventory as base data. This work accounted for the change in fleet composition, as well as the technological improvements industry anticipates are achievable from engine and airframe (weight and aerodynamics). The top line in Figure 3 identifies today's fleet plus traffic growth NO_x emissions assuming no improvements in fuel efficiency or combustion technology over today's fleet. The effect of fuel efficiency improvements (due to normal market pressure) will reduce the global NO_x emissions burden significantly. Small NOx reductions are possible with the introduction of combustor technology at 30% below CAEP2 NOx levels. Slowing the upward trend is conditional on the introduction of ultra-low NO_x concepts capable of reducing current best levels by 60% (entry into service (EIS) of 2035 assumed in the DTI study).

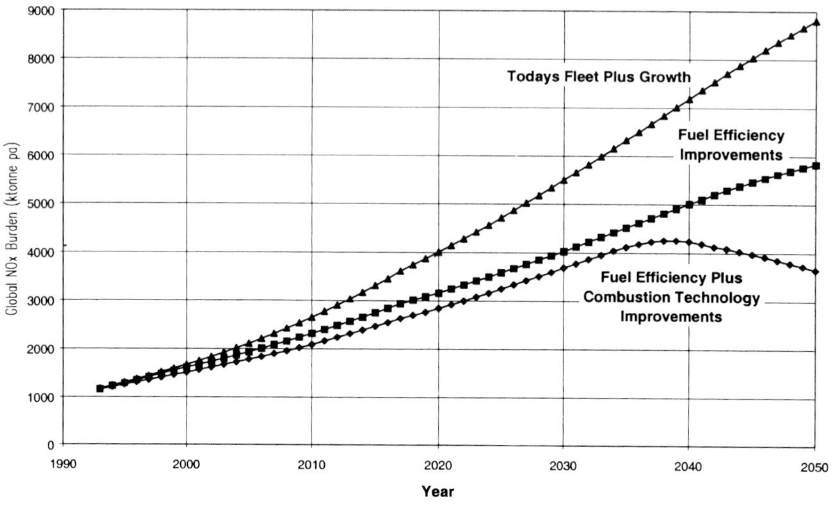

Figure 3 : Global Aircraft Cruise NO_x Emissions Burden Predictions

Until low-emissions technology capable of cutting NO_x by up to 60% over today's best levels is accepted by both airworthiness authorities and customers, aviation's emissions impact will become greater as air traffic grows and ground sources come under stricter control. As a result, while the need to control air quality around airports through the existing LTO cycle or with a new parameter including aircraft productivity will remain, there is significant pressure to create an additional regulatory regime that addresses the global implications of high altitude aircraft engine emissions ([8] & [9]). The choice of new legislation will depend on the level of certainty of scientific knowledge over global aircraft pollution issues especially from scientific assessment bodies such as UNEP/WMO (United Nations Environment Programme and World Meteorological Organisation) on ozone depletion and by the IPCC (Inter-governmental Panel on Climate Change). As understanding of the atmosphere improves clearly the rules should be evolved to legislate the most harmful species to both global and local environments, and hence any new rules should allow evolution.

One of the major resulting challenges is that the development of a future regulatory regime covering high altitude emissions, will require reference to the aircraft. The combination of engine, airframe and mission determines the flight emission levels. The same engine can power different airframes with quite different missions (eg. powerplants for the Boeing 767 two engine aircraft and the Boeing 747 four engine aircraft can be the same). Different flight missions result in different impacts, since the atmospheric implications of engine emissions such as NO_x and water vapour (H_2O) depend on the point of release.

3. REDUCING AIRCRAFT-GENERATED EMISSIONS THROUGH COMBUSTOR TECHNOLOGY

3.1 Single annular combustor technology

In response to emissions regulations, improved combustor technology has been the route to the reduction of smoke, CO, UHC and NO_x exhaust pollutants favoured by engine manufacturers. In reducing smoke, CO, UHC and NO_x emissions, a technical compromise has to be found, not only between one specie and another, but also between one specie and the many other combustor performance requirements such as size, low pressure drop, low wall temperatures, adequate ignition, altitude relight, blow out limits and outlet temperature distribution (pattern factor). It is possible to generalise and state that, with single annular combustor technology (SAC), any change in operating conditions or configuration that reduces NO_x tends to exacerbate the problems of low power emissions (CO and UHC) and smoke at high power, and vice versa.

During the past two decades or so, the levels of smoke, CO and UHC produced by SACs have been successfully cut. The efficiency of modern SACs lies at around 99.5% at idle, the engine operating mode at which the combustion efficiency is lowest. Further reductions in CO, UHC, H_2O, CO_2 and SO_2 (sulphur dioxide) mission emissions are possible with improved engine s.f.c. and aircraft fuel efficiency. However, NO_x has proved far more intractable in spite of the various research efforts that have taken place since the early 1970s. This is because there has been an increasing demand for more powerful and fuel-efficient powerplants for high-capacity aircraft, which have led to increasingly adverse inlet combustor operating conditions.

Recent high overall pressure ratio civil aero engine developments have proved that advanced single annular combustor technology can comply with the latest 1998 CAEP/4 ICAO standards with margin. In the case of the Rolls-Royce Trent 800 powering the Boeing 777, this has been made possible through such measures as the use of advanced airblast fuel atomisation, advanced cooling technologies, and combustor shortening (reduced residence time).

3.2 Future low NO_x combustor concepts

In anticipation of further tightening of the NO_x standards, manufacturers are continuously investigating a range of advanced combustor concepts which have the potential to reduce NO_x by greater than 60% relative to the current technology. Some of these combustors designed specifically for low NO_x first appeared in the 1970s, but remained on the shelf for another decade due to a greater interest during that time in reducing CO, UHC and smoke.

New techniques employ one of two principal methods to reduce flame temperature and hence NO_x without unacceptable penalties in CO, UHC and smoke emissions as well as in combustion performance at low-power conditions : Burning either fuel-rich or fuel-lean, (Figure 4 shows the various technical paths to achieve this). While the former rich/quench/lean (RQL) concept has the advantage of offering a wide range of stable combustion operations, there is considerable difficulty in maintaining low levels of smoke. Furthermore, as excess air, not actively involved in the early reaction process, is added to the rich burning products, stoichiometric conditions are bound to be obtained that will lead to high NO_x. Fuel-lean combustion avoids these difficulties but can pose combustion stability and/or durability problems. Fuel-lean combustion would have the big advantage that minimal NOx would be produced at the lower conditions found at engine cruise.

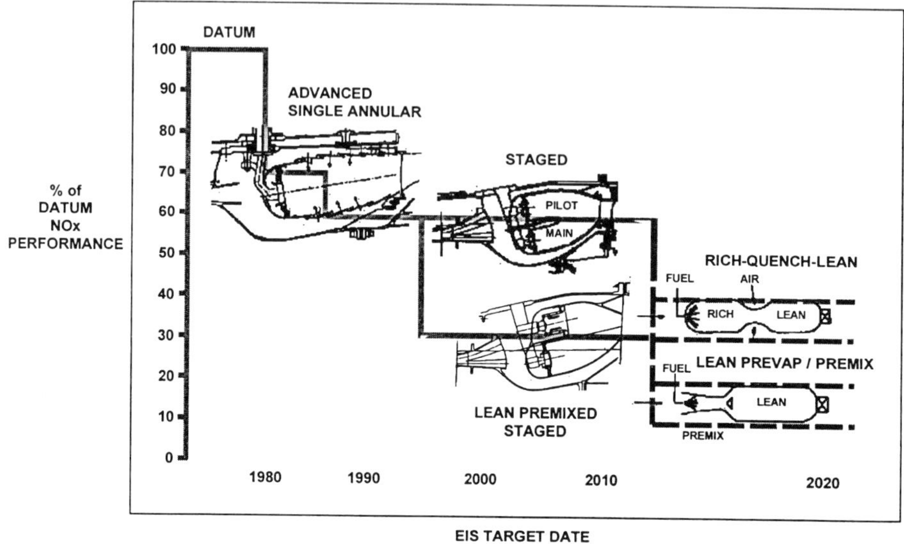

Figure 4. Low NOx Combustor Development

A further step to reducing NOx would be to run fuel lean and with prevapourised fuel. Lean premixed prevapourised combustion (LPP) appears to be the best technology to achieve ultra low NO_x emissions from practical combustors [10]. Low levels of smoke are also realised due to lean burning. However, the problems of spontaneous ignition and flashback as well as acoustic resonance need to be fully addressed before the concept can be applied with confidence. Doubts will remain whether they will be able to operate in very high pressure ratio engines due to limitations of auto ignition delay time. This may force the industry to choose between further fuel efficiency and NO_x control. Clearly the decision at CAEP4 to adjust the slope of the NOx rule back to 2 was sensible in light of the difficulties in limiting NOx at high engine pressure ratios. The engine manufacturers will clearly have to consider a compromise on the LPP combustors design for these high pressure ratio applications, and may not be able to run such a system prevapourised at take-off and climb conditions. If a combustor can be run at LPP during cruise conditions then hardly any high altitude NOx will be produced.

4. REDUCING AIRCRAFT-GENERATED EMISSIONS THROUGH AERO ENGINE DESIGN

4.1 Aero engine operation and it's effect on emissions

Measures that could contribute to reducing the atmospheric impact of aircraft emissions fall into two distinct categories: Those designed to reduce the flight emissions output by technical improvements and those designed to reduce the flight emissions output by operational improvements. The present paper focuses primarily on technical emissions reduction measures through aero engine cycle optimisation and the use of low-NO_x combustor technology. Potential operational improvements not covered by this paper include the use of alternative fuels, the chemical treatment of current aviation fuels, the optimisation of airframe design, and the better scheduling of flight operations. Most fuels (including hydrogen) will still produce NO_x emissions due to the high flame temperatures promoting the oxidation of atmospheric nitrogen. Importantly, the current LTO rules do not differentiate between the operations of cargo carriers, business jets or passenger carriers, and it is believed that any future rules should maintain this uniformity of standard.

Low emissions requirements are not normally considered during the initial stages of the aero engine design process. This is because SAC technology has proved up to the present time that it could meet on its own the low emissions LTO standards set out by EPA and ICAO. Furthermore, there is a strong commercial incentive to minimise specific fuel consumption (sfc) which inherently helps to limit mission emissions.

The choice of engine cycle to meet the performance and s.f.c. requirements will affect the combustor inlet thermodynamic conditions and as a result CO, UHC and NO_x emission indices. Owing to the high efficiency of current combustors, the amount of CO_2, SO_2, and H_2O emissions are a linear function of fuel flow, and are fairly insensitive to the main engine cycle parameters and the combustor inlet thermodynamic conditions. As a result, fuel efficiency goes hand in hand with low CO_2, SO_2 and H_2O emissions. NO_x is more strongly affected by the choice of engine cycle, and the anticipated evolution of the engine cycle in the future means that a range of ultra low-NO_x technologies are being actively investigated. Low-power emissions are also influenced to some extent by the cycle with a high pressure ratio constituting favourable conditions to maintain low CO and UHC levels at idle.

4.2 Optimising for low specific fuel consumption

Since the beginning of the jet era in the 1950's, there have been strong economical and operational incentives to improve aircraft fuel burn. Figure 6 shows the trends in engine fuel consumption and fuel burn per passenger over a 1000 nm mission relative to a De-Havilland Comet 4 powered by Rolls-Royce Avon engines. The latest generation of aircraft have a fuel burn per passenger level approximately 30% of the Comet. This improvement can be equally attributed to engine and airframe advances. The emissions of CO_2 and H_2O will be reduced by a corresponding amount.

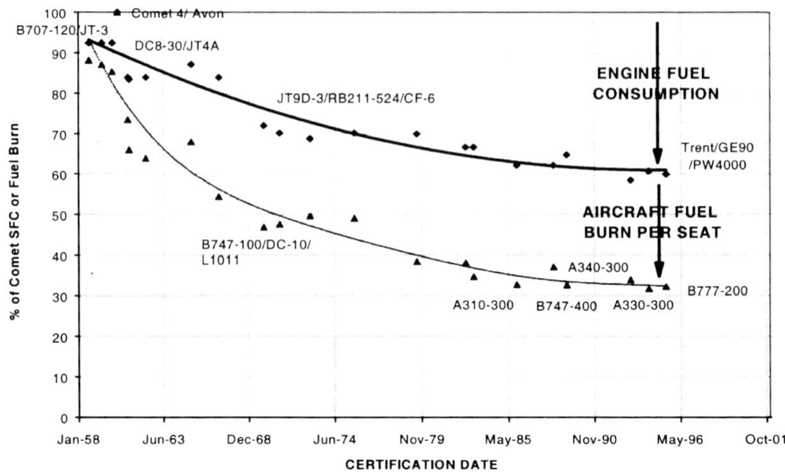

Figure 5 : Fuel Efficiency Improvement of Long Range Transport Aircraft

The significant improvements in engine fuel consumption have been made possible by advances in engine overall efficiency and reduced specific fuel consumption through increases in both thermal and propulsive efficiencies. The early turbojets achieved overall efficiencies of about 20%. Low bypass ratio engines made available in the early 1960s offered overall efficiencies of 25% thanks to improved propulsive efficiency. Current high bypass ratio turbofans have achieved substantial improvements in both thermal and propulsive efficiencies and offer about 35% overall efficiency. Even higher overall efficiency engines, perhaps in the region of 50%, may be possible with ultra-high bypass ratio turbofans, such as propfans and advanced ducted propulsion systems.

Improved thermal efficiency comes from better thermodynamic cycles (Figure 6).

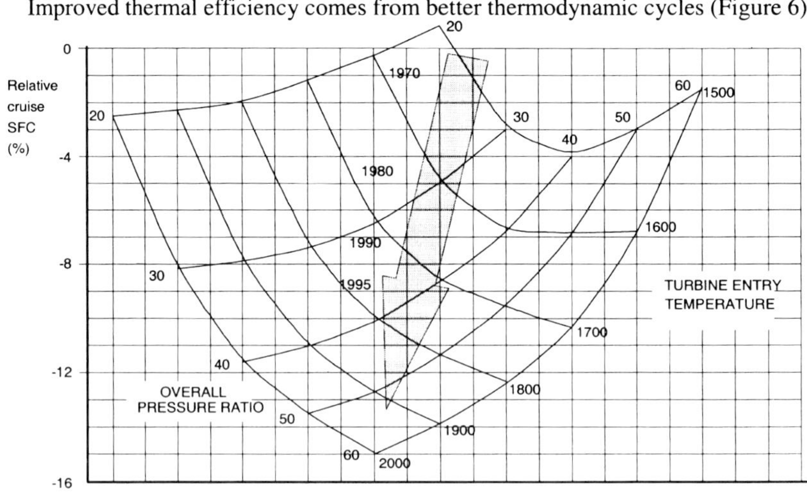

Figure 6 : Engine Cruise s.f.c. vs. Overall Pressure Ratio and Temperature at T/O

Improvements in thermal efficiency are caused by advances in the engine thermodynamic cycle, in particular increased overall pressure ratio (OPR), turbine entry temperature (TET) and component efficiencies. The arrow in Figure 6 indicates the trend of OPR and TET with time for large commercial turbofan engines since about 1970. The s.f.c. improvements are solely due to increased thermal efficiency. It reflects largely the progress made in materials and cooling technology, particularly in the high pressure turbine.

A high bypass ratio (BPR) combined with a corresponding reduction in fan pressure ratio (FPR) is necessary to produce a high propulsive efficiency. The lower fan pressure ratio of higher BPR engines implies a lower exhaust jet velocity hence a reduced jet noise level. This is possible because there is the same amount of energy in the jets leaving the fan from a larger mass of slower air.

A high thermal efficiency requires increased work per unit gas flow through higher OPR, higher TET and higher component efficiencies.

Whilst absolute levels of water vapour, carbon dioxide and SO_2 emissions fall almost linearly with reductions in s.f.c., low NO_x requirements necessitate low combustion temperatures i.e. low OPRs and TETs. Until now, advances in single annular combustion technology have prevented NO_x emission indices from rising when TET and OPR have been increased. However, further increases in OPR and TET necessitate development of low NOx combustor concepts.

4.3 Optimising for low LTO NO_x emissions

The challenge with LTO NO_x is to keep flame temperature down at high-power conditions, against the historical background of increasing inlet cycle temperatures and without incurring unacceptable penalties in terms of low power emissions, durability, flame stability and exit temperature distribution.

For a given combustor design and fuel injection system (i.e. a given fuel/air mixing quality), NO_x is primarily a function of combustor residence time and flame temperature. The flame temperature is mainly determined by inlet combustor temperature and pressure, and zonal fuel-to-air ratio. The NO_x emissions index increases as engine thrust increases, because combustor inlet temperature and pressure increase and combustor overall flame temperature becomes hotter (Figure 7 shows a typical NOx curve against combustor inlet temperature).

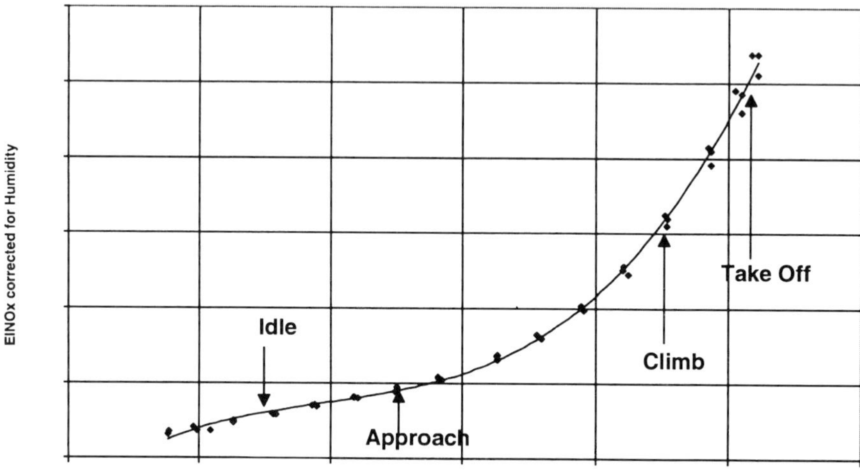

Figure 7 : Typical EI NO_x Emissions Characteristic

- Emissions measured at stabilised conditions
- Curve fitted to measured data
- Emissions evaluated at LTO mode points

The engine manufacturer, may choose to optimise his combustor for the higher conditions of climb and take-off because they dominate the NOx LTO cycle summation. The combustor stoichiometry would be optimised such that the region where stoichiometric conditions are achieved is small in volume and the gases have a short residence time in such a region. This optimisation may ignore cruise conditions which generally occur at combustor inlet temperatures between approach and climb conditions, and at weaker combustor conditions (higher AFRs).

4.4 Optimising for low cruise NO_x emissions

The LTO emissions as a percentage of total flight cycle emissions are small and for long haul flights insignificant as shown in Table 3.

	LTO as % of Total		
	Fuel	CO_2	NO_x
Short Haul / B757 London-Glasgow (283 nm)	37	37	32
Long Haul / DC10-30 Los Angeles-Tokyo (4725 nm)	3	3	2

Table 3 : LTO Emissions as a Proportion of Total Flight Emissions

In this respect, the question arises whether the current LTO standard is adapted to limit aviation's impact at high altitude. The new flight cycle emissions parameter that is under development by CAEP will enable the regulatory bodies to target cruise operation. This will help place the onus on the engine manufacturer to limit the NOx emissions at cruise operations aswell as over the LTO cycle.

Figure 8 shows the LTO NO_x characteristics of two different combustors. Engine B would produce much higher LTO NO_x than Engine A. However, at cruise, where a long range aircraft would burn over 85% of its fuel, the relative NO_x performance is reversed. With the current LTO based regulations, Engine B with the lower cruise NO_x emissions could fail the CAEP/4 recommendations, even though it could be argued that it has a better global environmental performance.

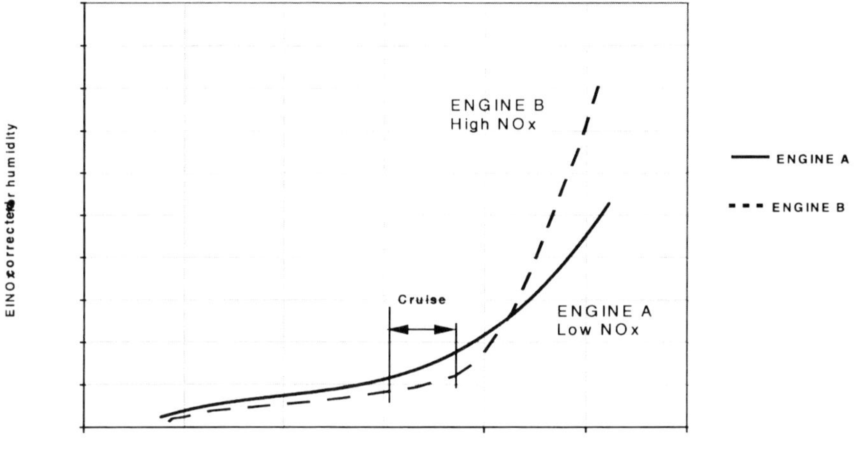

Figure 8 : Comparison of NO_x Characteristics from 2 Combustors

Future emissions regulations must address such areas where the compliance with airport air quality is realised at the expense of high altitude environmental protection. This is not to say that all combustors with low LTO NO_x have high cruise NO_x. The majority of cases where new technology combustors have been introduced into engines have resulted in better NO_x performance at both high power conditions and during cruise.

4.5 Effect of choice of engine BPR

Manufacturers choice of engine bypass ratio (BPR) has seldom taken into account the affect of BPR on engine emissions. Engine cycles and core size are determined by the maximum climb requirement. The engine is sized to an overall pressure ratio and turbine entry temperature, and to a certain extent there is then a choice of BPR. The choice of BPR can be dictated by sfc, noise levels, maintenance cost, weight for the application, core through-put, and fan Mach number. Traditionally engine manufacturers have chosen a BPR where the thermal and propulsive efficiencies can be matched. BPR's have grown over time because the thermal efficiencies have improved enabling larger fans to be driven. Noise rules, like those at Heathrow, are tending to drive engine designs to higher BPRs above the matched thermal to propulsive efficiency value, and it is likely that this trend will continue.

Recent studies have considered the effect of bypass ratio on NOx performance, and all these studies have shown the same generic effect. Figure 9 shows the result of one of these engine preliminary design studies. This study looked at the engine performance (including NO_x emissions) of engines with a bypass ratio ranging from 6.5 to 13.5. The engines incorporated similar levels of combustor technology. All the engines were designed to give equivalent thrust at top of climb conditions as well as the same rated take-off thrust. Figure 9 shows that higher BPR engines give higher NOx at cruise conditions, but they look much better than low BPR engines over the LTO emissions cycle.

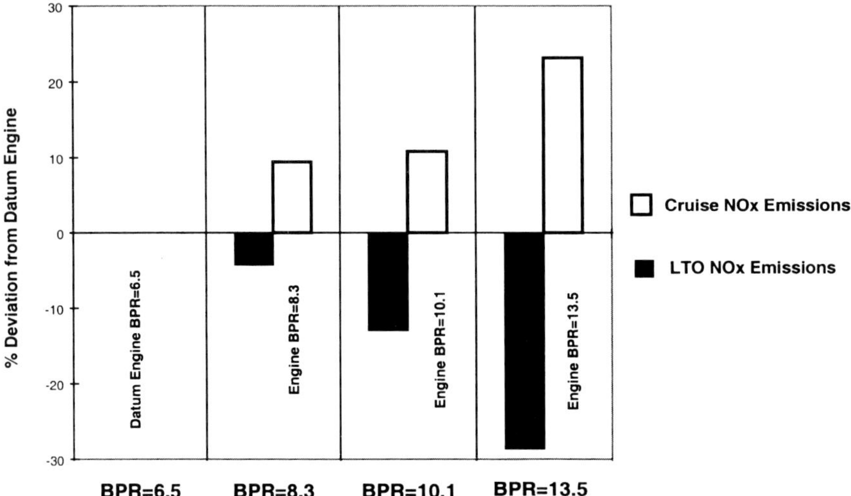

Figure 9 : Deviation of LTO and Cruise NO_x Emissions with Engine Bypass Ratio

The generic understanding of this BPR effect on emissions is detailed in the following text:-

(i) Considering the effect of BPR on NOx emissions during the LTO cycle:
At take-off (T/O), the lapse rate with forward speed is significant (see Figure 10). The available thrust of higher BPR engines becomes relatively higher as forward speed reduces (TET fixed), because of the differences in propulsive efficiency. Therefore, at a fixed T/O thrust for a fixed take-off field length the higher BPR engine will be more throttled back. The better propulsive efficiency allows the manufacturer to run the core cooler to get a given thrust. The higher BPR engine will therefore give lower OPR, lower TET and lower NOx at T/O conditions. The differences will be further enhanced at static conditions.

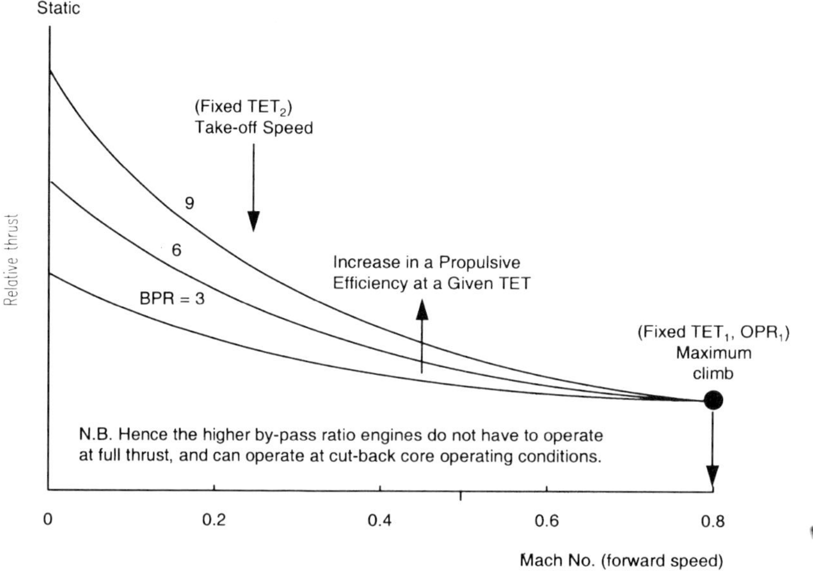

Figure 10. Propulsive Efficiency Differences at Varying Forward Speeds.

(ii) Considering the effect of BPR on NOx emissions during cruise:
For a given engine designed to a maximum climb condition the thermal efficiency will be independent of the BPR. The choice of BPR dictates the propulsive efficiency and how the core works at cruise conditions. The high BPR engines have a big propulsive efficiency advantage at high specific thrust levels (specific thrust is thrust/inlet mass flow), and at these high specific thrust conditions they have a significant s.f.c. advantage. At lower specific thrust levels the propulsive efficiency tends to a 100% so the bigger BPR engines lose a lot of their propulsive efficiency advantage. Conversely the lower BPR engines have a greater propulsive efficiency gain as they are throttled back from the climb to cruise rating. Clearly the harder core operating conditions outweigh the sfc advantage of high BPR engines, in terms of cruise and flight cycle NOx production. In summary as you throttle back to cruise conditions, the propulsive efficiency advantages of the higher BPR engines is lost, and therefore to maintain the thrust you need to run the core harder.

Irrespective of the NO_x issue, with today's technology levels, BPRs greater than 9 are unlikely to deliver lower mission fuel burn for a given design maximum take-off weight (MTOW) due to higher weight and drag. At today's levels of technology, the optimum bypass ratio with respect to mission fuel burn lies between 6 and 9 depending on the installation. Noise requirements may dictate the final choice of BPR.

4.6 **Effect of engine and airframe performance deterioration and variability**

The extent to which the fuel burn and hence flight cycle emissions levels will be maintained over the years depends on the initial choice of cycle and materials for the engine as well as the airframe aerodynamic and structural design. The engine performance deterioration will typically account for 80% of the total aircraft fuel burn deterioration, with the remainder attributable to the airframe. The s.f.c. deterioration stems from the necessity to maintain the aircraft take-off and climb performance with degraded engine components.

Although the fuel burn deterioration will translate to an equal deterioration of the flight emissions levels, the increased fuel flow to compensate for the loss in component efficiency will affect NO_x emissions indices. The increased combustor inlet pressure, temperature, and overall flame temperature will lead to a higher NO_x emissions index.

The effect of deterioration on low power emissions of CO and UHC should not be detrimental since the degradation of non-combustor components effectively result in hotter combustion at higher pressure. At these higher conditions less unburnt and partially burnt emissions species will be produced. The changes in the trace species indices remain small relatively to H_2O and CO_2 levels. The deterioration of CO_2, H_2O and SO_2 emissions will be in line with that of fuel burn.

During the development of the LTO emissions regulations a variety of engine types of different ages were tested to understand the effect of engine to engine variability and the effect of engine deterioration [11]. These 1960's and 1970's engines showed considerable variability in engine emissions, and so a statistical compliance factor was added to the rules (i.e. engine certification measurements had to show large emissions margins to the rules to prove that the majority of the fleet met the rules). The development of any new emissions parameter will have to revisit the engine to engine variability and engine deterioration aspect of a new fleet wide rule. It should be hoped that the improved performance retention of modern engines would enable the statistical compliance factors of future rules to be reduced.

5.0 **DEVELOPMENT OF AN EMISSIONS REGULATION METHODOLOGY FOR THE ENTIRE FLIGHT CYCLE**

One of the major problems with developing a new cruise/climb methodology will be the difficulty in deciding the conditions to regulate at. This is partly due to the different aircraft flight requirements, and partly due to the fact that the engine/airframe operation will change as the fuel is burnt off (Figure 11 shows a typical flight). Half the weight of the aircraft may be fuel on a long haul flight, so the operators may even step up the cruise altitude mid-way through the flight (eg. B747 often cruises at 33Kft initially and then climbs to 35Kft).

There are potentially a number of alternative methods for achieving a new parameter, but whichever method is chosen, the cruise phase of the engine operation will obviously become very important for long haul operations. Engine cycles are seldom chosen for emissions reasons and yet the choice of cycle effects the emissions. Clearly the development of a new rule will emphasise the need for low emissions at cruise as well as over the LTO cycle.

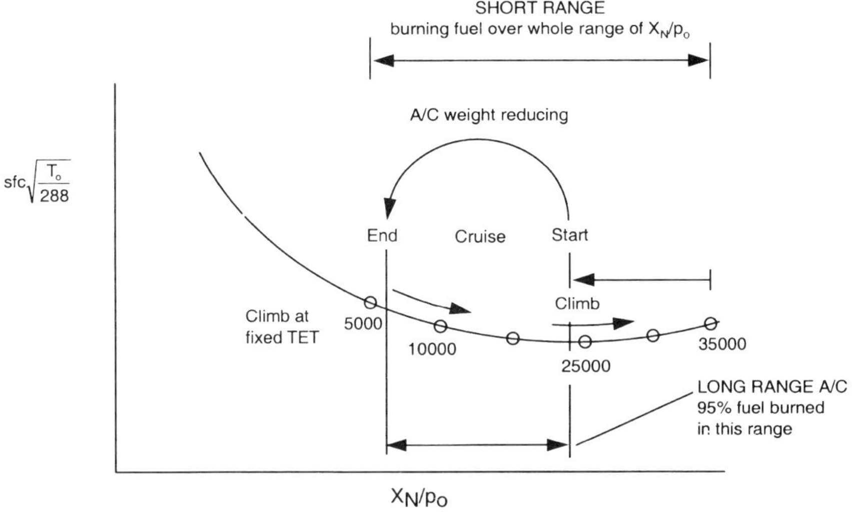

X_N is specific thrust, P_o is ambient pressure, T_o is ambient temperature.

Figure 11. Typical Flight Profile For Subsonic Aircraft

The new emissions parameter will need to consider many of the points stated previously within this paper:-

(i) Current scientific understanding of the atmosphere suggests that the critical species at cruise are NOx and CO_2. Water vapour and particulates may also become important, but can only be dealt with when more data is available. CO and UHC emissions are significant within the LTO cycle, but may be emitted in negligible amounts at cruise.

(ii) Subsonic aircraft will produce the majority of emissions, and no supersonic fleet is envisaged in the immediate future. The rules will still have to be fleet based and allow statistical compliance for engine to engine variability and engine deterioration.

(iii) The new rule will have to be aircraft based, but engine based certification with a link to the airframe via MTOW and airframe/engine flight cycle performance is preferable. This will enable equal weighting for passenger flights, business flights, and cargo carriers.

(iv) A certification parameter is desirable which prevents a double standard or anomalies between cruise and LTO performance.

(v) The focus of the new rule should be provision for protection of the environment, both in terms of local health and the global climate.

The conclusions from the recent post-CAEP4 WG3 meeting (12) were similar to those above. This WG3 meeting suggested that engine and airframe manufacturers report the parameter below for discussion at their next WG3 meeting:-

$$X_I = \frac{Y_{LTO}\ x\ fn(time) x\ fn(alt) + Y_{climb}\ x\ fn(time) x\ fn(alt) + Y_{cruise}\ x\ fn(time) x\ fn(alt) + Y_{take\text{-}off}\ x\ fn(time) x\ fn(alt)}{MTOW}$$

Where: X_I, is the notional value specific for each emission species (ie. the total emissions for comparison with any regulations)

Y is the mode specific value (in the initial studies this will be values for CO_2 and NOx at the different conditions).

fn(time is the time in each mode (using typical flight cycle data).

fn(alt) is a weighting for the effect of the species at different altitudes (allows evolution dependant on the scientific assessment of each specie effect at its place of deposition).

MTOW is the aircraft maximum certificated take-off weight (this allows a flight cycle duty and length to be encompassed (ie large aircraft fly further)).

The main advantage of the parameter, as written, is that it may be applied both locally and globally by using different altitude weightings fn(alt). Evolution of a future flight cycle emissions methodology has begun with the intended formation of a new rule by the end of 2001.

6.0 CONCLUSIONS

Since the beginning of the jet era in the 1950's, aviation industry has made significant fuel efficiency improvements. Market forces will continue to drive for further fuel efficiency improvements with their corresponding benefits in terms of CO_2 and water vapour emissions. Although the law of diminishing returns will apply, lower s.f.c. levels are possible with higher bypass ratio engines, advanced thermal cycles, and further advances in component efficiencies.

Higher cycle temperatures are the price to pay for engine growth, improvements in cruise s.f.c., and increases in the top of climb thrust. Radically different combustor concepts may be required in the future, if they are to generate low NO_x emissions in the face of increasingly adverse combustor conditions. New LTO regulatory standards made stricter than the 1998 CAEP/4 recommendations, and new cruise emissions regulations will force engine manufacturers to minimise emissions over the entire flight cycle..

Future emissions regulations must ensure that airport air quality and the long term global atmospheric environment are given equal attention. Effort must be focused on research of the atmospheric impact of aviation emissions, hence future rules should allow evolution as scientific understanding of the atmosphere develops. The best way to proceed towards new legislation would appear to be to continue with engine based certification with a link to the airframe via MTOW and airframe/engine flight cycle performance. The emissions results would have to be reported over the entire flight cycle including cruise conditions. Evolution of a future flight cycle emissions methodology has begun with the intended formation of a new CAEP rule by the end of 2001.

REFERENCES

[1] ICAO, International Standards and Recommended Practices, Annex 16 to the Convention on International Civil Aviation, Volume II, First Edition, 1981. Superseded by the Second Edition in 1993.
[2] ICAO, Conclusions of the CAEP Working Group 3 Meeting, Berne 1997.
[3] J. Sanborn, Feasibility Paper for NOx Stringency Increase Considerations for Engines in the 6,000 to 20,000 Rated Thrust Range, 1998.
[4] ICAO, Report of the 4th Meeting of the Environmental Protection Montreal 6-8 April 1998.
[5] Schumann U, Impact of Emissions from Aircraft and Spacecraft Upon the Atmosphere - An Introduction, In U. Schumann and D. Wurzel (eds.) : '*Impact of Emissions from Aircraft and Spacecraft Upon the Atmosphere*', Proceedings of an International Scientific Colloquium, Cologne, Germany, April 18-20, 1994, pp. 8-14.
[6] Boeing, 1997 Current Market Outlook, World Airplane Demand and Airplane Supply Requirements, Boeing Commercial Airplane Group Marketing.
[7] Rolls-Royce, Market Outlook 1997-2016.
[8] Wright M, Future Trends in Environmental Legislation, Additional Paper, Seminar 23 : Environmental Issues (1), Aerotech'95, 17-19 October 1995.
[9] Ralph M O and Newton P J, Experts Consider Operational Measures as Means to Reduce Emissions and their Environmental Impact, *ICAO Journal*, March 1996, Volume 51, No 2.
[10] Lefebvre A H, The Role of Fuel Preparation in Low-Emission Combustion, *Journal of Engineering for Gas Turbines and Power*, October 1995, Vol. 117.
[11] M.Platt & E.R.Norster, FAA, Time Degradation Factors for Turbine Engine Exhaust Emissions, April 1979.
[12] ICAO, CAEP Working Group 3 Task Group Meeting, Geneva, 11 June 1998.

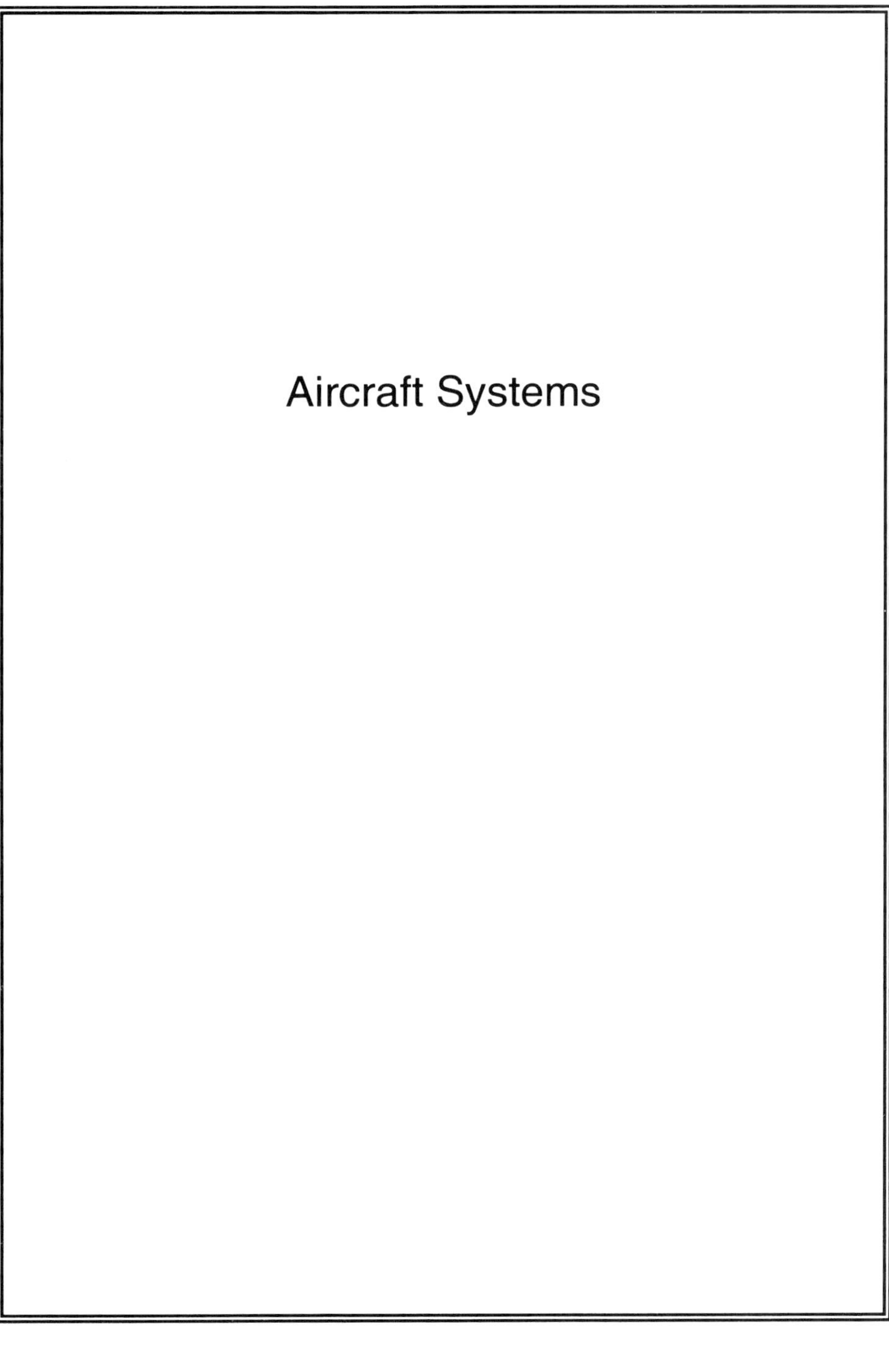

Aircraft Systems

C545/030/98

A systems approach to flight controls

R McKAY MIMechE and **A McLOUGHLIN** BEng, MEng, MIEE
Lucas Aerospace, Wolverhampton, UK

SYNOPSIS

In previously developed aircraft flight control systems, it has been the airframers responsibility to fully define the system prior to it being put out to tender. However, in more recent flight control system developments, there has been a move towards an open period of system definition, where the system supplier has liaised closely with the customer to jointly develop the system definition. This has largely been driven by the reduction in time to market for new aircraft.

This paper uses the experience gained by Lucas Aerospace in developing the N250 flight control system to show how the development of such a system, and the timescales involved in its production can be improved by adopting a collaborative systems approach to its engineering.

1. INTRODUCTION

The historic approach to the provision of flight control systems for aircraft has been for the aircraft manufacturer to produce a written specification for the units which comprise the whole system. This specification, often produced by the airframer in isolation, without consultation with the equipment supplier, can lead to repetition of work by both customer and supplier, and a set of requirements which may not be wholly understood by the supplier leading to a less than optimal system implementation.

Lucas Aerospace provides the global aerospace industry with high integrity systems in flight controls, engine controls, electrical power generation and management and cargo handling, all backed by a worldwide customer support operation.

LucasVarity designs, manufactures and supplies advanced technology systems, products and services in the world's automotive and aerospace industries.

Lucas Aerospace were chosen as contract leaders for the Indonesian IPTN N250 aircraft's Flight Control System, with complete responsibility for the control of the three axis fly-by-wire actuation system. Lucas Aerospace worked closely with IPTN at the start of the project to jointly define the requirements for the flight control system. Drawing on aircraft level and equipment level expertise in this way produced a more complete system specification.

Benefits of this approach include all the requirements being understood by both customer and supplier; a cleaner customer/supplier interface; optimised system components, all leading to technologically advanced, but cost effective solution being engineered.

A number of lessons have also been learned from this project, mainly in the areas of customer/supplier management and in dealing with the difficulties in managing such a complex system. In future projects, effective requirements management will be instrumental in controlling systems which are becoming ever more complex.

2. HISTORICAL BACKGROUND TO FLIGHT CONTROL SYSTEM DEVELOPMENT

The provision of flight control systems has undergone a slow but steady development over the years.

2.1 Up to 1940's

In the early days systems were invariably mechanical. The pilot was directly linked to the control surfaces such that he could feel what was happening resulting in very simple systems. The aircraft manufacturer took responsibility for the design and manufacture of all the systems within the aircraft allowing an optimised and high integrated design against their own requirements.

2.2 1950's to 1960's

The increase in aircraft speeds and size resulted in the requirement for power flying controls. The power flying controls were complex, redundancy was required to achieve safety, hydraulic power sources had to be provided and artificial feel systems required to give the pilot tactile feedback he was used to. The huge increase in both complexity and technology forced a change of approach. Typically the aircraft manufacturer continued to be responsible for the system configuration and developed expertise in all areas of system design allowing them to define the configuration required and the major features of the component units. Equipment suppliers were used to design the equipment against the requirements and were only given responsibility for the detailed definition of the component units. The Victor and Lightning aircraft are typical of this engineering approach. Units from these aircraft are depicted in Figure 2-1 and 2-2.

Figure 2-1 Victor Elevator Control Actuator

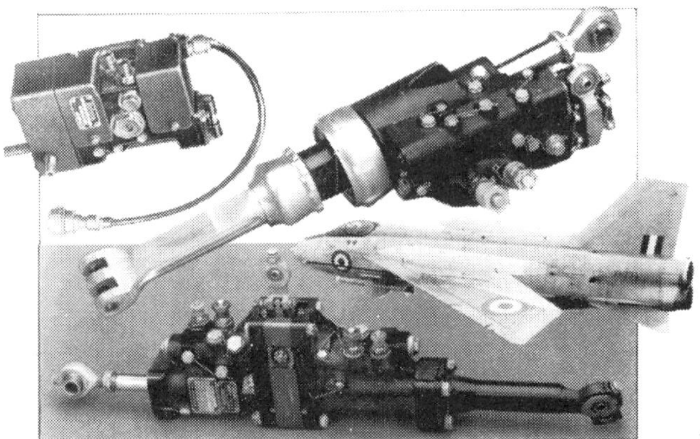

Figure 2-2 Lightning Flight Control Actuators

2.3 1970's to 1980's

System complexity continued to increase driven by the need to reduce aircraft weight and operating costs. Suppliers were given responsibility for larger systems e.g. secondary flight controls and increased responsibility for the configuration definition of other flight control units. This allowed a steady reduction in the level of aircraft manufacturer support necessary in the base technology areas.

Compressed project time scales started to drive the need for concurrent working in order to allow the airframe and equipment supplier to work together to jointly define the system configuration. The A300 Flap and Slat system are typical of the systems developed around this time, as depicted in Figure 2-3.

Figure 2-3 A300 Flap and Slat System

3. CURRENT APPROACH TO FLIGHT CONTROL SYSTEM DEVELOPMENT

The responsibility for the flight control system configuration is still usually with the airframer. The typical process followed to define and procure the flight control system is detailed below, and shown in Figure 3-1.

- an initial baseline system configuration is arrived at by iterative systems studies examining the parameters and benefits of alternative configurations.
- expert support is solicited from flight control suppliers through the issue of formal RFI's on particular system features resulting in the optimisation of the baseline system.
- a solution is reached which is believed to be optimum and consistent with the top level system requirements.
- the requirements for the component units or sub system elements are then specified and a formal bid phase entered.
- suppliers selected
- a joint definition phase is entered where the supplier and customer jointly optimise the system requirements and solution. At this point the airframer and the supplier start working concurrently.
- component units are designed, tested and certified against the defined requirements.

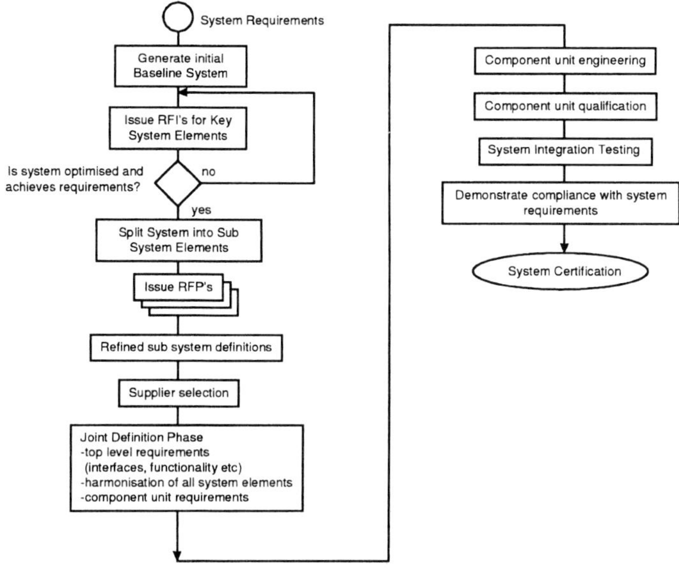

Figure 3-1 Flow Chart of Typical Current Approach

3.1 Benefits

- clear contractual responsibilities at proposal
- a high level of control of the system configuration by the airframer

3.2 Problems

- insufficient recognition of actuation supplier expertise in the initial system definition leading to a low level of optimisation.
- Pre contract system evolution is sequential resulting in extended definition times.
- Requires high level actuation system expertise by the airframer which is duplicated at the supplier leading to an increase in overall system costs
- Very inefficient for complex systems.

It is clear that significant overall improvements can be implemented by improving the supplier / airframer interaction during the pre-contact system definition allowing an earlier optimisation of the system configuration resulting in a major reduction in the overall system development timescales.

4. N250 FLIGHT CONTROLS - A SYSTEMS APPROACH

The N250 is a Regional Twin Turboprop aircraft which was conceived by Indonesia's IPTN (Industri Pesawat Terbang Nusantara), primarily to meet the needs of its own domestic travel market.

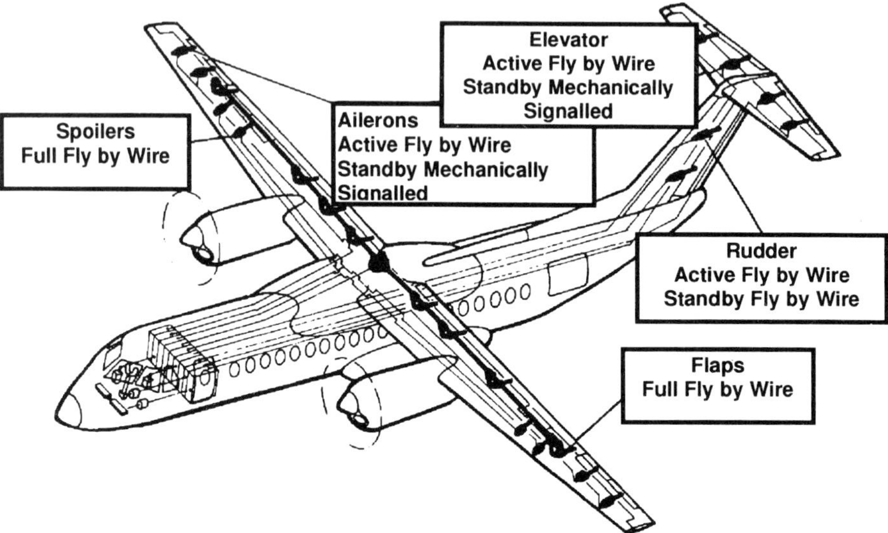

Figure 4-1 N250 Flight Control System Layout

Traditionally, this size of aircraft has used mechanical signalling of hydraulically powered actuators, or direct pilot control of the flight control surfaces to provide the most cost effective flight control system solution. IPTN wanted N250 to be a state of the art aircraft, and therefore pushed hard for a fly-by-wire flight control system. These are typically used in larger more expensive aircraft which can take advantage of the improvements in cost of ownership offered by such a system. There was therefore a major challenge to meet with the N250 flight control system in terms of providing a state of the art system at an affordable original equipment cost.

N250 was the first aircraft for which IPTN had held sole design responsibility and therefore they were new to the definition of the systems necessary to support such an aircraft. IPTN relied heavily on a number of system suppliers to provide the expertise to bring together such a high technology aircraft.

In order to best meet the needs of this aircraft a systems approach was adopted at a very early stage in the definition of the entire flight control system. This approach required a significant commitment from both customer and supplier in order to develop a system definition from which the flight control system could be designed. Much of the work undertaken for this activity was completed prior to contract award by the customer, and was used to develop a comprehensive system requirements definition document held by the customer.

4.1 System Definition

The key to making the system requirements definition meet the needs of the aircraft was that a dedicated team of supplier engineers were co-located at IPTN's Indonesian facility for an extended period of time. These engineers assisted directly in the development of the system requirements definition, working closely with the customer's flight control system engineers, structural engineers and avionic system engineers. This gave the opportunity to compile the requirements documentation with a high level of completeness.

Since there was representation from both customer and supplier working together in the system requirements development team, the requirements were more complete due to the specialist input from both sides; airframe level expertise from IPTN and actuation system expertise from Lucas Aerospace. This led to a system which was optimised for the application, offering a cost effective, technically sophisticated system for the aircraft.

4.2 Requirements flowdown

Following the system requirements definition at a flight control system level, these top level system requirements were broken down into a suite of sub-system specifications. From these, each part of the flight control system could be defined. Again, a system approach was adopted in order to optimise the benefits from working on the complete flight control system.

The flight control system on N250 comprises a number of flight control surfaces controlled by a number of electronic control units:-

- 2 x Aileron surfaces (2 ECUs)
- 2 x Elevator surfaces (2 ECUs)
- 1 x Rudder surface (2 ECUs)
- 4 x spoiler surfaces (split into 2 pairs) (2 ECUs)
- 1 x Flap system (1 ECU)

The use of a systems approach in determining the functionality required in each ECU meant that the ECUs could be designed as identical units; their ultimate function being determined by their position in the avionics rack. Since a single supplier had responsibility for the entire flight control suite, a number of decisions were made to ensure that a suite of identical control units was a viable solution.

The Flap system can be used to illustrate the type of decisions which were taken in order to design for common controllers. In a conventional flap system, control of the flow of hydraulic fluid is directed by a number of solenoid valves which perform an on-off operation. In the N250 system, the control of this function is via a proportional servo valve. This decision has the system level effect of allowing the use of a common ECU type since the majority of the actuators which make up the flight control system are based on the proportional servo valve technology. Hence, a significant benefit can be achieved to the overall system by making such decisions at the highest possible level.

In order to facilitate a controlled and methodical breakdown of requirements into the individual surface control systems, a suite of documents was derived as shown in Figure 4-2. At the top level was a hardware specification detailing all of the hardware specific requirements. This referenced a set of software requirements documents, one each for the five sub-systems comprising the flight control system; i.e. aileron, elevator, rudder, spoiler and flap. These gave details of the specific requirements for each of the sub-systems.

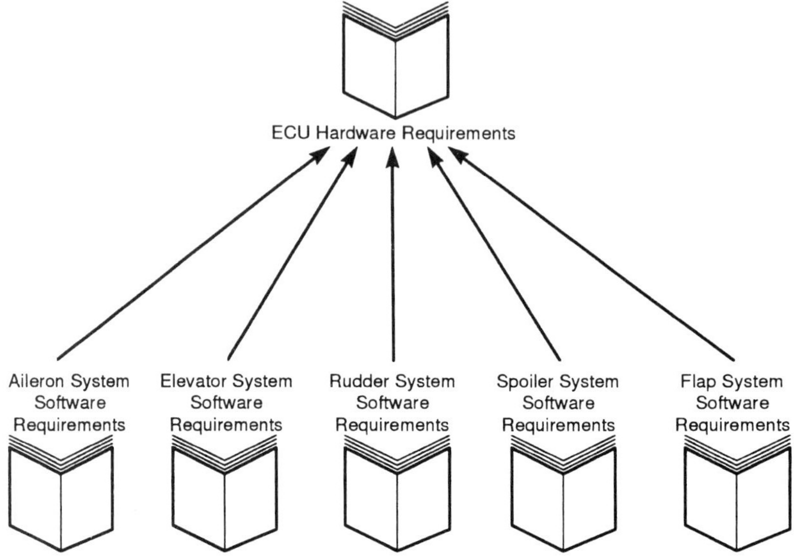

Figure 4-2 System Requirements Document Breakdown

Further benefits were possible due to operating initially at a flight control system level. There were various aspects of the software system design, and subsequently software coding and testing which could be common across the sub-systems, hence reducing the overall workload necessary to complete the flight control system.

4.3 Tracking of requirement changes

In a complex system such as that implemented on the N250 and also as a consequence of the early requirements definition, it is almost certain that the initial requirements which are laid down to define the system are subject to a number of later revisions. In defining such systems it is unlikely that all issues will be correctly defined in the initial documentation suite.

Fundamental to the successful implementation of the sub-systems of the N250 flight control system was the tracking of changes to requirements, and technical discussions which may have relevance to requirements. In order to keep track of these, three systems were put in place, the operation of which was agreed and followed by both customer and supplier.

- Specification Change Memo (SCM) system - put in place to facilitate changes made by the customer to their specification
- ECU Change Request (ECR) system - put in place to control changes made by the supplier to the various sub-system requirements documents.
- Co-ordination Memo (CM) system - put in place to cover items of discussion between the customer and supplier, or supplier and customer.

In order to maintain the specification of such complex systems in a condition which is understood by customer and supplier, use of tracking schemes such as these is vital. Much confusion can arise out of customer/supplier discussions based on fluid requirements where both parties may not be talking from the same baseline set of requirements.

4.4 Lessons Learnt

A large proportion of the benefits achieved by using the systems approach were from the close customer/supplier working relationship. Customer and supplier worked as a joint team in the areas of

- system définition, allowing an optimum flight control system for the aircraft
- system test on Iron Bird, using the expertise of both customer and supplier to trace and correct any test discrepancies
- certification support offered as required by supplier

One of the major challenges in fulfilling the objectives of the N250 programme was in meeting the tight timescales which were imposed in order to meet first flight. The Request for Proposal (RFP) was received from the customer late in 1991. First Flight of the aircraft was mandated as August 1995, which allowed approximately 3 years for the development and qualification of the aircraft systems with a major element of new technology for this type of aircraft.

If a traditional end-on-end engineering approach had been adopted for this programme, then these timescales would not have been met. However, due to the approach taken for the project, the first flight date was successfully met.

Using a systems approach provided a number of benefits including

- on-time flight clearance of flight control system
- single ECU type across all flight control applications

5. Future Trends

The aircraft industry is continuing to strive for improvements in terms of timescales, cost and quality. This is driving a continued change in the way systems are managed. A few of the underlying future industry trends are listed below.

- Move of the role of smaller Airframers away from component manufacture
    - Procurement of all aircraft subsystems
    - Airframer acts as systems integrator controlling system requirements and system integration
    - Systems supplier responsible for the subsystem configuration
- Larger airframers looking to procure fewer and larger systems packages
    - results in increased responsibility of systems suppliers
    - requires concurrent working between the airframer and supplier to optimise the integration of sub systems and the parent aircraft systems
- Greater risk sharing

- supplier are being asked to invest in the aircraft becoming a risk sharing partner
- Continued compression of time scales beyond those thought possible just a few years ago.
- Increased system complexity
  - increases the complexity of interfaces
  - increased integration between systems to reduce the total hardware requirements.
  - requirements management between systems suppliers (no customer in the loop)

6. Conclusions

We have come along way from the early design and supply of complex systems. The move of responsibility for systems design from the prime to the supplier has allowed major savings. Reductions in duplicated expertise has produced major overall cost savings. Early requirements definition has allowed major reductions in timescales. The increased understanding of the overall requirement by the supplier has also lead to a greater level of requirements understanding resulting in a greater level of optimisation of the system design.

Major challenges still exist in the provision increasingly complex systems, in ever decreasing time scales, at reduced cost and with increased technological challenge.

Systems engineering is the key to success with requirements management an essential element.

The system process must continue develop with the market.

C545/039/98

Fuel systems as an aircraft utility

T TULLY BSEE, MSAE
Smiths Industries Aerospace, Malvern, PA, USA

With ever increasing pressure to enhance functionality while at the same time reducing the cost and weight of aircraft systems, the trend towards integrated utilities architectures is likely to expand as future aircraft programs seek to exploit these benefits. This paper describes the utilities integration of the Smiths Industries fuel system on Raytheon Aircraft Company's new business jet, the Hawker Horizon.

This paper presents the architecture of the Horizon integrated fuel system which includes the fuel quantity gauging and fuel control functions. In addition, the role suppliers as system integrators in today's aircraft development environment is also discussed.

System Overview

The fuel quantity indicating and fuel control functions of the Hawker Horizon are integrated into the aircraft power and utility management system. The Hawker Horizon is a new mid size business jet being developed by the Raytheon Aircraft Company (RAC) of Wichita Kansas. RAC selected a utilities architecture for the airplane integrating various avionics, data acquisition and vehicle management functions into four (4) modular avionics units (MAUs). The MAUs illustrated in Figure 1 are being supplied by Honeywell's Business and Commuter Aviation Systems Division as part of their Primus Epic System . The fuel quantity indicating system comprising the fuel quantity measurement and level sensing functions is implemented as two circuit card assemblies each residing in a separate MAU. The fuel control system comprising the engine & APU fuel feed and transfer functions is integrated into the aircraft power management system. The electrical load controllers provide 3 phase variable frequency voltage for the AC standby boost pumps and 28VDC for the DC start pump and fuel system valves. The electrical load controllers use solid state power controllers with both programmable and fixed current outputs which provide on/off power control and protection. The thread that ties the power management and utilities systems together is the Avionics Standard Communications Bus (ASCB). The ASCB is a high speed serial ethernet type bus used to transfer data between the airframe systems and the flight deck displays. At the MAU the serial data is converted by a Network Interface Controller (NIC) into 32 bit data words for use by the modules in the MAU rack.

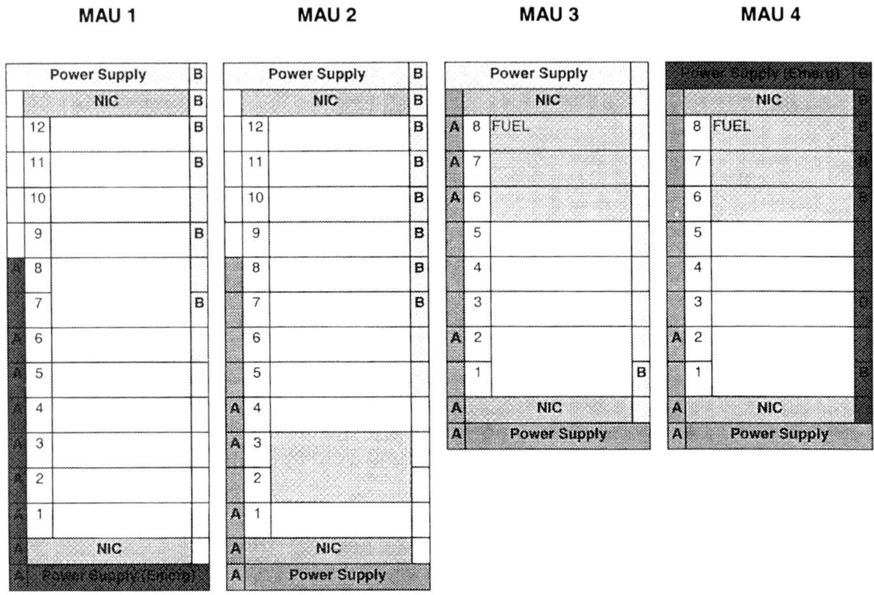

Figure 1 Hawker Horizon MAU Partitioning

2. Fuel Quantity Indicating System

The Horizon Fuel Quantity Indicating System is illustrated in Figure. The Fuel Quantity Processor Units (FQPUs) provide for the measurement of fuel quantity, level sensing and fuel tank temperature sensing. The system is configured such that each FQPU measures the fuel quantity in one wing and the level sensors of the opposite wing. This ensures that no single failure can result in the loss of fuel gauging and level sensing for a given wing. The probe and sensor interfaces are made through the FQPU front module. No fuel specific interfaces are supported through the MAU backplane. The backplane connectors are dedicated to the ASCB interface and for providing conditioned DC power supplies to the modules.

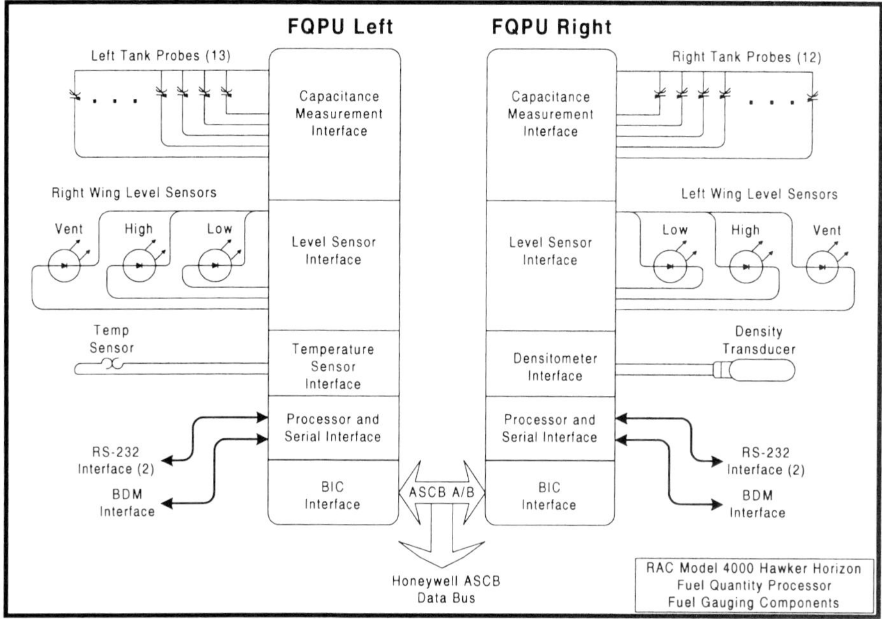

Figure 2 Fuel Quantity Indicating System Block Diagram

2.1 Fuel Quantity Measurement

The system uses twelve (12) linear, AC capacitance probes in each wing to determine the fuel volume for that wing tank. Each probe is multiplexed and measured individually to permit continued gauging in the event of a probe failure. The height volume profiles for each probe are stored in FQP memory. The height volume profiles are determined from a proprietary solids modeling tool that analyzes the response of each probe as the fuel height is simulated from empty to full in small increments. This analysis is performed at various aircraft pitch and roll angles to generate a set of coefficients representing the height - volume relationship of each probe over the entire flight envelope. These coefficients are stored in FQP memory and applied to the individual probe measurements in accordance with the aircraft attitude as received over the ASCB.

The fuel quantity densitometer installed in the left wing measures the fuel density with a vibrating cylinder transducer. The cylinder is driven by the FQP into resonant vibration at a frequency dependent upon the fuel in which it is immersed. The period of oscillation is measured by and converted into density by the FQP. The density is transmitted across the ASCB so it is available to both FQPUs. The fuel volume as determined by the capacitance probe measurements is multiplied by the measured density to determine the fuel quantity in pounds (or kilograms).

A compensator probe installed in the left tank allows for the calculation of back up fuel density in the event of a densitometer failure.

The direct measurement of fuel density and the application of attitude correction coefficients enables the system to achieve fuel measurement accuracy approaching 1% of Full Scale.

The fuel quantity of each wing tank as well as the total aircraft fuel quantity is transmitted via the ASCB to the flight deck for display on the fuel page synoptic. In addition to the fuel quantities, fuel density and the individual probe capacitances are placed on the ASCB for display on the flight deck.

2.2 Level Sensing

Optical level sensors are installed in each tank to provide for low fuel warning, high level shut off and fuel in vent warning. The FQPU monitors the wet/dry status of each level sensor to generate the warnings as well as to initiate refueling valve open/closed commands. The high level sensor acts as a back up to the fuel quantity measurement function. If the gauging system fails to command the refueling valve closed at the preselected shut off point and/or at the maximum volumetric shut off point the fuel will submerge the high level sensor which will in turn initiate fuel valve closure.

Optical level sensors provide faster response time, improved reliability and enhanced Built In Test (BIT) capability over the float type or thermistor based level sensors in common use.

2.3 Temperature Sensing

A sleeve mounted Resistive Temperature Device (RTD) is installed on the left wing rear spar to provide for the measurement of fuel temperature. The FQPU measures the resistance of the RTD and converts this into temperature for transmittal via the ASCB to the flight deck. The fuel temperature is monitored by the flight crew to ensure that the fuel does not reach the waxing temperature ($\approx$ -40°C) at which point it cannot be pumped to the engines.

2.4 Ground Refuel Panel

A ground refuel panel (GRP) is located at the right forward fairing adjacent to the single point pressure refueling station. The GRP illustrated in Figure3 is used to monitor and control aircraft ground refueling/defueling operations. The GRP provides the means to preselect a desired fuel quantity load prior to the start of ground refueling. The desired fuel quantity load is divided equally between the two wings and the refueling valves are commanded closed as the fuel quantity approached the desired shut off target. Communication with the FQPU is accomplished through a generic I/O module resident in the MAU. The GRP encodes the front panel position switches and persecuted fuel quantity into ARINC 429 transmissions to the I/O module which translates the data into 32 bit ASCB format for transfer to the left FQPU. The left FQPU transmits the on/off commands for the GRP legends and the fuel quantity to be displayed via the ASCB to the generic I/O module. The I/O module translates the ASCB data into ARINC 429 transmissions to the GRP.

The refueling valve commands are under the control the FQPU. When the fuel quantity in a tank approaches the desired shut off point the FQPU issues a valve closed command to the appropriate load control via the ASCB. Sufficient anticipation is applied to the command to account for bus latency and valve response to ensure that shut off is within +/- 30 pounds of the target.

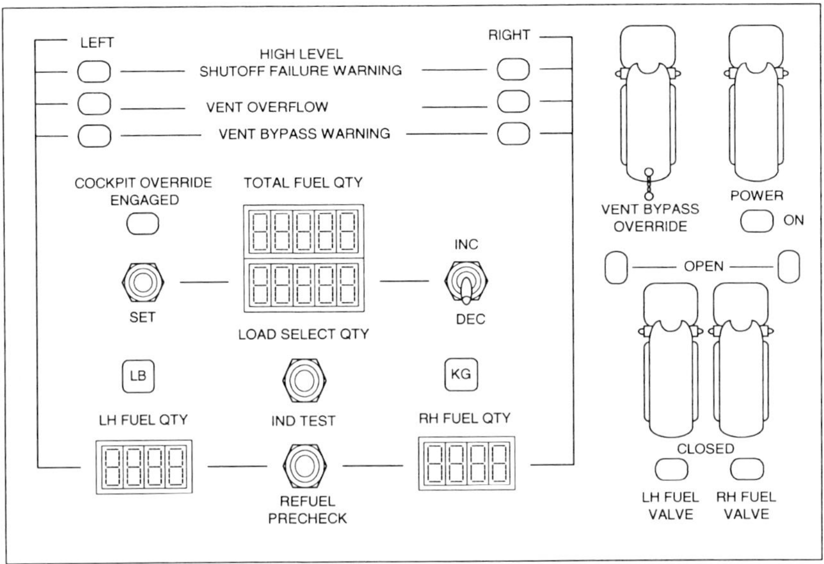

Figure 3 Ground Refuel Panel

2.4.1 Flight Deck Control of Refueling

The utilities based architecture provides for a simple means of controlling ground refueling operations from the flight deck with the Multifunction Control Display Units (MCDUs) . Figure 4 illustrates the Fuel Quantity Indicating System interfaces supporting ground refueling. When enabled by a switch in the overhead panel, the FQPUs take control of the MCDU and put a simulated fuel control panel on the MCDU screen. MCDU soft keys are programmed to emulate the functionality of the corresponding switches on the GRP. In this way a flight crew member or other authorized person can oversee the fueling operation. There are several operators that will receive their Horizon aircraft without the GRP even installed. The MCDU will be the sole means of controlling pressure refueling.

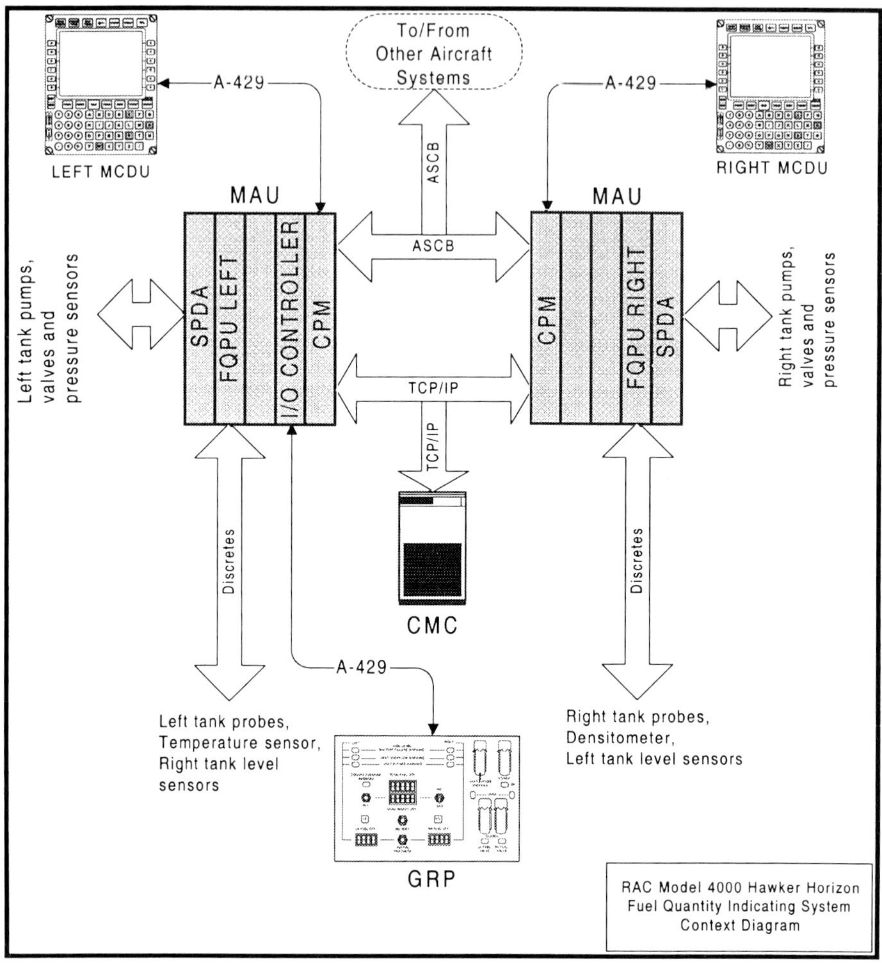

Figure 4 Ground Refueling Interfaces

2.5 Impact of Utilities Integration on FQIS Design

Integrating the FQIS into the aircraft utilities system had a number of significant impacts to the design. The foremost being the requirement to host the ASCB interface on the FQPU. This interface referred to as the Backplane Interface Controller or BIC is essentially a dual port RAM with sufficient arbitration and isolation logic to prevent any single module from taking down the entire bus. The BIC chipset is defined by Honeywell and each module in the MAU has the identical BIC circuitry installed.

The MAU rack provides conditioned power supplies of +3.3VDC, +5 VDC and +/- 14VDC. This eliminates the requirement of each module to derive analog and digital supplies from the aircraft 28 VDC bus. However, modules requiring other voltages must derive them from these MAU provided supplies or have the 28VDC supply routed through the module front cover. The FQPU requires a minimum of 15VDC for the probe drive signals and is deriving these supplies from the +5VDC supply with onboard DC-DC converters.

Since all FQPU communications with the airframe occurs via the ASCB there are no onboard bus drivers such as RS422 or ARINC 429 transceivers. In addition to reduced circuitry requirements on the FQPU this also results in reduced system wiring weight since there is no requirement for separate busses to transmit the fuel quantity data to the flight deck. In a stand alone or non integrated architecture it is not uncommon for a fuel quantity processor to support four (4) or more separate ARINC 429 data busses.

Another added benefit of the MAU based processor is the reduced & simplified mechanical packaging requirements. A stand alone fuel quantity processor would require a complete mechanical enclosure with internal circuit card holding provisions. As a circuit card assembly or module the FQPU mechanical design and hardware mounting provisions are significantly reduced.

3. Fuel Control System

The fuel system control function is completely integrated into the utilities system. There is no stand alone Fuel System Controller or Fuel Control Relay Panel.

Figure 5 is a block diagram of the fuel distribution and control logic. The pump on/off and valve open/closed commands are initiated by the load control modules in response to discrete switches on the overhead fuel control panel. Valve positions and system pressures are monitored by the MAU and SPDAs with all information made available on the ASCB for display on the flight deck.

The AC boost pumps are driven with 3ϕ variable frequency (375 to 700Hz) 115VRMS from the PDA. The use of variable frequency eliminates the requirement for (and corresponding weight of) a constant speed engine driven generator.

The DC start pump is a 28VDC canister type pump used to start the APU.

All electrical connections to the pumps are made external to the fuel tanks.

The engine firewall shut off valves are electric motor operated valves used to control the flow of fuel to the engines. The APU firewall shut off valve is a solenoid operated valve. The open/close commands from the load control module are hardwire or'ed with firehandle to ensure emergency valve control is available in the event of a load controller fault.

The crossfeed valve is an electric motor operated valve which permits feeding fuel to both engines from a single tank in the event of a pump failure or to correct a lateral imbalance.

The transfer (designated XFER) pump is a 28VDC bi-directional gear pump used to transfer fuel between wings for maintenance operations or to correct a lateral imbalance.

The refueling valve open closed commands are received by the load controllers from the FQP via the ASCB. The refueling valves are driven closed and disabled when the aircraft is in the air.

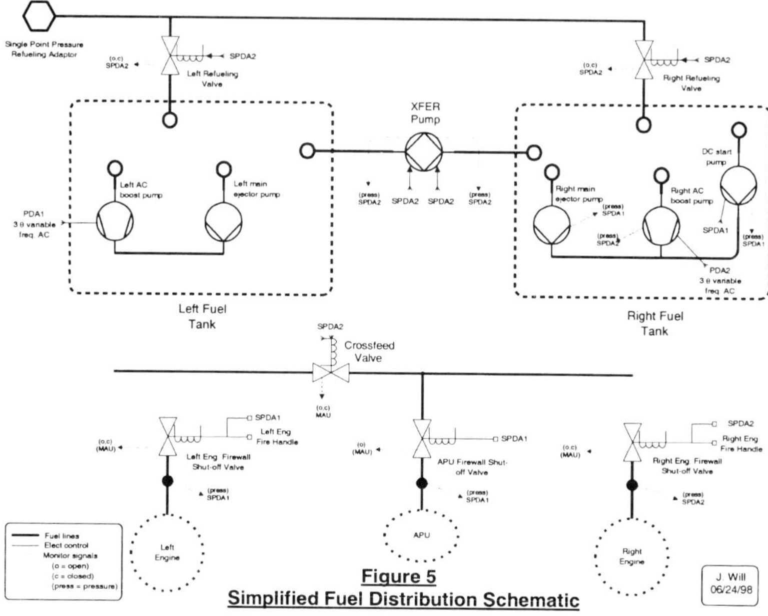

Figure 5
Simplified Fuel Distribution Schematic

3.1 Impact of Utilities Integration on Fuel System Control Design

The fuel system control function is well suited for integration into a utilities based system. The logic associated with the valve and pump controls is usually very straight forward and can easily be integrated within the power management functions.

By providing the switch and valve position information on the system bus (i.e. ASCB) all systems making decisions based on this data have access to it without having to wire the corresponding input(s) to each LRU. The Horizon implementation illustrates how the fuel system control function can be distributed between the Fuel Quantity Indicating System and Power Management System and eliminate the requirement for a stand alone fuel system controller.

This implementation requires careful coordination between the various system integrators to ensure the system failure modes and hazard assessments address all of the safety issues associated with fuel system design.

4. System Integration Responsibilities

The Horizon program continues the trend toward suppliers gaining a larger role in aircraft integration. It is becoming less common for an airframe manufacturer to design and build a new airplane by competitively bidding the individual components (pumps, valves, gauges, controllers etc.). It is more likely that complete systems will be competed with suppliers responsible for selecting the components that are most cost effective and best suited to ensuring the system achieves it performance requirements. With the large development costs as well as the substantial risks associated with the terms and conditions of most development contracts the term "risk sharing partner" has been come into vogue.

As a risk sharing partner on the Horizon program, Smiths is responsible for the design, development, integration and certification support of the complete fuel system

It is important to recognize that as suppliers are awarded complete systems there is a corresponding increase in responsibility for integrating those system into the overall airplane and supporting the airplane certification effort. Suppliers that are accustomed to supplying individual components to an Interface Control Document (ICD) or Source Control Drawing (SCD) might easily be overwhelmed as they try to fulfill the role as system integrator.

In recognition of the extensive coordination necessary to accomplish the integration smoothly and per schedule, Raytheon required that all Horizon's major system suppliers maintain an on site presence at their facility in Wichita. The extensive interaction and communications necessary to ensure successful systems integration could not be possible without having all the suppliers design representatives in a single location.

As system interfaces and installation designs are undertaken the inevitable interferences between system plumbing, wiring and structure must be resolved quickly. A strong on site representation is a must if a supplier expects to influence changes and compromises in their favor.

As the program advances into the system integration phase there will be extensive on site support required by the supplier's hardware and software teams at the Honeywell integration facility as well as Raytheon's new Integrated System Development Facility (ISDF) currently under construction. Demonstrating satisfactory FQPU performance in the MAU and verifying that the fuel system control logic has been implemented correctly cannot be accomplished as stand alone activities. They must be conducted in the overall integrated system environment.

In support of the fuel system certification effort, Smiths has prepared an extensive FAR part 25 compliance document which identifies all of the applicable part 25 FARs and the intended means by which compliance will de demonstrated. We will be responsible for reviewing all fuel system related certification test procedures, compiling the results and verifying their compliance with the regulations.

The responsibility of a system integrator is not complete until the airplane is certified.

5. SUMMARY

While this paper does not provide an overall analysis of the benefits achieved by utilities integration nor does it present all of the utilities architectures that are available, it does present an overview of the integrated fuel system on the Hawker Horizon program. The benefits achieved by this integration as they pertain to the fuel system include

- elimination of onboard power supplies from the Fuel Quantity Processor
- simplified and lighter mechanical packaging requirements
- reduced wiring due to the elimination of separate busses between the Fuel Quantity Processor and the flight deck display system
- elimination of a stand alone fuel system controller

C545/034/98

Dual use of variable speed constant frequency (VSCF) cyclo-converter technology

V BONNEAU BSEE, MSEE
Smiths Industries Leland Electrosystems, USA

SYNOPSIS

The conventional means of generating 115 VAC, 3-phase electrical power on civil aircraft is to use mechanical Integrated Drive Generators (IDGs) to generate constant frequency (CF) 400 Hz. This method has finite limitations on reliability based on needs for periodic maintenance and mechanical wear-out. Recently a few civil applications have begun using all electronic DC link/inverter Constant Frequency Generator (CFG) electronic approaches for 400 Hz power. While this technology offers little or no periodic maintenance, it still suffers from a reliability standpoint due to low field numbers and lack of product maturity.

The Smiths/Leland CFG Cycloconverter is an alternative all electronic method of generating 400 Hz electrical power which has wide usage within the US military community. Over 4,000 Cycloconverter generation systems are in service with over 6 million flying hours.

This paper reviews the Cycloconverter principle of operation, contrasts DC-link failure modes and shows how the cycloconverter technology may be adapted for civil use with resultant improved reliability expectations.

1.0 INTRODUCTION

Aircraft electrical power has been gradually evolving over the last 50 years. First propeller driven airframes used a very simple 6, 12 or later 28 VDC system. These arrangements were of low power (<5 KW) and almost always used a DC battery as back-up. In the 1940's and 1950's, aircraft power requirements increased in KVA to the point where 28 VDC systems were getting heavy from machine and feeder weights. A shift in electric power occurred from 28 VDC exclusively to a mixture of 28 VDC and 115 VAC / 400 Hz. The AC power was used for all non-critical loads such as pumps, heating, lighting etc. and the 28 VDC with its battery back-up was left for flight critical loads like flight controls, navigation, engine controls, etc. By using 400 Hz, 3-phase power, the feeder weights and prime generation machines would be smaller and lighter.

In the 1960's and later, jet engines were produced on larger aircraft and again the need for more KVA arose. The jet engine also created a demand for some type of constant speed device between the engine and generator in order to yield a constant frequency (400 Hz) from the variable speed jet engine. In the 1960's and 1970's this device was a mechanical transmission or constant speed drive (CSD). Being a mechanical, complex device, the CSD required regular maintenance for wear components and had a relatively low reliability (2000 to 5000 hour MTBF). During that period the CSD was the only way to make constant frequency from variable speed however, and almost all commercial and military aircraft used the CSD or integrated drive generator derivative (IDG).

The 1980's and 1990's saw a new area of activity in aircraft electrical power generator come about, initiated by the military side. Electronic frequency conversion from a variable frequency generator began to be developed and an alternative to the CSD/IDG was found in electronic variable frequency to constant frequency (VSCF) converters. The first two types of VSCF were cycloconverter or AC to AC, SCR-based techniques and DC link, AC to DC to AC power transistor-based techniques. Both conversion types were developed and operated in significant production quantities to yield accurate field reliability and fault mode data. In the 1990's several additional power generation methods were explored, again mostly on the military aircraft side. These are 270 VDC, switched reluctance and variable frequency. Of these only variable frequency is being pursued on the civil side of the aircraft market.

This paper attempts to explain both cycloconverter and DC-link VSCF system designs in a simple enough method so as to contrast normal and fault-mode operation. A reliability analysis is shown for the cycloconverter in a commercial transport, 60 KVA application.

2.0 TECHNICAL OVERVIEW FOR CONSTANT FREQUENCY POWER GENERATION

Constant Frequency Generator (CFG) can be accomplished on an aircraft, in general, two ways. One is to use a mechanical speed regulation device and a constant speed generator. This has been done over the years and is known as a CSD or IDG as previously described.

The alternative method is with a directly driven generator running at a variable speed and corresponding variable frequency with a constant frequency electronic converter. The CFG or variable speed constant frequency (VSCF) electrical conversion can be done several ways, but the two most prevalent approaches are the AC to AC cycloconverter and the AC to DC to AC DC-link inverter. This paper will focus on the electronic conversion method of CFG contrasting the DC link and cycloconverter topologies.

Figure 1.0 shows a block diagram of a simple VSCF function on an aircraft engine. The engine and power take-off deliver torque to an accessory drive gearbox. The gearbox in turn delivers engine torque at the correct speed to the generator. Often in VSCF applications, the accessory gearbox also provides oil to the generator for cooling and lubrication.

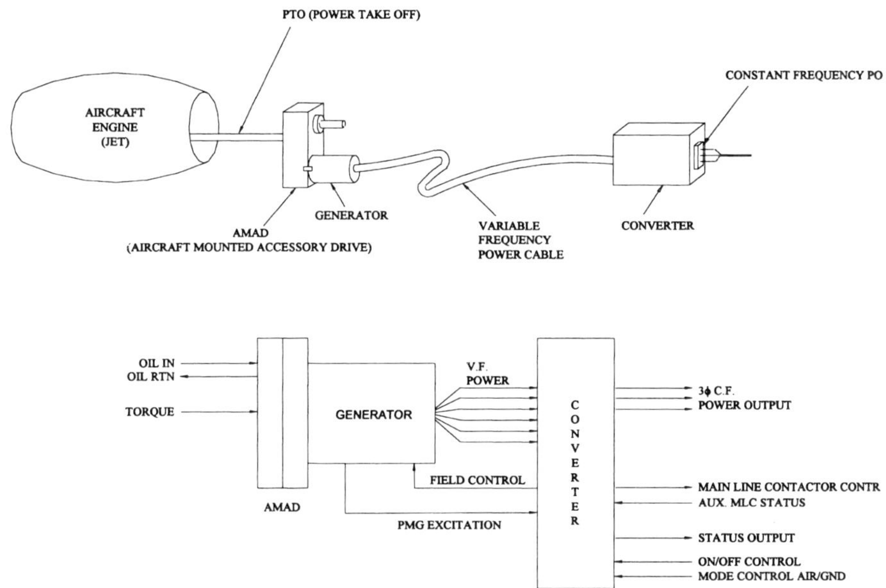

Figure 1.0: VSCF Function on an Aircraft Engine.

In general, VSCF generator shaft speeds are as high as is practical in order to have the lightest weight machine. Generator speeds will vary approximately 1.7 to 1 or 2.0 to 1 depending on the jet engine range of speed. Typical generator speeds are 12,000 to 24,000 rpm or 15,000 to 30,000 rpm. Generator output frequency is dependent on the converter requirements.

In Figure 1.0, the generator output frequency will vary proportionately with engine speed. The converter is required to take the variable frequency and convert it to a fixed frequency at 400 Hz while also maintaining a 115 VAC, 3 phase voltage regulation. The converter has sub-tier functions of generator exciter regulation, main AC line contactor control and aircraft control interfaces of on/off, mode and status. For the purposes of this paper, we will concentrate on the power stage differences between DC link and the cycloconverter since all other functions are common between technology approaches.

2.1 DC Link Converters

The DC link CFG system approach uses two power stage sections in order to convert the variable frequency input power to constant frequency AC. Figure 2.0 depicts the DC link power converter topology and shows these two major stages along with the key components. From left to right we see the variable frequency 3-phase AC voltage is converted to a DC link through a full wave rectifier bridge of six diodes. This DC voltage is filtered in bus capacitor C1. Next, a six switch, 3-phase inverter is used to "chop" the DC bus into a pulse width modulated output of 3 sine waves with harmonics. There are a multitude of switching schemes employed to produce the inverter output of 3-phase AC. They all use some form of pulse width modulation to synthesis the desired 400 Hz output. An example of a typical Pulse Width Modulation (PMW)[(1)] is shown in Figure 3.0. Most hard-switch, DC-link inverters follow a similar pattern. The output 3-phase waveforms are then filtered through the output choke/capacitor networks. An additional neutral forming output transformer will be necessary in this approach.

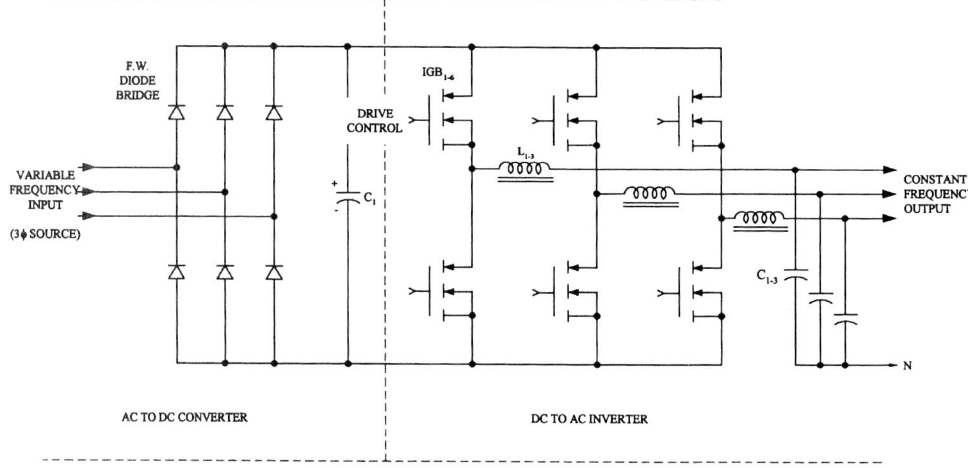

Figure 2.0: DC Link Power Converter

[(1)] PMW is 3-Phase Voltage Source Inverters; Power Electronics; Mohan, Undeland & Robbins

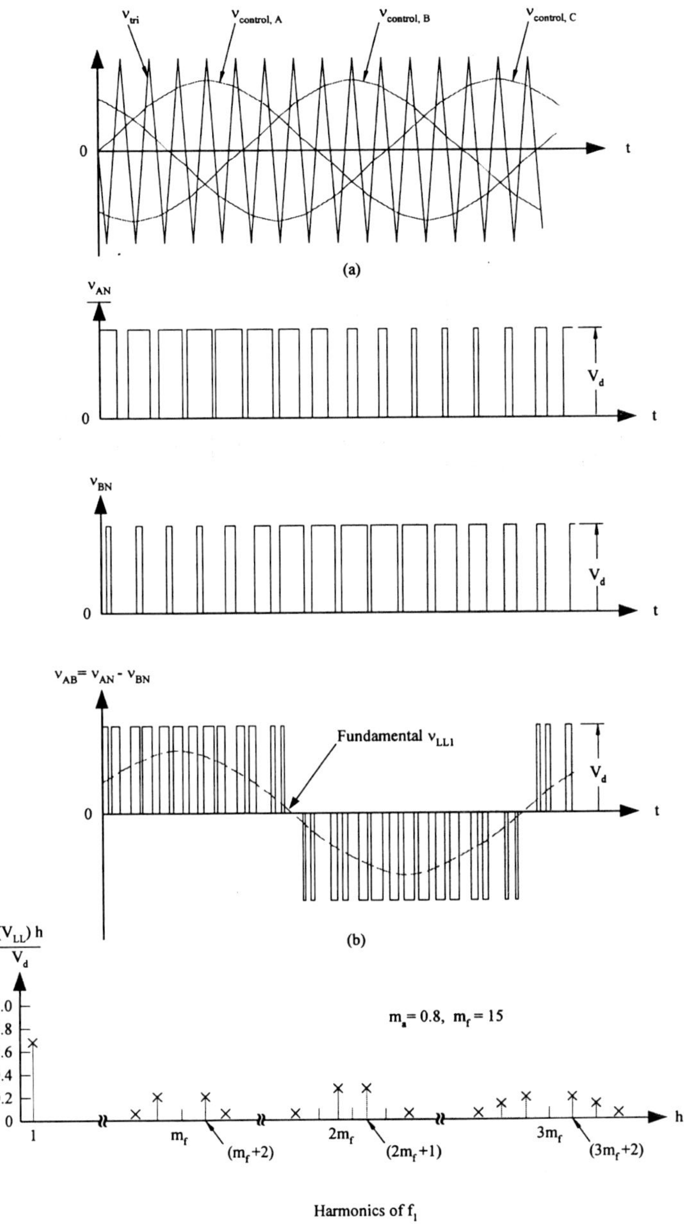

Figure 3.0: Three-Phase PWM Waveforms and Harmonics Spectrum for DC Link Inverter

Under normal operation, the DC link converter control functions to deliver 3 phase power in a continuous mode. Input AC power is rectified into DC and inverted into constant frequency AC, 400 Hz. Load transients are sensed and regulated to give a tightly controlled output with typical aircraft level power quality requirements of

- 5% Total Harmonic distortion or less
- ± 2.5 Vrms regulation max. with up to 1/3 unbalance loads
- No frequency transients
- Minimum voltage transients at step loading
- Controlled start-up and shutdown

Most operating requirements also specify some type of overload requirement such as 150% or 200% for short durations of time. Typical times of 5 minutes at 150% and 5 seconds at 200% of the full continuous load is specified to handle brief system overloads like motor starts or equipment bus transfers.

2.1.1 Fault Protection for the DC Link Converter

There are two types of fault protection needed for any CFG system. One is to protect load equipment from poor power quality or from extended fault current situations on a bus/load system. The second type of protection is for internal DC link converter protection, where internal faults may produce damage to the converter.

Typical external fault protection is listed as

- Overvoltage (inverse time delay)
- Undervoltage (5 seconds)
- Wrong frequency (immediate)
- Excess DC content (inverse time delay)
- Excess Zero sequence voltage (5 seconds)
- Feeder fault (immediate)
- Short circuit (5 seconds)
- Excess distortion (5 seconds)

Internal fault protection is converter type specific and for the DC link system has a typical list of

- DC link overvoltage
- IGBT gate misfire
- IGBT shoot thru current
- Input rectifier fault
- Internal feeder fault
- IGBT over temperature
- IGBT short circuit

The internal DC link protection circuits are generally centered around protecting against an IGBT misfire, overvoltage or short circuit condition. A failure to this device will quickly cascade into the other inverter components and is not recoverable.

2.2 Cycloconverter Based CFG

The cycloconverter is an AC to AC converter without an intermediary DC-link section. The cycloconverter in effect switches sections of the input variable frequency AC voltage into a constant AC output. The cycloconverter topology requires that the input frequency be at least 3 times higher than the minimum output frequency for high quality power required by aircraft loads.

Unlike the DC-link, the cycloconverter maintains the same neutral point between input and output AC so an isolation transformer is not needed.

A single phase diagram is shown in Figure 4.0 for a six phase input to 3 phase output cycloconverter. A 3 phase input cycloconverter could also be used for discussion purposes, but a six phase input is more typical for a high power unit. The term "Cyclo" was coined years past from the observation that the power switch tends to be gated sequentially and "cycle" through the high frequency input to synthesize a desired output (400 Hz). In Figure 4.0 we see 12 SCR's arranged to provide the positive and negative half of the reference 400 Hz wave by rectifying or inverting the high frequency wave. As a rule of thumb, the generator frequency needs to be 3 times the output frequency minimum to yield less than 5% THD after the L/C filter. From the waveform of Figure 5.0, we see how each SCR (thyristor) is gated after an appropriate delay angle to provide the correct output "slice" to the 400 Hz waveform. An output L/C filter reduces the excess harmonic energy in a similar fashion as the DC-link output filter.

The cycloconverter power state topology has only one conversion stage as shown. This means that it is inherently more efficient with lower power component losses to shed than the previous DC link approach. The amount of silicon area in the cycloconverter is also less than the DC link based with one power stage. Silicon total area is the measurement technique commonly used today between power conversion topologies as power switch count varies with the package choice for the power device. This is easily illustrated by showing that today's 60 KVA cycloconverter consists of nine power SCR hybrids while a 60 KVA DC link can use two parallel IGBT's for each switch at a total power IGBT hybrid count of twelve.

Figure 4.0: Cycloconverter Power Stage (One Phase of Three)

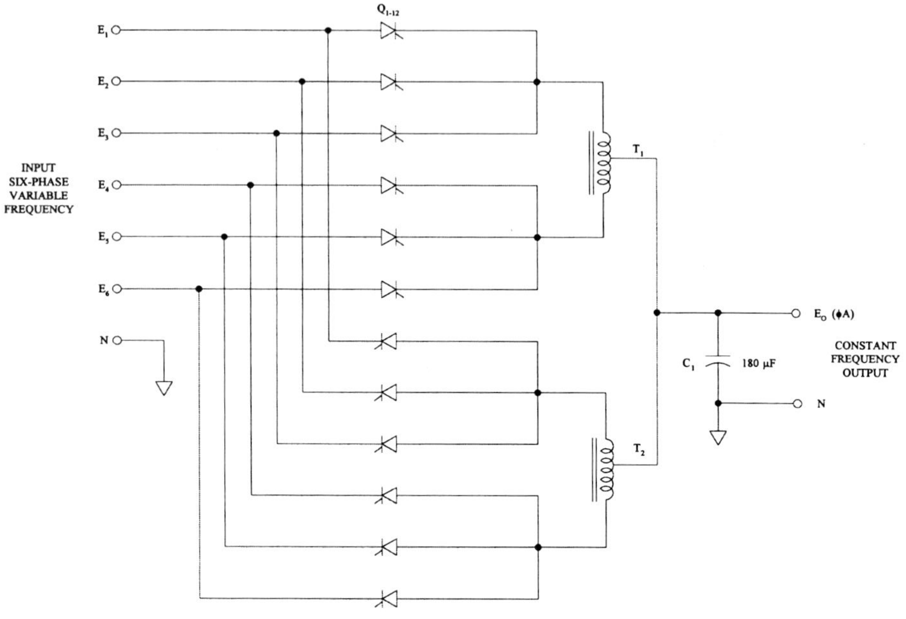

Figure 5.0: Cycloconverter Waveforms.

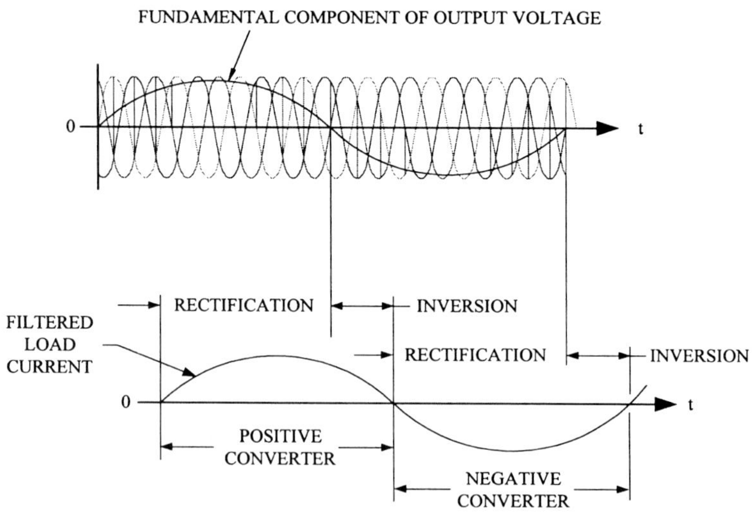

In the cycloconverter topology, the generator input power source is connected to the output at the appropriate time to reproduce the reference AC waveform. Frequency conversion is through a naturally commutated SCR bridge made up of a positive and negative bank of power devices. Harmonic energy is filtered using interphase transformers and AC output capacitors. Control of the cycloconverter senses several parameters.

- AC input voltage
- AC reference
- Output AC voltage

These signals are used to establish thyristor delay angles and to obtain input waveform knowledge about voltage and time. A firing signal is modulated with the 400 Hz reference to gate each successive SCR device. Feedback provides regulation of voltage and some form of non-linear load correction.

As in the DC-link VSCF, the cycloconverter VSCF controls its 400 Hz waveform to yield the same quality requirements of harmonic reduction to 5%, voltage regulation to ± 1.5 Vrms etc.

2.2.1 Fault Protection for the Cycloconverter

The same situation for protection is present in the cycloconverter system as was needed in the DC link. Typical fault protection is for things like overvoltage, undervoltage, feeder fault etc. Where the cycloconverter differs, however is in the internal fault protection design. Here the only significant fault protection needed is for internal feeder faults such as wiring shorts or SCR shorts to the chassis. Unlike the DC link IGBT, the SCR can misfire without catastrophic results. The misfire of an SCR or miscommutation results in a momentary output waveform disturbance until the fault current goes through zero and the device turns-off naturally. The SCR i^2t rating is always sized to handle repeated miscommutations without failure. Internal cycloconverter fault protection is usually:

- Internal feeder fault
- AC generator overvoltage
- Excessive SCR miscommutations.

3.0 RELIABILITY DISCUSSIONS

3.1 Failure Mode Comparisons

The key to high reliability is to have a robust, balanced design where no components are over-stressed. Using high reliability components does not guarantee the product will have high reliability or fail free life. High reliability is a result from several factors such as

- Normal and abnormal fault mode protection
- Identified user applications and operational scenarios

- Component reliability match to application needs
- Protection for common fault propagation
- Protection for uncommon faults
- Weed out of infant mortality prior to use

While this list is not all inclusive, it does point out several recurring themes concerning reliability. They are simply stated as:

1) Fully understand the user application and match the design accordingly.

2) Plan for the abnormal problem or occasional user scenario to occur without having a failed product as a result.

We can try to review the DC link and cycloconverter VSCF approaches from these standpoints. A reliable VSCF converter needs to be able to handle all aspects of the end item use as well as have good dependable protection for occasionally occurring issues. The best analysis approach for a fault mode review is to list a common review criteria. One such list is seen as

A) Noise / control problems
B) Load issues of hard, soft and intermittent shorts
C) Over temperatures.

3.1.1 Noise / Control Problems

Noise and/or control circuit problems always seem to be present in power electronic converters. Most products require efforts by designers to shield or filter control lines internal and external to the converter in order to reduce any susceptibility that may make an undesired state of operation. In general, these problems cause the main power devices to switch at the wrong time or not switch when commanded. No design is foolproof. There will always be a time, however infrequent, when a switch will not work as desired. In the case of the DC link VSCF, an uncontrolled IGBT gate command usually results in a shorted DC link bus and at least two failed IGBT devices. Current cannot always or easily be limited fast enough to save the devices from exceeding their safe operating area (SOA) limit and the devices will fail. A well designed DC link converter must find a way to protect from a hard bus fault or short-thru current by adding fault detect protection circuits. These circuits have to be very fast acting and will cause an immediate shutdown of the entire converter to "save" the IGBT switches from moving beyond the SOA.

Noise and/or control problems tend to act differently in the cycloconverter. A misgated SCR will result in a brief short circuit of the input generator phase as the misgated SCR goes through the miscommutation event and the current moves back to zero at AC zero voltage crossing. Protection of the SCR's from failure is accomplished by choosing the appropriate i^2t rating of the device and matching it to be larger than the output generator inductance di/dt rating. A miscommutation will not usually see a converter shutdown as the output waveform from the cycloconverter will be disturbed by the misgating event, but not result in an immediate need to terminate operation. The SCR is an i^2t device with an inherent current overload capability as a function of time as compared to the IGBT.

3.1.2 Load Issues of Hard, Soft and Intermittent Shorts

Short circuits and overloads are handled by the DC link VSCF through design margin in the power components. Most often, the short circuit current and overload power are able to be provided without any radical changes to the full load, continuous power design points. The 5 second overload and short circuit requirements are designed around faulted load equipment circuit breaker clearing times which do not cause enough of a thermal time constant to be a problem for the magnetics and, in most cases, the IGBT's and diodes. The 5 minute overload is reviewed in detail and very often is a design corner point for the power semiconductors.

The cycloconverter tends to handle the heavy overloads and short circuits a little easier than the DC link in that there is no DC link voltage to depend on. During a heavy load, the DC link filter capacitor (C1) will need to hold-up or the output inverter bridge will not produce enough voltage. The cycloconverter has no internal bus sag during these situations and only the input generator response to heavy loads comes into play. Both converter types do produce more harmonic distortion during shorts and overloads.

Soft or intermittent short circuits is a real sticky point for the DC link VSCF. Repeated, intermittent shorts will push the IGBT devices out of the SOA curve and eventually cause a converter failure. The cycloconverter, without an SOA limitation will tend to simple recover or miscommutate and then recover from the intermittent short circuit. It is important to note that most aircraft faults tend to be intermittent as they often are a result of chaffed wires around bulkheads, or vibration/acceleration/position sensitive failures and tend to be short in time duration.

3.1.3 Over-Temperatures

Over temperature conditions occasionally occur in electronic systems. Some scenarios that lead to this in air or liquid cooled applications can be listed as:

- Brief, abnormal ground operating times
- Hot taxi after aborting a take-off
- Heat soak back from cooling fan failure (if equipped)
- Low fuel conditions after a long hot flight (if equipped with a fuel to air heat exchanger)
- Equipment failures causing intermittent faults or cycling operation.
- Adjacent equipment failures causing close proximity thermal increases.
- Unspecified operating conditions during ground maintenance.

The result on a VSCF converter from any of these conditions is higher temperatures on the power and control components. High temperatures, in general, lower safe operating current and voltage capabilities. In case of all semiconductors, internal operational parameters are usually degraded. For the IGBT, the device will catastrophically fail if high temperatures cause uncontrolled latch-up. SCR's will also be affected by high die temperatures. Turn-off time (Trr) will lengthen beyond specified times and the device may miscommutate. Again, as in the previous discussions, the thyristors will usually recover and will continue to function while IGBT based DC link inverters will fail if not equipped with fast thermal shutdown protection.

3.2 Cycloconverter Reliability Analysis for Dual-Use Civil Applications

Smith Industries, Leland Division has manufactured a 40 KVA and 60 KVA cycloconverter CFG for various CFG military aircraft applications. In this analysis the unit reliability history and MIL-HDBK-217 prediction methods are used to develop a projected civil transport reliability mean time between failure (MTBF) for these types of equipment in civil transport applications.

Figures 6 and 7 show the basic cycloconverter CFG hardware with and without the covers. The CFG shown is an integral package unit with both rotating generator and static cycloconverter housed in a single, QAD band attached structure. Shop replaceable assemblies are used for a very maintainable, module design.

3.2.1 CFG Reliability Model

Figure 6. Basic Cycloconverter CFG with Cover

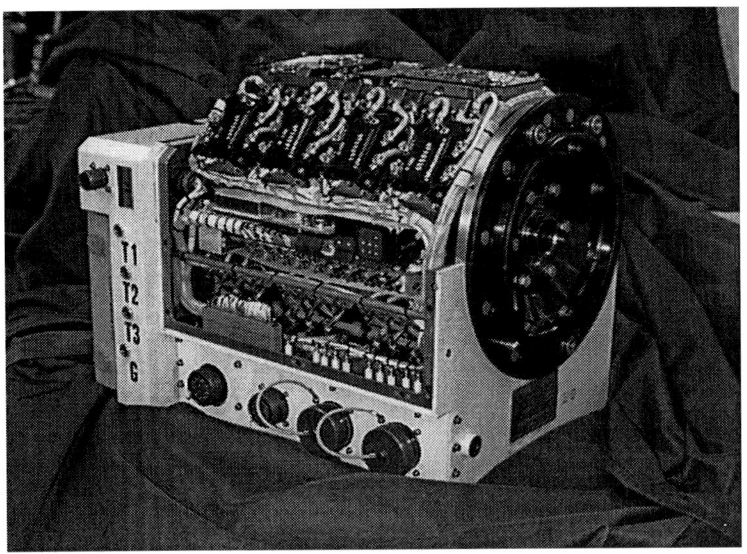

Figure 7. Basic Cycloconverter CFG without Cover

Two critical parameters of the reliability model must be considered in order to analyze a fighter aircraft environment vs. a transport or cargo aircraft installation. These parameters are: temperature (π_t) and environment (π_E). Final configuration of the system must also be considered in the analysis; specifically, what sections of the hardware remain in an engine area and what sections go into an equipment bay. For this analysis, the baseline configuration of the hardware changes. In the fighter application, a single, integral generator/converter unit is located in the engine area. The transport application leaves the generator on the engine gearbox, but moves a separately packaged converter to an equipment bay. Both inhabited and uninhabited environments are examined for the converter in the equipment bay.

The assumption to "split" the CFG electronic converter away from the engine area is used since this reduces typical (average) semiconductor operating temperatures from 95°C down to 40°C. The existing engine area location is not a reliability issue to the generator, because the generator easily accepts 80°C cooling oil. However, the location is a reliability issue to the converter, because the low power control, protection and interface electronics should operate with minimum junction temperatures for highest reliability. Given that the parts selections are held constant (high reliability, stress screened and tested components are used in both approaches), temperature reduction of the electronics yields a potentially large reliability improvement. The following assumptions are made for this reliability analysis:

A) The transport aircraft configuration is a separate generator & converter unit with the generator as-is (in the engine area) and the converter moved to an air-cooled equipment bay. (Liquid cooling, which is more efficient, would possibly enable higher reliability figures and a lighter/smaller electronics package).

B) Since the generator section of the CFG remains in a relatively harsh engine area environment, its failure rate will not be adjusted when analyzed in a transport application. This is a more conservative assumption but is deemed reasonable based on actual reliability rates of engine-mounted equipment.

C) Power electronic components of the Cycloconverter (such as IPT's, SCR's & filter capacitors) will not be adjusted in reliability for any temperature improvement, because these parts produce heat during operation (losses) anyway. These components may be subjected to similar junction temperatures whether in an air-cooled equipment bay or in the liquid-cooled engine area. Again, this relatively conservative assumption is based on the efficiency differences between air-cooling and liquid-cooling. Usually, liquid cooling is not available in an equipment bay without special provisions.

D) Another failure rate improvement in the power electronics of the Cycloconverter is in the environmental factor (π_E) moving from a fighter to a transport aircraft. Low level electronics will see failure rate improvements due to both π_E and π_T, since there is no appreciable self heating in the signal-level circuitry.

3.2.2 Reliability Estimates

Research was conducted using typical CFG Cycloconverter component types and MIL-HDBK-217E to develop an average π_E and π_T factor. In general, for a given semiconductor component failure rate (λ_P), a sum of the environmental factor (π_E) and temperature factor (π_T) is multiplied through other factors and constants. The resultant failure rate gives the predicted rate of component failure per one million operating hours. This is illustrated in Equation 1.0 below:

$$\lambda_P = \pi_Q \bullet ({}_{K1}\,\pi_T + {}_{K2}\,\pi_E) \bullet \pi_L$$ **Equation 1.0**

π_Q is a general device quality factor and π_L is a maturity factor. Both are assumed held constant for this analysis. K1 and K2 are other factors of the component temperature and environment also held constant.

The information in Table 1.0 shows device π_E and π_T parameters from typical device data. This data results in the factor improvements illustrated in Table 2.0.

This data translates into the average π_E and π_T improvement factors shown.

Table 1.0
π_E and π_T Factors for Cycloconverter Components

Device Type	AUF		AUC		AIC	
	π_E	π_T	π_E	π_T	π_E	π_T
Monolithic IC	3.0	1.9	2.6	0.2	2.5	0.2
Hybrid Device	4.0	0.0002	1.7	0.0001	1.5	0.0001
Transistors	65	0.0018	10.5	0.00098	9.5	0.00093
Diodes	25	0.00082	16.5	0.00041	15	0.00034
Zeners	70	0.0011	5.0	0.00072	4.5	0.00068
SCR's	65	0.0065	10.4	0.00019	9.5	.00017
Resistors	15	0.0027	3.2	0.00035	3.0	0.00022
Capacitors	30	0.0095	4.0	0.00063	2.5	0.00052

Table 2.0
Resultant π_E and π_T Improvement Factors.

Device Type	Δ AUF to AUC		Δ AUF to AIC	
	π_E	π_T	π_E	π_T
Monolithic IC	1.15	9.5	1.2	9.5
Hybrid Device	2.35	2.0	2.67	2.0
Transistors	6.19	1.83	6.80	1.93
Diodes	1.51	2.0	1.67	2.41
Zeners	14	1.53	15.55	1.61
SCR's	6.25	3.42	6.84	3.82
Resistors	4.68	7.71	5.0	12.27
Capacitors	7.50	15.07	12.0	18.27
Average Factor	5.4	5.3	6.4	6.4

3.2.3 Cycloconverter Civil Application Prediction

The present SI/Leland integrated CFG unit reliability model is illustrated in Table 3.0 which contrasts both uninhabited and inhabited environments in transport applications. The model is partitioned in Shop Replaceable Assemblies (SRA's). Descriptions of each SRA are listed in the following data:

SRA No.	Name	Description / Function
A1	Control Module	SCR, current limit and power stage control. Gives each SCR hybrid device a preprogrammed phase retard corresponding to the overall AC output voltage error level. Low level circuits.
A2	Interface, Protection and Regulation	AC control loop, interface and protection circuitry. Functions to control and monitor the 3-phase AC output. Low level circuits.
A4	Signal and Feedback	AC voltage feedback and other divider circuits used to monitor the output regulation, low level circuits with high voltage spacing.
A5	Transient Suppressor	Generator line-to-line snubber R/C networks. Filters the generator variable frequency waveform. Power circuitry.
A8	Sense/Link PWB	Small PWB to interface generator variable frequency circuits with the A4 and other modules. Power circuitry.
A9	Generator Voltage Regulator	(GVR) regulates the main generator exciter field and controls variable frequency voltage. Power circuits.
C1-3	120μf AC Capacitors	Three hermetic, 200 Vrms AC filter capacitors. Filters each AC output phase for <4% THD.
CTA	Current Transformer Assembly	Used at the aircraft point-of-regulation to monitor for differential line protection.
PS1	Bias Power Supply	Internal VSCF bias power supply. 28VDC, ±12VDC and +5VDC @ 200 watts.
T1-6	Inter Phase Transformers	IPT's link SCR phase groups to filter the 400 Hz AC phases. These inductors (6) are power magnetics.
T7	LFCT & Mechanical	Low frequency current transformer for VSCF AC

		output current monitoring. Functions as differential protection with the CTA and output current limit.
Z1-9	SCR Module	Power silicon controlled rectifier assembly. Functions as the main switch.
G1	Generator Assembly	Main generator rotor, stator and frame assembly.
Chassis	Main and Rear Chassis	Main, rear and misc. chassis metal (aluminum) parts.

Table 3.0 CFG Failure Rate Matrix

SRA	AUF	AUC	AIC
	F/ 10^6 Hrs.	F/ 10^6 Hrs.	F/ 10^6 Hrs.
A1	36.79	3.44	2.87
A2	15.63	1.46	1.22
A4	22.82	2.13	1.78
A5	1.43	0.26	0.22
A8	0.65	0.11	0.10
A9	1.34	0.25	0.21
C1-3	17.4	3.22	2.72
CTA	0.24	0.04	0.04
PS1	9.26	1.71	1.45
T1-6	3.79	0.70	0.59
T7	1.12	0.20	0.18
Z1-9	80.92	14.98	12.64
G1	27.79	27.79	27.79
Chassis	0.91	0.16	0.14
Failure Rate Total	220.09 F/10^6 Hrs	56.45 F/10^6 Hrs	51.95 F/10^6 Hrs
Predicted Mean Time Between Failure (MTBF)	4543 Hrs.	17,714 Hrs.	19,246 Hrs.
Adjustment factor (based on in-service actual vs. predicted Cycloconverter history)	**21.3515%**	**21.3513%**	**21.3513%**
Realistic (adjusted) MTBF	**3573 Hrs.**	**13,932 Hrs.**	**15,137 Hrs.**

The matrix in Table 3.0 illustrates both failure rate and corresponding mean time between failure (MTBF) calculations of the CFG Cycloconverter system for fighter and transport environments. The summary values of MTBF, with factored adjustments for actual

VSCF field history, yield an excellent, realistic transport environment reliability in the 14,000 hours range.

4.0 SUMMARY DESCRIPTION

This paper has presented comparison information in some detail concerning a DC link and cycloconverter variable speed constant frequency system for 400 Hz aircraft power. Contrasts of major failure modes and reliability assessments have been discussed based on those failure modes in general terms. The results show that the hard switch, DC link approach tends to be less reliable and have more catastrophic fault modes than the naturally commutated cycloconverter. Key reasons are the inherent IGBT switch safe operating area limitations as compared to the robust I^2t capability of the cycloconverter thyristor.

From a power conversion topology viewpoint, the single stage cycloconverter has a higher reliability from greater efficiency and less total power silicon area. A two stage DC-link has a greater high stress component parts count in both power devices and power device support circuity

Creating a reliable unit costs the manufacturer in several ways.

1) Lot's of development time and materials for design improvements, elaborate testing to find the problems and long enough development schedules or prior R&D time to understand the needs exactly.

2) Unit protection complexity tends to increase exponentially with the addition of power semiconductors or power handling stages.

Both of these points are common sense and relate to the initial ideas presented about making a reliable VSCF product. The points state to

- understand the requirements fully and completely test the product
- make the product as simple as possible with lot's of stress margin.

If we assume that a CFG design approach of either the cycloconverter or DC-link system is done well . . . i.e. point number 1 is accomplished, then the only comparison issue has to do with complexity.

Clearly, to protect an IGBT hard switch device for all normal and induced (external) failure modes requires lots of monitor and fast acting circuits . . . much more than the cycloconverter since the thyristor has fewer failure modes. These extra, complex circuits will lower the reliability of the DC link given the same factors of application environment with the cycloconverter. Intelligent devices with large scale integration such as DSP's, microprocessors, PLD's and FPGA's do tend to help this issue of protection circuit complexity. Over time we are seeing an improvement in reliability predictions from the use of these devices. This trend should continue.

A CFG Cycloconverter used for demanding Military aircraft applications has been reviewed for civil transport aircraft use. Mean time between failure rates are projected with conservative factors based on actual field data for the civil use in the 14,000 hour area. It is believed that this level of reliability will be met in the civil environment and will be the highest, fault tolerant CFG system available for jet engine aircraft.

C545/031/98

Developments in variable frequency electrical systems

P MILLER BSc
Lucas Aerospace, Hemel Hemstead, UK

SYNOPSIS

Civil aircraft electrical power over the last forty years has been 400 Hz Constant Frequency (CF), traditionally using a complex and expensive hydro-mechanical Constant Speed Drive.

Industry interest is growing in Variable Frequency (VF) power, with similar power quality as CF, as the benefits in reliability and life cycle cost of VF are becoming better known. The effects of VF on electrical utilisation loads are now being quantified.

The power quality requirements for VF are included in the latest update to the ISO 1540 aircraft electrical power quality document that is due for issue during 1998.

The first application of VF power generation on a civil turbofan aircraft is the Bombardier Global Express® long range executive jet.

WHY VARIABLE FREQUENCY?

For over forty years the primary source of electrical power on civil passenger transport aircraft has been 115 Vac, three phase at a constant frequency of 400 Hz. Traditionally the constant frequency is produced by driving the rotor of the generator at a constant speed. A Constant Speed Drive (CSD) produces this constant speed from the variable speed output of the main engine accessory gearbox.

This CSD is a complex hydro-mechanical device, consisting of many close tolerance components. The close tolerance components make it an expensive device to manufacture and maintain, it also requires a high number of maintenance inspections and activities. Average generator loading and cooling oil inlet temperature both impact the reliability of the CSD. A CSD run at or anywhere near it's maximum power will have a short operational life.

At present the only viable method of producing Constant Frequency (CF) power is the Integrated Drive Generator (IDG). The IDG combines the CSD and generator in one unit. It is therefore, smaller, lighter and has simpler maintenance schedules than a system with separate CSD and generator.

Over the last 20 years an alternative method of producing CF power has been tried in a few civil applications. CF power is produced by electronic conversion from variable frequency power produced by a generator directly driven from the variable speed engine accessory gearbox. This is known as Variable Speed Constant Frequency (VSCF). But after early promise, this method has not lived up to reliability predictions and is no longer viewed as viable in the civil aircraft arena in the near future.

Variable frequency electrical power has been used on turboprop aircraft for many years. The generator is directly driven from the main engine accessory gearbox, so the frequency of the generated power is directly proportional to the speed of the engine. The power quality of such systems has not usually been an issue as the power is used mainly for general heating and anti-icing purposes also the engine speed range is relatively narrow. For the wider speed range turbofan engines, in applications where most of the loads are supplied from the ac power, it is necessary to address the subject of power quality.

Since 1991 Lucas Aerospace has developed Variable Frequency (VF) power systems with the same power quality as present CF systems for operation with the wider speed range turbofan aircraft. These VF power systems deliver the same power quality without the additional maintenance costs and lower reliability of the CSD.

VARIABLE FREQUENCY POWER QUALITY

Lucas Aerospace's aim for the VF power system was to supply power over the full VF range to the same quality as present CF system requirements i.e. MIL-STD-704E, DO-160C/D, etc. This includes all parameters (voltage transients, etc.) except the frequency range. Therefore, the only difference between a CF system and a VF system is the wider operating frequency range of the VF system. This was done to minimise the effect on electrical utilisation equipment.

This aim was achieved by taking account of the following higher frequency characteristics of generators during the design:

- A VF generator must be designed with lower reactances than an equivalent CF generator to counter the effects of higher frequencies on the power quality. This causes a VF generator to be approximately 10% heavier than an equivalent CF generator.
- A fast acting voltage regulator is required to limit the load switching voltage transients to the specification levels. This is a function of the Generator Control Unit (GCU).
- Fast acting overvoltage protection is required to limit the maximum voltage following a shorted field condition at high frequency to the specified level. This is a function of the GCU.
- On large aircraft with long feeders, the feeder impedance has more influence on the phase voltage unbalance than does the generator impedance. This affects both CF and VF systems, but the VF system is more affected because of the higher maximum frequency (approximately 700 Hz opposed to 420 Hz). There are cable installation concepts which can dramatically reduce the reactive component of these feeder impedances, these would increase the system's resistance to phase voltage unbalances. These concepts should be seriously considered for future large aircraft.

ISO 1540 - CHARACTERISTICS OF AIRCRAFT ELECTRICAL SYSTEMS

The Society of Automotive Engineers (SAE) AE-7 committee 'Aerospace Electrical Power & Equipment' has been working to produce a VF power quality requirements document for the past few years. At the same time the International Standards Organisation (ISO) and the SAE have identified the need to co-ordinate the requirements of the separate aerospace standards for all equipment concerned with aircraft electrical power. It was agreed between ISO and SAE that the VF power quality requirements would be incorporated into the next update of the ISO 1540 'Aerospace - Characteristics of Aircraft Electrical Systems' document. This update would be used to standardise the requirements for all equipment concerned with aircraft electrical power including generation, distribution, utilisation equipment and ground power.

Airframe, generation and utilisation equipment manufacturers are all working to expedite the issue of this update, which should be during 1998. Presently the draft is out with industry for comment.

The VF power maximum and minimum frequency limits were agreed after consideration of many conflicting needs and impacts, including:

- The minimum frequency limit must be set to include the 380 Hz minimum frequency of external power supplies.
- Present equipment can operate at the emergency minimum frequency of 360 Hz.
- Generator and electromagnetic load power density increases at higher frequencies, weight and volume can therefore be reduced.
- Feeder voltage drops are greater at higher frequencies, due to feeder reactances.
- Current and voltage distortion requirements are more difficult to achieve at higher frequencies.
- This document should set standard ranges to promote the design of common utilisation equipment, as opposed to designing aircraft specific equipment with aircraft specific frequency ranges.

The minimum frequency limit was set at 360 Hz because this includes the external power minimum frequency and is the minimum abnormal condition frequency set by the present DO-160D utilisation equipment requirements document. Therefore, utilisation equipment will not require any additional heavy electromagnetic core material and the maximum frequency is restricted to manageable levels.

To cater for the difference in speed range of turboprop and turbofan engines and for fitment to aircraft that can differ greatly in size, the requirements for VF power quality have been split into two sections:

- Narrow range of 360 Hz to 600 Hz, covers turboprops and other turbofans with narrow speed ranges (up to 1.66/1). Narrow range VF has exactly the same power quality limits as CF.
- Wide range of 360 Hz to 800 Hz, covers all present and foreseen turbofans. This range also covers the effects of future large passenger aircraft where the feeder impedance has a far greater effect than the generator impedance on power quality. The wide range VF power quality limits are slightly wider than the CF limits for some characteristics, while the rest are the same as the CF limits.

IMPACTS OF VARIABLE FREQUENCY ON UTILISATION EQUIPMENT

Lucas Aerospace has studied the operation of utilisation equipment on VF power with utilisation equipment manufacturers. The equipments most affected by VF power are asynchronous motors and fluorescent lighting. The effects on these and other equipment is discussed in this section.

Hydraulic power

Electro-hydraulic pumps (EHP) are usually variable displacement type pumps driven by an asynchronous motor. The variable displacement pump is designed to deliver a variable flow rate at a constant pressure.

For a pump driven by a constant voltage, variable frequency supplied motor, the pump will normally operate at maximum displacement to deliver rated flow when driven at minimum speed. The maximum flow capability of the pump will increase linearly with speed/frequency without additional flow limits. This pump would require a larger motor capable of delivering maximum flow at maximum speed at maximum displacement, than the equivalent CF motor. At low power ratings (<4 kW), the weight increase is small and the fact that the starting current is higher than it's CF equivalent can be tolerated. However, as the rating of the hydraulic system increases and becomes a major load on the electrical system, the situation needs to be addressed more carefully.

Starting can be ensured by preventing the pump from immediately going to maximum displacement, which limits the torque demand during run up, by incorporating a depressurisation valve. Addition of a flow limiter will limit the flow rate to the rated level and so provide a maximum output power that is constant with speed and equal to that of the CF pump. This causes the torque requirement to decrease with speed and allow the VF asynchronous motor to operate within its stable region. This motor would need to be approximately 15% larger and heavier than the equivalent CF motor, to cater for the motor torque reduction at higher frequencies.

Fuel system

The only fuel system equipment affected by VF operation are fuel pressure boost and transfer pumps which are usually fixed displacement types. These pumps produce a motor load torque that is almost constant with speed, this translates to a power output that is proportional to speed. This allows a VF asynchronous motor to be designed to operate within its stable region.

The pump is designed to deliver the rated pressure and flow rate at minimum steady state frequency. Thus the CF and VF powered pumps will be of the same capacity and weight. The critical design point for the VF asynchronous motor is starting at maximum frequency due to the lower torque of the motor at high frequency. Therefore the VF motor needs to be larger than an equivalent CF motor, resulting in an estimated overall weight increase of 10-15% for the motor pump unit.

The engine fuel pressure boost system design is not detrimentally affected by VF operation. The design of the fuel line between the boost pump and the engine is not affected by the range of pressures present over a flight profile.

The fuel transfer system is not detrimentally affected by VF operation. Transfer pumps are designed to deliver the minimum desired pressure and flow rate at the minimum steady state frequency. Control of the transfer system uses quantity measurement. Therefore, the only effect of VF will be on the time taken to transfer a set quantity of fuel.

Air conditioning system

High power air conditioning fans, such as recirculation and avionics cooling fans, which are designed for 400 Hz operation will not operate satisfactorily over the whole VF range. These fans are either axial or centrifugal types which exhibit torque demands that are proportional to the square of the speed. The decreasing torque of the motor, combined with the increasing torque requirement of the fan as the motor speed increases, will push the motor into its unstable operating region. A replacement for the basic asynchronous ac motor is therefore required.

A popular motor technology is the 'brushless' dc motor with power and control electronics to convert the input power into the motor drive current waveform. These power and control electronics can be designed to operate from VF, CF and dc power. The speed of the motor is not a function of the supply frequency but of the electronics controlled drive waveform. Thus the speed of this motor can be set as a function of a system parameter e.g. air temperature. The main benefits of 'brushless' dc motors with electronic speed control are:

- More efficient than asynchronous motors - power is used close to unity pf opposed to the 0.7 to 0.8 pf lagging for the asynchronous motor.
- Lighter than asynchronous motors - rotor is non-solid permanent magnet, unlike the solid, laminated soft iron core of the asynchronous motor rotor.
- Similar volume, when including speed control and Built In Test electronics.
- No in-rush current with 'soft start'.

Fan manufacturers are already using this type of motor for both variable and constant frequency systems. e.g. VF powered Technofan (France) units on the Saab 2000 and VF powered Sundstrand Aerospace recirculation fan on the Bombardier Global Express®.

Lighting

External lighting that is ac powered either transforms the 115 Vac down to 28 Vac or uses the 115 Vac directly. This lighting is almost exclusively incandescent lighting, which is not affected by this range of variation in frequency. Tests have shown that the 115 Vac to 28 Vac step-down transformer designed for 400 Hz operation behaves in the same manner over the full VF frequency range as at 400 Hz . Thus the operation of the external incandescent lights is unaffected by operation on VF power. Only the anti-collision strobe light power supply unit would need a minor component modification to operate on VF power.

Electroluminescent lamps are widely used for panel illumination on flight decks. This type of lighting operates on ac which may be at supply frequency but is normally provided by a small inverter supplied from dc. The associated inverters are not affected by VF as they are dc supplied, but they could simply be modified to operate from VF power. Therefore this type of panel lighting can easily operate from VF power.

Reading lights are incandescent and as described above are unaffected by VF power. If these lights are powered by dc power from seat mounted control units, the power supply of the control unit might need a minor component modification to operate on VF power.

The main internal lighting is fluorescent, which has historically been excited by using ballasts working at the supply frequency of 400 Hz. Lighting systems using these ballasts will not give a constant light intensity over the VF frequency range and so are not compatible with VF power. For the past 10 years, high frequency ballasts and fluorescent tubes have been available. These fluorescent lighting systems work in the kHz range, so produce the exciter frequency internally to the ballast. Thus the light intensity is independent of the input power frequency.

These high frequency systems also offer the following advantages over the old 400 Hz systems:

- Higher ballast and tube efficiencies reduce the number of tubes and ballasts required to light the cabin with the same light level.
- The reduced number of tubes and ballasts, reduces the power requirement of the high frequency lighting system compared to the 400 Hz system.
- High frequency ballasts incorporate pf correction and so cause much less power supply distortion than the basic 400 Hz ballasts.

Avionics

Avionic equipment that at present is supplied from 115 Vac power converts this input power to other voltage levels (e.g. 12 Vdc; +15 Vdc; -15 Vdc; etc.). These other voltage levels are used by internal electronic circuitry and for system communication signalling. Thus, it is only this power conversion circuitry that may require modification to operate with VF power.

Electrical power

Transformer Rectifier Units (TRU) are major users of ac power, as they convert the ac power to dc power to supply the aircraft dc buses. Unregulated types of TRUs consist of an input transformer with wye/delta secondary windings supplying two full wave rectifier bridges. The outputs of these bridges are connected via an interphase reactor to provide the required quality of 28 Vdc power. The only component in the TRU that is frequency sensitive is the transformer. A transformer designed to operate at 400 Hz will operate satisfactorily at higher frequencies, within the ranges of interest. Tests have shown that, over the VF range, a TRU designed for 400 Hz operation performs very close to it's performance at 400 Hz. Including the efficiency that was within 0.25% and the harmonic content of the generated dc power remained within MIL-STD-704E limits. The TRU cooling fan motor is usually a small asynchronous motor operating at high slip. These small asynchronous motors are much more tolerant of supply frequency variations because the speed does not increase greatly with increasing frequency. Thus no modifications are needed to these TRU cooling fans.

Galleys

Galley ovens consist of the main heating element (approximately 1 to 4 kW) and the associated electrical or electronic control circuit. Additionally, a fan assisted oven will incorporate an asynchronous motor driven fan. This fan motor rating is normally about 200 Watts. The heating element is resistive and is therefore not sensitive to frequency variations. The control electronics are not directly powered by the VF supply but by power converted from the VF supply. Thus, it is only this power conversion circuitry that may require modification to operate with VF power. This oven fan is very similar to the TRU cooling fan. They are both small asynchronous motors operating at high slip and as such are tolerant of frequency variation over the VF range. Thus no modifications are needed to the oven fans. Tests at Lucas Aerospace have confirmed the satisfactory operation of a standard fan assisted galley oven over the VF frequency range.

BOMBARDIER GLOBAL EXPRESS® - 1ST VF CIVIL TURBOJET AIRCRAFT

In 1994, Lucas Aerospace was selected by Bombardier to be the prime contractor responsible for the complete electrical power generation and distribution system for the new Global Express® aircraft.

The Global Express® is a medium to long range luxury executive jet aircraft with a maximum range in excess of 6000 nautical miles and endurance of approximately 14 flight hours. The design aims for this aircraft included the requirements for very high despatch reliability. The aircraft primary electrical power was specified to be Variable Frequency from program launch. This was the first application of variable frequency power generation on a civil turbofan aircraft.

Primary electrical power is 115/200 Vac, 3 phase, variable frequency between 324 Hz and 596 Hz (engine speed range 1.84/1). There is no CF power on the aircraft. 28 Vdc power is converted by TRUs directly from the primary VF ac power. All electrical loads have been successfully integrated with the electrical power generation and distribution system.

THE FUTURE - LARGE GENERATORS AND AIRCRAFT

Lucas Aerospace are developing the technology for large VF generators of greater than 100 kVA, which would be aimed at future high capacity aircraft. These large VF generators would exhibit the same power quality characteristics (ISO 1540) and high reliability of the present smaller generators.

As the lengths of the electrical power feeder cables and their associated impedance increases, their detrimental effect on power quality increases. Lucas Aerospace believes that the lengths envisioned on future large aircraft will require action to reduce feeder impedance. Investigation is presently being carried out on methods to reduce the impedance effects on power quality.

As a consequence of the increases in the number of electrical loads and the relative number of power electronics loads (e.g. In-flight entertainment), the harmonic distortion levels on the electrical power supplies have been increasing. This increasing trend can only continue if nothing is done to limit the harmonic distortion reflected back onto the power supplies from the utilisation equipment. Therefore, future utilisation equipment design should include functions to reduce these harmonic distortion effects on the power supplies.

Utilisation equipment manufacturers are designing more VF compatible equipment, so the choice of utilisation equipment is increasing.

SUMMARY

The benefits of VF power are becoming more widely recognised and the utilisation equipment affected by VF power have been identified and solutions developed. More VF compatible utilisation equipment is becoming available hence increasing the choice available.

The first application of VF power on a turbofan aircraft, the Bombardier Global Express® executive jet, is now flying

For the future large aircraft, high power VF generators are being developed and solutions are also being developed for the installation challenges posed by the size and power requirements of these future aircraft.

CNS/ATM

C545/050/98

A partnership for prosperity

T LAVEN
IATA Asia Pacific, Singapore

INTRODUCTION

The air transport industry plays a major role in world economic activity. In every region of the world, countries large and small depend on the aviation industry to fuel their economic growth and their financial strength. Air transport drives economic progress, and in turn benefits from it. It acts as an economic catalyst, promoting business and leisure activities, contributing to growth and increasing efficiency. In an increasingly global society, the contribution of the airline industry to the world's economies is growing more important every day. International Civil Aviation will have a major role to play in ensuring the economic recovery of Asia. International air routes must become the highways of the sky and support the development of commerce and tourism, in the same way that the inter-state highways did in the United States decades ago.

Unprecedented and sustained growth in air traffic is straining the Asia/Pacific Region's ATC infrastructure. In the period between 1991–1995 traffic at the five leading International Airports in the region (Seoul/Kimpo, Hong Kong/Kai Tak; Tokyo/Narita, Singapore/Changi and Bangkok/Don Muang) grew by some 39 million passengers. This growth at these five airports generated nearly 250,000 additional flights. In the next period 1995-1999 similar growth is expected. In the last eight years air traffic has doubled in the Asia/Pacific Region and it is poised to do so again within the next six years. Throughout the region there is a growing divergence between the demand for passenger and freight services and the capacity of airports and airspace.

- *Region-wide traffic grew at an average of 10.3% per annum between 1985 and 1993;*
- *It is forecast to grow by an average 8.6% per annum between 1993 and the year 2000;*
- *Beyond the year 2000 it is expected to grow at some 7% per annum until the year 2010.*

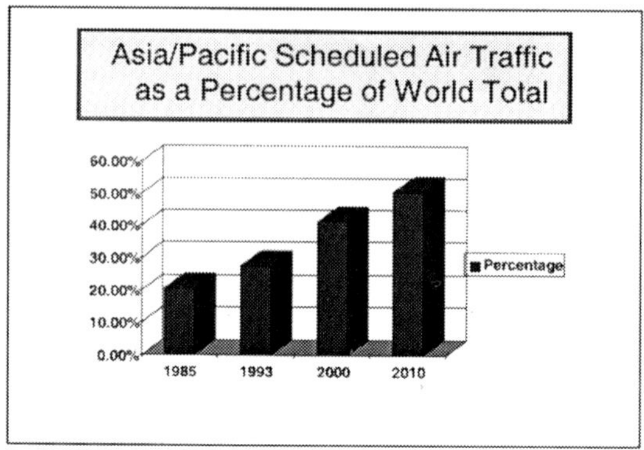

The Region's share of world-wide international scheduled passenger traffic has grown substantially;

- *It was only 26.4% in 1985;*
- *It increased to 35.3% in 1993;*
- *It is forecast to be 41.2% by the year 2000, and*
- *It is expected to reach 50.5% by 2010.*

The Region's share of domestic and international scheduled passenger traffic is considered it is found that;

- *It was only 16.5% in 1985;*
- *It increased to 29.9% in 1995;*
- *It is forecast to be 34.0% by the year 2000, and*
- *It is expected to reach 42.9% by 2010.*

As the consequence of this growth and the forecasts, the Asia-Pacific region has put far more investment into airports in the past ten years, than any other region of the world. A substantial number of major new airports have been constructed, or are being built. Many of these could be termed 'mega-airports', with design capacities in excess of 45 million passengers per year. In Europe, one the world's most congested aviation areas, the only major new airport to open in recent years has been Munich 2, and that took over 30 years from inception to completion. Only four new airports, in Athens, Berlin, Milan and Oslo, are currently being planned or built. Similarly, the recently inaugurated Denver International Airport is the only major 'green-field' international airport to open in the United States over the past twenty-five years and no new airports are currently under construction.

However, unlike Europe, airspace capacity improvement in Asia is falling behind airport capacity development. Unlike the aggressive airport development, development of ATC facilities and the whole air navigation system has not kept pace with the growing demand. It must be recognised that much of the new demand is concentrated in peak periods - an average growth rate of 8% may be in fact a 12% growth during the peak hours. Serious capacity problems are being experienced by long-haul operators on the trunk routes between North America and the Asia and between Asia and Europe.

The North Pacific (NOPAC) route network linking North America and Asia is the most densely travelled airspace in the Pacific. The passenger traffic between North America and North East Asia totalled 11.9 million passengers in 1995. This high volume of passengers is reflected in the number of aircraft movements. In 1992 the Japan Civil Aviation Bureau (JCAB) estimated that the NOPAC by 2000 to reach 66,000 movements a year and would be tripled by 2010. Yet by 1997 the NOPAC actual traffic count was 60,417 movements (16.45% over 1996 and some 1,230 flights/week) with an additional 4,074 movements (nearly 80 movements/week) siphoned off the NOPAC and placed into the Russian Far East (RFE).

The current volume of traffic in the NOPAC—as many as 40 aircraft in a single sector at any one time, already. places high workloads on air traffic controllers. The lack of capacity adversely affects the airline operators, departure delays are extensive, many flights operate at low and uneconomic cruising levels and often on routes which are not the optimum for the prevailing winds. The FAA is very aware realises that current procedures and current automation cannot efficiently handle any further increase in traffic. The application of the Future Air Navigation System (FANS) and its CNS/ATM technology could resolve the issues of safety, capacity and efficiency.

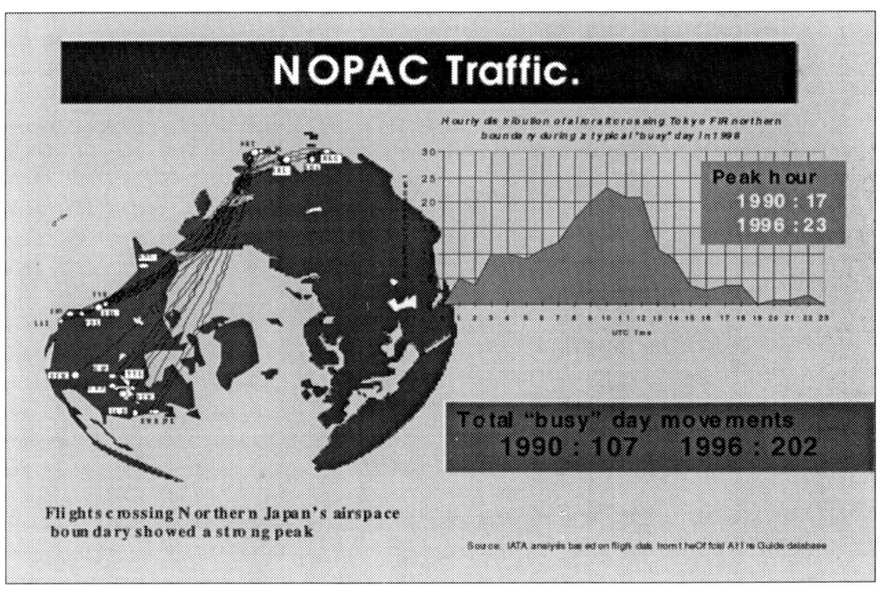

As far back as 1989, IATA's Asia/Pacific Regional Technical Office developed substantial documented and fully justified proposals for the establishment of a route network linking North America and the Orient over the Russian Far East (RFE) to support the future growth of traffic between Asia and North America. This network was seen to be the key to providing alternative more direct, fuel-efficient routes than those available in the North Pacific (NOPAC) area. These routes, termed Siberia 1, 2 & 3 and Kamchatka 1, 2 & 3 were initially proposed to the Russian Government on behalf of IATA, by the US State Department.

However, the entire route structure as envisaged by IATA has never been implemented and therefore the immense operational benefits that could have been derived from having the full number of routes available have never been achieved. Only the two southern most routes have been opened to international air traffic, albeit with significant limitations on their time of availability, their air traffic capacity and thus limiting their economic viability. Some of these routes have now been re-opened in October 1997, and IATA continues to press for more of them to be implemented. A new route A218 was promulgated in the AIP-Russia in June 1997 and became available for international operations in early 1998. This route is in effect the Siberia 1 route proposed by IATA in 1989. It has the potential to provide a near optimum route for operations between the East Coast of the United States and Japan.

The reasons for all of these limitations are based upon political, economic and technical considerations. The capacity constraints and high en-route charges almost make the routes uneconomical for daily use by the airlines. This in turn deprives the air traffic service providers in the RFE the financial opportunity to develop their air traffic management systems including increased personnel benefits to properly accommodate the rapidly growing demand for their services. Airlines have often stated that less than one fifth of their current flights that could have benefited from the use of the RFE routes actually fly the routes.

There is a substantial volume of traffic on the 'Trans-Asia' routes; used by non-stop traffic between Europe and airports in Thailand, India, Indonesia, the Philippines, Pakistan, Malaysia, Singapore and China including Hong Kong. The following chart, (Asia-Europe & Asia-Mid East) shows substantial peaks in the traffic pattern. These have grown dramatically. In 1990 the peak hour had 23 aircraft, by 1996 it had increased to 38. These traffic peaks arise from 'scheduling windows" caused by restrictions on departure and arrival times as the result of the night-time closures of airports in Europe. This peak traffic creates major problems for traffic departing from Asia to Europe.

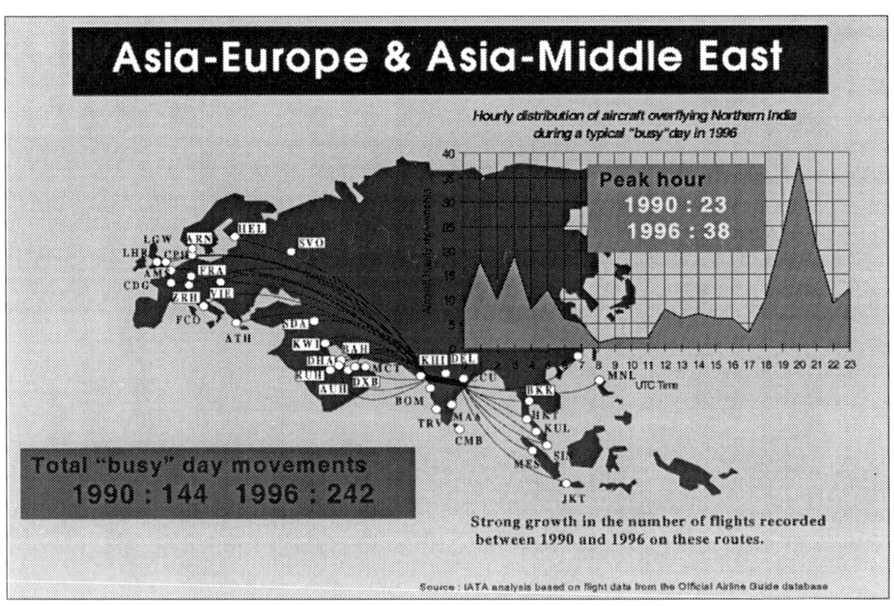

The total number of flights in the busy day also showed a substantial increase from 1990, where there were 144 flights to 1996 where there were 242 flights. Examination of airline schedules for the winter season 1997/1998 reveals that during night time peak hours 366 weekly flights operate over the Indian Sub-continent from Bangkok (160), Singapore (130), Kuala Lumpur (54), Hong Kong (15) and Manila (7) to the Middle East and Europe.

Compared with the Summer 1997 schedule it reveals that the departures per week during the peak hours from Bangkok and Singapore has increased considerably with 14 % and 28 % respectively. The growth from Kuala Lumpur appears to be 6%.

However, as the result of increasing use of ATS route B330 over China, traffic from Hong Kong routing over the Indian subcontinent has decreased with approximately 50 %. With the opening of the new airport at Chek Lap Kok and the availability of shorter more direct routes from Hong Kong to Europe over China and Kazakhstan it is expected that most, if not all operations between Hong Kong and Europe will avoid the routes over the Indian Sub-continent.

The opening of new routes to support CNS/ATM (FANS) operations between Asia and Europe will reduce the volume of traffic operating over the Russian Federation. Planned FANS routes include; Singapore/Kuala Lumpur - Delhi, over the Bay of Bengal, and also Delhi – Trabazon (in Turkey) over Pakistan, Afghanistan and Iran, as well as new conventional routes over Pakistan, Afghanistan and Turkmenistan.

IATA's view is that unlike the other routes, between Japan and Europe, China and Europe, and South Korea and Europe, traffic between S. E. Asia and Europe has practical alternatives to routes over the Russian Federation. At least half of the some 750 flights a week do not enter the airspace of the Russian Federation. Instead many flights route over Iran and Turkey. This is determined by major factors in the flight planning process

- *The possibly more favourable winds over Iran and Turkey—especially for eastbound traffic to S. E. Asia to Europe, and the development of shorter cost-effective CNS/ATM routes over Pakistan, Afghanistan and Iran;*
- *The opening of new routes over Afghanistan and Turkmenistan will also draw traffic away from the Russian Federation, especially that destined for Southern and Central Europe.*
- *Operational difficulties operating North of the Himalayas over the high Tibetan plateau which will restrict the route to those aircraft with FANS-1/A equipment and the need to carry additional passenger oxygen.*

Operational inefficiencies such as departure delays, operations at low and uneconomic cruising levels, en-route holding, re-routing and en-route diversions are imposing a major economic penalty on the airline operators. Estimates of the costs of congestion measured by increased operating hours, additional fuel and reduced operating revenue, exceed US $250 million annually.

Despite substantial and effective action by ATS Providers to increase airspace and air route capacity by conventional means, demand is once again outstripping capacity. Anticipated traffic growth cannot be supported by the current procedural air traffic systems in both oceanic and low-density continental airspace. The possibility of enhancing capacity by the provision of radar services is in many areas impossible and in others either difficult or financially impractical.

Within the region airspace capacity problems are well recognised. The Malaysian Minister of Transport, in his speech, opening the 30th Annual Conference of the Directors General of Civil Aviation - Asia/Pacific in September 1994, referred to the growing problem of airspace congestion with these words;

"Another serious challenge to the Asia/Pacific Region is the problem of airspace congestion. Unprecedented growth in air traffic is straining the region's ATC infrastructure. Unlike Europe, airspace capacity improvement in Asia is falling behind airport capacity development. Asia could find its new airports, which can handle millions of passengers, congestion bound because of ATC infrastructure constraints. Some have described this as the "Looming Crisis" in the Asia/Pacific Region. If immediate and long-term solutions are not found this problem of capacity imbalance is expected to worsen and consequently stifle the growth in aviation."

"Due to the fact that the region is faced with a highly fragmented airspace network and a huge variance in the sophistication of ATC systems, there must be a regional approach to the problem. The future lies in the creation of an integrated air traffic infrastructure. It is in this context that the implementation of satellite based CNS/ATM (Communications, Navigation and Surveillance/Air Traffic Management) systems must gain greater momentum."

CURRENT ATS SITUATION

The major challenge to addressing the increasing number of flights is the provision of adequate aeronautical communications. Procedures to be applied, by both controllers and pilots, must compensate for shortcomings in point-to-point communications systems. Some of these shortcomings encountered in the Asia/Pacific Region and other regions include the following:

Inadequate and partial implementation of Aeronautical Fixed Communications Services; which are not able to meet the operational requirements, especially for;

- *ATS voice communication (ATS direct speech circuits for inter-centre communications), and*
- *Aeronautical Fixed Telecommunications Network (AFTN) for data communications.*

Inadequate and partial implementation of Aeronautical Mobile Communications Services and operational problems;

- *Large parts of the area are outside of the range of coastal VHF aeronautical mobile stations leading to extensive reliance on HF for air/ground. communications through a third party (radio operator).*
- *Highly congested HF frequencies. HF Congestion, in some areas is aggravated by the misuse of HF aeronautical mobile facilities to supplement the un-implemented or unreliable point-to-point fixed service circuits established for ATS co-ordination, further adding to the congestion.*

The poor state of aeronautical communications places a high and unsatisfactory work load on air traffic controllers and aircrew.

- *Pilots are frequently required to establish contact with the ATS units 15 to 20 minutes in advance of entering its area of jurisdiction to deliver flight plan data or entry point notification.*
- *Lack of area control service in some FIRs and an inability to accept reduced separation minima, which would provide additional capacity.*
- *An Air Traffic Advisory Service is provided by some ATS Units on a number of trunk routes because of poor aeronautical communications, even though the provision of an Air Traffic Advisory Services was only intended to be considered as a temporary form of ATS.*

During the peak periods, the demand for operations at economic cruising levels on the trunk routes is greater than the capacity of the routes. This creates significant high workloads on the controllers (who have to resolve the situation) and problems for the pilots, as the flights frequently have to operate at low and uneconomic cruising levels, accept en-route holding, or in some cases re-routing on to longer and more circuitous routes. Instances have occurred where the combined effects of these constraints have led to unscheduled and expensive en-route technical landings at intermediate aerodromes.

While efforts are being made to improve the reliability and the range of Remote Control Air Ground (RCAG) VHF sites, the integrity, reliability and performance of the HF network still remains the main obstacle to providing full air traffic control service. HF air/ground communications could be improved, however, this would take considerable time and expense and would still not meet the operational requirement for the provision of area control service for which Direct Controller Pilot Communications (DCPC) is a key feature. Also, improvement in HF communications would do little to reduce the workload of the Air Traffic Controller as communications through an intermediary would still be used.

Urgent steps are need to increase efficiency, reduce airspace complexity, reduce ATC provider workload and provide additional benefits to users. The possibility of an early transition to the new Communications, Navigation and Surveillance/Air Traffic Management (CNS/ATM) system—the technologies of which will provide the controller with a visual display of aircraft positions using Automatic Dependent Surveillance ADS, where airborne navigational data is made available to ATC by data link techniques and which Direct Controller Pilot Communications (DCPC). Through fast reliable Two-Way Data Link (TWDL), will provide substantial enhancements to the existing procedural control procedures.

In the last few years there has been a "Sea-Change" in aviation. Many airlines are no longer "State-owned" and have to ensure their profitability to survive. Many ATS providers have also been privatised and together they are welding a partnership in which both stakeholders—user and providers are striving to work together to provide optimum levels of service in the most efficient and economical manner.

The implementation of CNS/ATM with its enhanced communications and surveillance capability, can not only be used to enhance capacity but also to permit the use of airspace which has not previously been available for civil air operations because of the lack of an adequate communications, navigation and surveillance infrastructure. CNS/ATM enables new Required Navigation Performance (RNP) routes to be developed, which have greater capacity than conventional routes, are more direct and operationally beneficial. At the same time non-equipped aircraft will also benefit as traffic demand on existing and congested trunk routes will diminish.

The ICAO Asia/Pacific Air Navigation Planning and Implementation Regional Group (APANPIRG) has recognised that the growth of international civil; aviation would require additional airspace capacity and that the capability of RNAV equipped aircraft should be exploited to provide both increased capacity and enhanced operational efficiency. It was noted that there were significant advantages in adopting the concept of Required Navigation Performance (RNP) in the Asia/Pacific Region as a means of achieving the necessary reduced separation minima.

The support of APANPIRG for the initial use of RNP type 10 in the Asia/Pacific Region provides a bench mark for new horizontal air traffic separation minima now being introduced by ICAO Annex 11. With the approval of RNP for aircraft navigation systems, airspace planners are being freed from the constraints of 100 NM lateral and 15 minutes longitudinal time separation in procedural airspace.

The initial application of RNP 10 took place in a number of Pacific Flight Information Regions (FIRs) on the 23rd of April 1998. This enabled the introduction of 50 NM lateral separation and will enable a change from 10 minutes longitudinal time separation (approximately 80 NM) to 50 NM distance separation. The initial use of RNP is the first step towards further reductions in procedural separation. The use of 30 NM Lateral and 30 NM Longitudinal separation associated with RNP 4 *(The RNP value for which the Boeing FANS-1 Package is certified for en-route operations)* can provide a dramatic improvement in capacity over currently applied standards.

CNS/ATM CONCEPTUAL DEVELOPMENT

It must be emphasised that the initial implementation of CNS/ATM in the Asia/Pacific Region is directed at providing an ***'enhanced'*** form of procedural air traffic control that will provide significant improvements in both capacity and operational efficiency. Substantial reduction in the workload of air traffic controllers is also expected as automation assistance is developed

Communications: Improved and reliable communications in the form of two-way data link will provide the basis for an improved Air Traffic Management (ATM) system.

- *Controller Pilot Data Link Communications (CPDLC) enabling the air traffic controller to have access to the navigation data in the aircraft's Flight Management System (FMS) and provide error free automatic position reports (initially these will be primarily "event-driven" but "periodic" and "on-demand" reports can also be used). the use of specific message sets will also reduce the workload on the controller and pilot;*
- *The introduction of ATC Interfacility Data Communications (AIDC) will permit the automation of the many co-ordination tasks that create a high workload on the controller*
- *Enhanced communications will, at last, enable the full exploitation of on-board navigation and flight management systems and permit significant reductions in the separation minima applied between aircraft. This will give significant improvements in air route capacity and operational efficiency while maintaining and also enhancing current levels of safety.*

Navigation: Improved navigation by the use of Global Navigation Satellite Systems (GNSS) provides greater accuracy and allow the certification of aircraft equipped with FANS-1/A to be certified as meeting RNP Type 4, thus permitting substantial reductions in separation minima in current procedural airspace, while maintaining or possibly improving safety.

Surveillance: The addition of Automatic Dependent Surveillance (ADS) will provide not only position information but also information on intended flightpath needed for automated air traffic management. This will enable the controller to determine flight plan conformance and provide the basis for automatic conflict resolution with other traffic. Display systems will also improve the situational awareness of the controllers.

Air Traffic Management: The new CNS facilities and the new ATM workstations will give the controller the tools to provide a safer more efficient and Enhanced Procedural Air Traffic Management Environment.

THE FANS 1 PACKAGE

The Boeing Commercial Airplane Group had a commitment to provide avionics functionality on the Boeing 747-400 aircraft for it to operate in a full CNS/ATM environment. This functionality was planned to be achieved through their offering of a modification to the aircraft Flight Management Computer System (FMCS) known as "Package B." This offering provided for more than 50 functional upgrades including;

- *Communications: the ability to use Controller-Pilot Data Link Communications (CPDLC) over the Aeronautical Mobile Satellite Service (AMSS) and over VHF data link services and for them to be relayed to ATS units over the planned bit-oriented Aeronautical Telecommunications Network (ATN).*
- *Navigation: the integration of the satellite based Global Positioning System (GPS) into the aircraft's FMCS for primary navigation.*
- *Surveillance: the provision of data for Automatic Dependent Surveillance (ADS).*

It was envisaged that this functionality would enable the Asia/Pacific Region to rapidly move ahead with CNS/ATM implementation. However, Boeing's offer was withdrawn in January 1992, because:

- *there was little prospect of the ATN being developed and becoming operational in the near-term;*
- *the specifications for CPDLC were as then undefined;*
- *GPS certification requirements were not known;*
- *in the absence of compatible ground system infrastructure there could be no operational benefits to provide the necessary incentives for the airlines to purchase the proposed package; and*
- *most importantly, nor could the aircraft systems be certified without a corresponding and operational ground system.*

Despite this setback, a core group of airlines (CX, NZ, QF & UA) who had participated in the Pacific Engineering Trials (PET) determined that significant reductions in their operating costs were achievable by flying optimised flexible CNS/ATM tracks on their Trans-Pacific routes. This group of airlines met with the Boeing Airplane Company and its avionics suppliers in June 1993 to seek alternative ways of providing an interim CNS/ATM capability in the near term. They reduced the number of required upgrade functions from over 50 down to five. It was recognised, at an early stage, that a single step replacement of existing systems with fully ATN compliant systems was not achievable.

- *Limited memory capacity and processor capability of existing avionics equipment was not capable of incorporating the complex protocols associated with the envisaged ATN.*
- *Substantial investments in existing systems (both in the air and on the ground) and limited resources drove the need to use existing avionics and air/ground communications systems with minimal modification.*

With little prospect of ATN capability being realised for operational use, certainly until much later in the decade; a strategy was developed which permitted the use of existing communications infrastructure and enabled the support of bit-oriented applications (new CNS/ATM functionality) over existing character-oriented networks while ensuring that current and future communications networks could coexist during the transition to ATN;

- *The existing aeronautical data (character-oriented) network—the Aircraft Communications and Reporting System (ACARS) could therefore be used without waiting for region-wide implementation of the proposed ATN. ACARS was*
- *Capital investment in ground communications networks could be deferred until ATN came of age.*

The group decided to implement an interim FANS capability which would make maximum use of existing aircraft avionics capability (which developed into Boeing's FANS-1 package offer) and develop ground ATC systems with matching capability. Although the AFTN is a character oriented communications network it has neither the speed nor reliability to be used for CNS/ATM.

The specific airborne avionics elements of the interim FANS-1 system are:

- *Aircraft Control and Reporting System (ACARS) data link over VHF and SATCOM using Data Level 1 and 2;*
- *ARINC 622 ACARS protocols which enables FANS compatible bit-oriented functions (applications) to be transmitted over the ACARS character oriented data link, and the ATC Facilities Notification (AFN) function to enable aircraft to establish Two-Way Data Link (TWDL) communications with Air Traffic Control Centres;*
- *Integration of the Global Positioning System (GPS) into the Area Navigation (RNAV) function of the Flight Management System (FMS)* GPS is used in the RNAV solution as an input to the RNAV solution for oceanic, remote, en-route, terminal and instrument approach procedures;*
- *Controller Pilot Data Link Communications (CPDLC) function implementing the message sets included in RTCA DO219.*
- *Automatic Dependent Surveillance (ADS) in accordance with RTCA DO 212 and ARINC 745;*

- *Note* FANS-1 uses Aircraft Autonomous Integrity Monitoring (AAIM). An Inertial Reference System is a very powerful means of augmentation for GPS. It can be used during short periods when the satellite* *navigation antennas are shadowed by aircraft manoeuvres, or during periods when insufficient satellites are in view. FANS-1 is certified to RNP-4, an accuracy of 4.0 NM (95% probability). This will be achieved 99.999% of the time with 23 healthy GPS satellites and provide a probability level of 1 x 10 $^{-7}$ that the actual position is outside the containment radius of 8 NM*

This interim FANS-1 implementation, is designed to evolve gracefully into the final FANS CNS/ATM implementation, after that has been designed and requirements have been validated. The objective was to have a minimum of throw-away design. Unlike Package "B" a corresponding and operational ground system existed to enable the aircraft systems be certified. Figure 1 (below) illustrates some of the components in the **end to end** data link system.

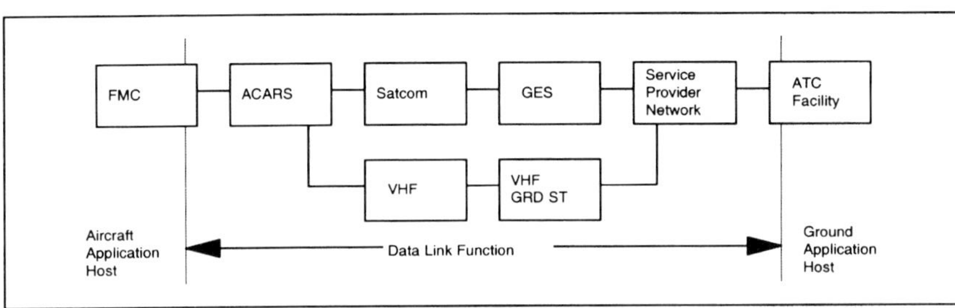

Figure 1

An agreement, in principle, to develop FANS-1 was of no practical value unless ATS providers were ready to support CNS/ATM operations by installing compatible ground system elements which matched peer functions with the aircraft avionics and provided some additional unique ground functions to support:

- *ARINC 622 ACARS protocols and ATS Facility Notification (AFN) function;*
- *RTCA DO219 CPDLC & DO 212 ADS*
- *Flight Plan generation and display;*
- *Traffic Situation Display;*
- *Controller CPDLC workstation;*
- *Ground-Ground ATC communication: and*
- *ATS unit/ACARS Service Provider interface.*

The Informal South Pacific ATS Co-ordinating Group (ISPACG) is an action-oriented group, initially established by Australia, New Zealand and the United States and later joined by Fiji, Papua New Guinea and French Polynesia. The initial purpose of this group was to explore solutions to the near-term problems that limit the capacity and efficiency of the air traffic services in the South Pacific. However, it took over the responsibility for the development of operational CNS/ATM system implementation plans in the South Pacific from the Tri-lateral (Australia, Japan and United States) Pacific Engineering Trials (PET) group which commenced in 1990 and itself was expanded to include a substantial number of States from the Asia/Pacific Region.

ISPACG/5 in October 1993 adopted an IATA proposal for an aggressive and expedited implementation of CNS/ATM, using the Boeing FANS-1 package in the South Pacific. Four States in the South Pacific (United States, Australia, New Zealand and Fiji) agreed in principle to implement the ground facilities, compatible with the FANS I package for the Boeing 747-400 aircraft, for operational use in April 1995.

While this was highly satisfactory, the airlines were unwilling to place any firm orders for the equipment and Boeing themselves unwilling to fund any further development unless there were formal agreements in place to confirm the decisions of ISPACG. The Annual meeting of the Asia/Pacific Directors General of Civil Aviation in November 1993 being held in Los Angeles provided an opportunity for IATA to organise a meeting with the Administrator of the United States Federal Aviation Administration, the Chief Executives of the Civil Aviation Authorities of Australia, Fiji and the Airways Corporation of New Zealand.

This meeting provided the necessary commitments to procure and put into operational use, in the time frame envisaged by ISPACG, the air traffic end-systems compatible with the FANS 1 system. With this support Boeing was able to put FANS-1 into production and the airlines were able to place orders for the equipment with confidence. The primary certification of the FANS-1 package, was jointly done by Boeing, QANTAS and the CAA Australia on 17 June 1995 and the first truly operational flight across the Pacific on a scheduled passenger flight by QANTAS took place on 22 June 1995, *the significance of this date* ***'6/22'*** *should not be overlooked.*

Working groups of ISPACG focused on the overall aspects of CNS/ATM implementation and specific areas such as data link communications. The Pacific was an ideal testing area for CNS/ATM, because;

- *The population of aircraft equipped with Satellite Communications for Controller Pilot Data Link Communications (CPDLC) was significant.*
- *The enthusiasm of the airlines for CNS/ATM was matched by the States, particularly in the South Pacific, where Fiji, a small island State, has been at the forefront of the implementation process.*
- *ISPACG/4, in May 1993 developed a phased timetable for full inter-networking by July 1994, to enable CPDLC to be used in lieu of HF in the Auckland, Nadi, Oakland, Sydney and Tahiti FIRs. The first step was for waypoint progress reports and acknowledgements, to be initially made by CPDLC in parallel with the normal HF reports and then to dispense with HF reports entirely.*

Over the past four years, extensive trials and demonstrations of ATC datalink by Australia, New Zealand and the Pacific States based on the FANS concept have led to the development and certification of avionics providing air traffic control communications capability over today's aeronautical communications networks. This capability has been matched by the development of Air Traffic Management end systems and procedures that now support an operational service.

The introduction of CNS/ATM compatible avionics, ATM workstations and the upgrading of aeronautical communications networks ensure the highest possible integrity can be met for ATC datalink services. Several other States have now implemented, or are in the process of implementing operational services using both ATC datalink and ADS..

The following Table shows the Status of implementation of CNS/ATM workstations in the South Pacific (ISPACG States) to support CPDLC, and in some cases ADS operations by FANS-1/A aircraft.

State	FIR	Availability	Capability
New Zealand	Auckland	Sept. 95	CPDLC + ADS
United States	Oakland	Oct. 95	CPDLC only
"	Anchorage	April 96	CPDLC only
French Polynesia	Tahiti	Oct. 95	CPDLC + ADS
Australia	Brisbane	July 98	CPDLC + ADS
Fiji	Nadi	2 Q 99	CPDLC + ADS

Most of these systems have the capability of Controller-Pilot Data Link Communications (CPDLC) system connection and maintenance, the establishment of ADS contracts and the processing and display of ADS position reports. Typical equipment hardware includes;

- *an IBM RS/6000 computer with 64 MB of RAM, 1 MB of Cache Memory and 2 GB Hard Disc Drive.*
- *Single or Dual 20 inch 1280 x1024 resolution colour monitors with a keyboard and mouse.*
- *an X.25 interface for connection to the aeronautical services supplier*
- *a printer to produce a log of incoming and outgoing messages and a hardcopy of either of the screen displays and a 4mm magnetic tape for message archiving.*

User interfaces are based upon windows technology and powerful workstations. Typical windows available to the controller are listed below. Most of them are available on request and can be positioned on the screen and even overlaid on other displays.

The following table provides an overview of current planning in the Asia & Middle East Regions for CNS/ATM transition to satellite based controller pilot datalink communications (CPDLC) and automatic dependent surveillance (ADS), in addition to the FIRs in the Table above;

Implementation in 14 Asian & Middle Eastern States

State	FIR	Availability	Capability
Singapore	Singapore	July 96	CPDLC + ADS
Hong Kong	Hong Kong	Aug. 96	CPDLC + ADS
Thailand	Bangkok	June 97	CPDLC + ADS
Malaysia	Kuala Lumpur	Jun 97	CPDLC + ADS
India	Calcutta	Mar 97	CPDLC + ADS
India	Delhi	Jul 98	CPDLC + ADS
India	Madras	Sep 96	CPDLC + ADS
India	Bombay	Oct 98	CPDLC + ADS
Japan	Tokyo	Sep 97	CPDLC + ADS
China	Urumqi,	1 Q 99	CPDLC + ADS
China	Chengdu	1 Q 99	CPDLC + ADS
China	Kunming	1 Q 99	CPDLC + ADS
China	Lanzhou	1 Q 99	CPDLC + ADS
Mongolia	Ulan Bator	4 Q 98	CPDLC + ADS
D. P. R Korea	Pyongyang	4 Q 99	CPDLC + ADS
Sri Lanka	Colombo	3 Q 99	CPDLC + ADS

Afghanistan	**Kabul**	**4 Q 99**	**CPDLC + ADS**
Myanmar	**Yangon**	**Sep 98**	**CPDLC + ADS**
Russian Federation	**Magadan**	**3 Q 98**	**CPDLC + ADS**
Iran	**Tehran**	**4 Q 98**	**CPDLC + ADS**
Pakistan	**Karachi/Lahore**	**4 Q 98**	**CPDLC + ADS**
Turkmenistan	**Ashkhabad**	**4 Q 99**	**CPDLC + ADS**
Uzbekistan	**Tashkent**	**4 Q 99**	**CPDLC + ADS**
Taiwan	**Taipei**	**4 Q 99**	**CPDLC + ADS**

The following table provides an overview of other States in the Asia & Middle East Regions for CNS/ATM, where States are evaluating potential implementation of CNS/ATM and the purchase of workstations.

Possible Implementation in Asia/Pacific, Middle East and African States

State	FIR	Availability
Republic of Korea	**Taegu**	**Under evaluation**
Indonesia	**Jakarta**	**Under evaluation**
Nepal	**Kathmandu**	**Under evaluation**
Bahrain	**Bahrain**	**Under evaluation**
South Africa	**Johannesburg**	**Tender from Thomson Accepted**
Mauritius	**Port Pleasance**	**Under evaluation**
Male	**Male**	**Under evaluation**
Bahrain	**Bahrain**	**Under evaluation**

In addition, China is in the process of implementing a VHF air/ground data link network. Some 22 Remote Ground Stations (RGSs)have already been installed and cover the densely used air route in eastern China with stations from Haikou in the South all the way to Harbin in the north. The next series of RGSs will cover the Asia-Europe air routes over Urumqi and then those over Kunming

Airlines will only invest in equipping aircraft for datalink services in circumstances where they are likely to realise financial or other benefits. Airlines will also wish to maximise the utilisation of existing investments. Many airlines, particularly those operating in the Pacific, perceive that there is an urgent requirement for the provision of ADS and CPDLC to get en-route clearances, and that ATN-based services cannot be developed within short enough timescales to meet that present demand. These airlines are interested in implementing FANS-1 based ADS and CPDLC because many of them already have Satellite Communication (SATCOM) equipment and carry the ACARS system which, with relatively low cost modifications, can provide the services.

Currently, Boeing has received, from 25 customers, some 500 orders for the FANS package for the B747-400. This figure is expected to grow dramatically within the next 18 months. With the development of FANS routes in Asia, providing alternative, shorter, less congested routes, a number of other airlines, including some major European airlines, are actively looking at installing FANS-1 on their B747-400 aircraft.

Three hundred and fifty B747-400 and B777 aircraft are now operational with FANS-1. The airlines concerned have entered into the necessary commercial agreements with ARINC and SITA for access to ATS end users for CPDLC and ADS.

Boeing have indicated that some 40% of all Boeing 747-400 aircraft will be equipped with the FANS-1 Package by the end of 1997. They confidently believe that 'market penetration' of the FANS-1 Package will reach 80%. Boeing are also providing the FANS-1 Package on their new B777 (Market B) aircraft All B777 (Market B) aircraft delivered during 1997, will be equipped with FANS-1 capability as standard—it is a software only upgrade for the Market A aircraft. The FANS-1 Package certification for the B777 will take place in the first quarter of 1997 and delivery will be a month behind the certification of each engine variant. Fifty-three B777 aircraft will be delivered by end 1996; 195 by end 1998.

Implementation of FANS-1 on the Boeing 737, Boeing 757 and the Boeing 767 is under study. In regard to the B757 & B767, Boeing has developed a preliminary design that provided for a complete upgrade of the FMC/EICAS to support FMC hosted FANS applications. It will also support growth to the new CMU interface for CMU hosted FANS applications (when that specification is completed) It also has the capacity for growth to ATN (for both solutions. Certification of the FANS-1 Package for their B757/767 "Pegasus" is targeted for mid December 1997. Boeing has received order for 50 packages for B757 aircraft and at least 2 airlines are near to making decisions on the package for the B767.

Other manufacturers are intending to provide systems with similar functionality to the Boeing FANS 1 package; Airbus is developing the FANS-A/AIMS for CNS/ATM for the Airbus family of aircraft. This has been 're-baselined' to use the new Flight Management System (FMS) as a platform. The FANS-A Package is expected to be certified and delivered in 1998 for the Airbus A330 & A340 aircraft. Currently Airbus has some 40 orders for the FANS-A package.. Airbus has designed the package so that it will support the implementation of the CNS/ATM-1 package during 1999, without further hardware changes. The same hardware platforms will go into the A319, 320 & 321 in September 1998. For these aircraft, the ACARS functionality will be housed in an ATSU rather than a separate MU enabling ARINC 622, or ATN functionality, or both. Airbus is also looking at the most cost-effective means for providing CNS/ATM capability on the A310 and A300-600 aircraft. McDonnell Douglas is also developing FANS functionality for its MD-11. The GPS package was due to be available late 1996 with a FANS-1 package ready by late 1997 or early 1998.

IATA has been carrying out a survey on airlines plans for CNS/ATM equipment installation—this supports the contention that by the end of 1998 there will be some 400 aircraft equipped with FANS-1.

Further equipage will be dependent on obtainable benefits from city pair air routes rather than on an FIR by FIR basis, thus the concerted action by groups of States in the region will play a significant role in the encouragement of FANS technology and provide operational benefits both to the users and providers. The ICAO APANPIRG CNS/ATM Transition and Implementation Plan endorses this approach.

There are a number of areas in the Asia/Pacific Region where the early implementation of CNS/ATM technology could significantly increase airspace capacity, reduce operational penalties being imposed by operations at low and uneconomic cruising levels and en-route holding, and departure delays. They are:

- ***the North & Central North Pacific (including the Russian Far East);*** *used by traffic between North America and the Orient;*
- ***the 'Europe-Asia' routes;*** *used by non-stop traffic between Europe and airports in China & Hong Kong, and possibly also to and from airports in Thailand, Malaysia and Singapore;*
- ***the 'Trans-Asia' routes;*** *used by non-stop traffic between Europe and airports in Pakistan, India, Hong Kong, Thailand, Malaysia and Singapore; and*
- ***the Trans-Siberian routes****; used by non-stop traffic between Europe and airports in China, Mongolia, Japan and Korea.*
- ***the Cross-Polar routes****; used by non-stop traffic between North America and airports in Asia.*
- ***the South China Sea;*** *used by traffic between South East Asia and North Asia/.*
- ***South East Asia to Australasia****; used by traffic between South East Asia and Australia and New Zealand.*

CONCLUSION

Substantial progress in CNS/ATM implementation is taking place in the Asia/Pacific Region. It is 'spilling-over' into the Middle East and Africa. IATA, its member airlines and the ATS Providers are forging a Partnership for Prosperity. The Asia/Pacific Regional Office - Infrastructure, continues to work closely with the airlines, States, ATS Providers, Communications Service Providers, Aircraft and Avionics Manufacturers and ATS System Manufacturers to expedite CNS/ATM implementation in a cost effective manner.

C545/049/98

The application of space technology to CNS/ATM for civil aviation

D H FEATHERSTONE CEng, FIEE, FRAeS
Inmarsat, London, UK

1 INTRODUCTION

1.1 The ICAO FANS Committees

Although the feasibility of applying space technology to communications between air traffic controllers and the pilots of aircraft operating in oceanic and other remote airspace was first demonstrated more than thirty years ago, the initiative that resulted in the practical realisation of aeronautical satellite communications was taken in 1983 with the establishment by the International Civil Aviation Organisation (ICAO) of the Future Air Navigation Systems (FANS) Committee. This committee was charged with estimating the growth in global air traffic through the second decade of the new millennium and proposing how best to manage the use of airspace in order to accommodate it. The committee concluded that increasing the capacity of the airspace to handle the expected traffic safely and efficiently would be impossible using the communications, navigation and surveillance (CNS) systems then in service. It went on to examine the operational, technical and economic characteristics of a number of alternative CNS technologies and determined that, for oceanic and remote airspace at least, satellite technology offered the most cost-effective solutions.

The FANS Committee developed the baseline technical and operational concepts for the application of satellite technology to communications and surveillance which were subsequently adopted as guidance for ICAO specialist technical panels established to draft Standards and Recommended Practices (SARPS) in these areas. It also devised the concept of "Required Navigation Performance" (RNP) for aircraft desiring to operate in specific airspace and identified the use of satellite navigation systems as one way, but not necessarily the only way, to achieve such performance. The committee's recommendations, including one advising ICAO to establish a new committee to oversee the work of the technical panels and to ensure that the development and transition planning for the implementation and utilisation of the new systems proceeded on a global co-ordinated basis and in a cost-effective manner, were contained in the report of its final meeting, held in May 1988

The follow-on committee, known familiarly as the FANS II Committee, was established by the ICAO Council in July 1989 and met for the first time in May/June, 1990. In addition to the oversight function, its principal tasks were the development of a global co-ordinated implementation plan, preparing guidance on satellite navigation systems utilisation planning and research into air traffic management (ATM) issues raised by the use of the new technology. Early in its lifetime, a most significant event was the adoption of the original FANS Committee's CNS/ATM concepts by the ICAO 10th Air Navigation Conference in September, 1991. This put the international civil aviation community's seal of approval on the path for the future mapped out by that committee and gave added emphasis to the work of the FANS II Committee concerning systems implementation. The FANS II Committee concluded its work in September 1993. Since then, implementation planning has continued largely in ICAO planning and implementation regional groups (the "PIRGs") with assistance from the Regional Affairs Office at ICAO headquarters in Montreal.

1.2 Other Initiatives

Other initiatives proceeded in parallel with the ICAO FANS Committees' activities. The Airlines Electronic Engineering Committee (AEEC) established a subcommittee in 1985 to define aeronautical satellite communications system and equipment characteristics for the airline industry. Also, Inmarsat began work on the development of an aeronautical satcoms system based on its prior experience with similar systems for the maritime community. After some initial divergence of views among the various parties had been resolved, these activities proceeded synergistically and led to Inmarsat introducing into service in December, 1990 an essentially ICAO SARPs-compliant aeronautical satcoms system using AEEC-specified avionics. Subsequent in-service development of the system and parallel refinement of the SARPs have closed the small gaps in SARPs compliance and produced a mature system which fully meets the stringent operational requirements of air-ground-air communications services for safety-of-life applications, including ATM.

From the outset, the FANS Committee and the members of the technical panel established by ICAO to prepare Aeronautical Mobile Satellite Service (AMSS) SARPs recognised that an aeronautical satellite communications system designed to handle only safety communications would not be economically viable. Means must be found to share the system's facilities between safety and non-safety services so that revenue derived from the latter would be available to underpin the former. Also, the global, or at least regional, nature of satellite services necessitated a different approach to the institutional aspects of providing them. It seemed unlikely that individual States would procure, implement and operate their own satellite communications systems as they had always done for VHF and other terrestrially-based systems. Even if they did, they would probably choose not to engage in the business of providing non-safety communications services in order to acquire the safety service capability. Everything pointed to accepting a concept in which (i) the SARPs defined a system which could, if desired, provide non-safety services in addition to safety services and (ii) State civil aviation authorities and other air traffic service providers who so wished could purchase safety services from third party providers, who might also be offering non-safety services. This was done, and all subsequent system and service design and implementation decisions have been based on it.

2 THE INMARSAT AERONAUTICAL SATCOMS SYSTEM

2.1 Basic Design Considerations

Although by the mid-1980s Inmarsat had amassed considerable experience in the design, implementation and operation of satcoms systems for the maritime community, it rapidly became apparent that if the stringent operational requirements of civil aviation satcoms were to be met, new technological ground would have to be broken. Economic viability for the system dictated that it must share the satellites already placed in orbit to support the maritime systems. Thus, the aeronautical system would have to coexist with maritime (and, later, land mobile) systems without mutual interference. Consideration would also have to be given to the need to minimise the size, weight and power consumption of the equipment carried on board aircraft and to the physical constraints of size, weight and drag on aircraft antenna design.

Inmarsat's satellites are located in the geostationary orbit at longitudes chosen to provide aggregate global coverage (except for the polar caps) from earth disc coverage (global beam) antennas. In fact, the four locations provide considerable redundant coverage of the earth's surface, as figure 1 shows. The satellites act as simple relay devices, receiving signals from ground earth stations (GESs) on carrier frequencies in the C-band and then re-transmitting them to aircraft on carrier frequencies in the L-band. This process, which is known as "bent pipe" operation of the satellite (see figure 2), is reversed for signals received from aircraft. The satellites' transponders, which perform the receive, frequency changing and transmit functions, are wide-band devices each capable of handling numerous carriers simultaneously. The critical RF signal path is that between the aircraft and the satellite. Aircraft antenna limitations result in this link being characterised by low signal to thermal noise ratios and multipath fading, the latter becoming particularly significant over the oceans and at low satellite elevation angles from the aircraft. Closing this link cost-effectively is arguably the biggest challenge in the system design.

2.2 The Aircraft Antenna

A major factor in this exercise is the design of the aircraft antenna. For all services other than a low-speed data link (of which more later), an antenna is required which provides gain by forming a beam which continuously points to the satellite as the aircraft executes its flight path. Several design approaches are possible but the majority of antenna manufacturers have adopted one in which the beam is created by selectively energising different groups of elements in an array such that correct pointing is maintained as the flight proceeds. Information for the antenna's beam steering function concerning satellite location is provided by the satellite communications system, while instantaneous aircraft position data is supplied by the aircraft's navigation sensors. The challenges in the design of such an antenna are many. For example, coverage must extend below the aircraft by a few degrees to ensure that satellites at low elevation angles remain illuminated during aircraft manoeuvres, sidelobes must be small enough to preclude effective illumination of unwanted satellites and excessive sensitivity to reflections and multipath, while the impact on of performance of shadowing from aircraft structure such as wings and stabilisers must be negligibly small.

The net result of the many trade-off studies on antenna parameters for the Inmarsat aeronautical system was the choice of a nominal gain of 12dBic. This permitted the aircraft/satellite RF link to support a maximum RF channel data rate of 21 Kbps, which translated to a maximum service channel data rate of 10.5 Kbps after the application of rate one half forward error correction coding. This data rate was compatible with the use of 9.6 Kbps voice coding, then considered to be the minimum coding rate providing acceptable voice quality, in a channel also accommodating the associated circuit-mode service signalling.

2.3 Data Integrity Assurance

In accordance with the requirements set by the ICAO FANS Committee, the system provides packet-mode data services in addition to the circuit-mode capability just mentioned. Data services are available at several bit rates up to the maximum of 10.5 Kbps. Considerable pains were taken to ensure the integrity of the data carried by the system for both circuit-mode and packet-mode services. The use of rate one half forward error correction coding has already been mentioned. Forward error correction is best suited to dealing with bit errors which occur randomly. However, it is characteristic of the RF medium that errors occur in bursts rather than at random. A process called interleaving is applied to the data stream before it is modulated on to the RF carrier. This causes errors occurring in bursts in the RF medium to appear to the forward error correction process in the receiver as though they are random, thus maximising its effectiveness. Subsequent de-interleaving restores the original sequence of the bits in the data stream before it is presented to the application process. The resulting residual bit error rate is nominally 1 in 10^5. It should be noted that interleaving/de-interleaving as employed in the packet-mode data service introduces delays of the order of one second into the transmission process. Delays greater than 30 msec are not acceptable for voice transmission, so for this service an interleaver/de-interleaver having slightly reduced effectiveness is employed to ensure that this figure is not exceeded. The residual bit error rate, however, remains well within the handling tolerance of the voice codec.

Additionally, further protection from errors not detected and corrected by this process is afforded for the packet-mode service by the use of a 16-bit cyclic redundancy check. For transmission purposes user data is organised into packets known as signal units. The 16-bit sequence, which is transmitted as the last two bytes of each signal unit, is the remainder of the modulo two division of the user data contained in the signal unit by a standard polynomial function. At the receiver, the received signal unit is subjected to the same division process and the remainders compared. If they are identical, the signal unit is deemed to be error-free and is delivered to the application process. If they are not identical, the signal unit is deemed to contain errors and is discarded. The system automatically requests the retransmission of the signal unit, which is again subjected to the cyclic redundancy check upon receipt. Up to five retransmissions may be requested should the first or subsequent retransmissions also fail the check. The net result of all the error detection and correction processes incorporated into the system is that the probability of a packet-mode data service user receiving data containing errors is very low; typically less than 1 in 10^{10}.

Finally, the system incorporates adaptive RF power control in both the forward and return directions. This is designed to optimise the economics of system operation by ensuring that no more than the minimum RF power required to maintain link closure, defined as operation at or below a specified maximum bit error rate, is demanded from either satellites or aircraft equipment. In most cases, the bit error rate performance of the link can be maintained at or below this specified worst case figure with less gain in the system than is actually provided. The additional margin is needed to ensure this performance is achieved at the edges of satellite cover or in adverse multipath environments, etc. In appropriate circumstances therefore, RF power can be reduced to take advantage of this excess gain and the released resources used to increase the capacity of the system, thus contributing to reduced operating costs. As implied, the power control mechanism adjusts the RF power output as a function of the received signal bit error rate.

At the time the system was designed, a need for a version offering only low-speed (600 bps RF channel rate) packet-mode data service was identified. This retains all the integrity assurance features just described but operates through a much simpler and less expensive aircraft antenna which, although providing full coverage of the hemisphere above the aircraft, does not employ a steered beam. This version of the system is known as "Aero-L", while the more capable version based on the 12 dBic steered beam aircraft antenna is known as "Aero-H". Both versions are recognised in the ICAO AMSS SARPs.

To summarise to this point, the Inmarsat aeronautical system was designed in the mid-1980s to meet the satellite communications system operational requirements developed by the ICAO FANS Committee and other civil aviation standards-setting bodies active at that time. The starting point for the design was the performance of the "Inmarsat-2" geostationary satellites then shortly to be introduced into service for Inmarsat's maritime services as replacements for the leased satellite capacity used for this purpose so far. These incorporated transponders operating in the lower 3 MHz of the L-band spectrum allocated by the ITU to AMS(R)S and provided global coverage (except for the polar caps) through the use of earth disc coverage (global beam) antennas. The aeronautical system entered service in late 1990 offering a full range of safety and non-safety circuit-mode (voice) and packet-mode (data) services. At the present time, more than 50 airlines have installed Aero-H and Aero-L avionics on over 1,000 aircraft, in addition to which more than 500 corporate aircraft are equipped. The majority of these installations comprise multi-channel voice and data Aero-H equipment.

2.4 Priority for Safety Services

An important point not so far mentioned is the requirement for safety services always to be accorded precedence in the system. This has been accommodated by incorporating into the design a prioritisation and pre-emption process which ensures that whenever there is contention for system resources between safety and non-safety applications, it is resolved in favour of the safety service. For example, a typical Aero-H aircraft installation is capable of supporting six simultaneously operating communications channels. One of these might be a packet-mode data channel and the reminder circuit-mode channels. If the five circuit-mode channels happen to be all in use for non-safety communications, e.g., passenger telephone calls, when the aircraft captain needs to communicate with ATC or his company AOC office, he can pre-empt the use of one of these channels for this purpose. Many other less extreme examples of this principle can be cited but the result is the same in every case, resources are always made available for safety services. The operating rules for pre-emption and prioritisation are set forth in the ICAO AMSS SARPs and in the ITU Radio Regulations governing the use of the L-band MSS spectrum.

2.5 Channel Types

The system incorporates four channel types, each of which is optimised for a different communications function. The "P-Channel" is a time division multiplex channel used in the forward direction (ground-to-satellite-to-aircraft) to carry system management information and the packet-mode data service. Each GES originates at least one P-channel operating at the lowest system bit-rate (600 bps) which identifies its presence in the system to aircraft earth stations (AESs) listening on its carrier frequency. Having derived system information from this transmission, an AES desiring to log on to the system via this GES will send it a request to do so on an "R-Channel". This is a random access (contention) channel used in the return direction (aircraft-to-satellite-to-ground) to convey system management data and support the packet-mode data service for short messages. Longer packet-mode data messages are conveyed from the aircraft to the ground on a time division multiple access "T-Channel". An AES desiring a transmission slot in a T-channel sends its request to the GES on the R-channel and receives its slot assignment from the GES on the P-channel. The AES then transfers its transmit function to the T-channel frequency and transmits its data in the assigned slot. Circuit-mode services are handled on "C-channels". A C-channel provides the equivalent of a copper wire connection, maintained for the duration of the transaction, between a terminal on the aircraft and its peer on the ground. Pairs of C-channel frequencies, one for the forward direction and one for the return, are assigned to AESs in response to requests for them conveyed to GESs via R-channels and responded to on P-channels. As noted earlier, the signalling required to manage circuit-mode operations subsequent to setting up a C-channel pair is conducted on the C-channels themselves, using bandwidth additional to that required for the service communication. The use of separate forward and return frequencies for a C-channel pair enables full duplex circuit-mode operations to be conducted.

The system is inherently multi-channel. As has been noted, Aero-H AESs typically support simultaneous operations on six channels, of which one is a packet-mode data (R- or T-) channel and the other five are C-channels. Aero-L AESs support one low speed packet-mode data channel only. In the case of the packet-mode service, different bit rates are handled on separate channels. A GES offering this service at a higher bit rate than the 600bps used for the system management channel will radiate the higher bit rate service on a separate P-channel. The carrier frequency assigned to the higher rate service will be part of the information conveyed to AESs on the GESs 600 bps P-channel. Suitably equipped AESs will then transfer to this frequency for the service. The same principle applies to R- and T-channels.

2.6 System Applications

Initially, the system was used primarily for passenger voice services (air-ground telephone calls) and aircraft operational control data communications. Quite soon, however, several oceanic air traffic service providers began to use the data link for technical and operational evaluations of the air traffic management concepts developed by the FANS Committee and the development of practical oceanic air traffic management systems based on them. A co-operative effort by the air traffic service providers, airlines, communications service providers and aircraft and equipment manufacturers concerned with operations in South Pacific airspace produced commitments by all parties to proceed with the operational implementation of such a system based on the use of controller-pilot data link communications (CPDLC) via the Inmarsat aeronautical system for primary communications, with satellite voice as back-up, and automatic dependent surveillance (ADS) as the airspace surveillance tool.

From the satellite communications service provider's point of view, ADS is simply a specialised application of the data link which enables position and other operationally significant information derived on board the aircraft to be transmitted to the ground without pilot involvement. There, it is used by air traffic management automation systems to generate information for the controllers who are managing the airspace. The information may be displayed to them on geographically-referenced airspace situation monitors (pseudo-radar) or in tabular form as appropriate. Controllers are able to vary the content and delivery rate of ADS messages from each aircraft to match their operational needs. They do this by defining their requirements in an "ADS contract" with each aircraft which is set up in a special data link message transaction. These contracts can be amended at any time. The ADS function runs in parallel with the CPDLC function.

Operational CPDLC and ADS services via Inmarsat aeronautical satcoms are now available to suitably equipped aircraft in a number of oceanic airspace sectors in the Asia Pacific region. Because, at the moment, such aircraft form only the minority of the total population operating in this airspace, and the number of suitably equipped sector control positions is also small, overall management of the airspace cannot be predicated upon these services. However, a significant start has been made and, in the fullness of time, the FANS Committee's vision will be realised.

3 NEW SATELLITES - NEW AERO SYSTEM DEVELOPMENTS

3.1 Spot Beams and Aero-I

In the last eighteen months the Inmarsat-2 satellites have been superseded by a new constellation of Inmarsat-3 satellites, also geostationary and located close to the Inmarsat-2s over the world's major oceans. The new satellites offer eight times greater communications capacity than those they have replaced, which remain in service as back-ups. They also incorporate spot beam antennas in addition to global beam antennas. Each satellite can generate up to five spot beams, each of which focuses satellite power on a selected area of the earth's disc illuminated by the global beam antenna. Figure 3 shows the spot beam coverage concept and figure 4 the aggregate spot beam coverage of the four operational Inmarsat-3 satellites. This has the effect of introducing gain into the aircraft-satellite communications link which can be exploited to reduce the RF power required to be generated per channel on the aircraft. This, in turn, translates into reduced weight, smaller size and, above all, lower cost for the aircraft earth station compared to the global beam-only Aero-H system.

The AES savings are, in fact, shared between the avionics black boxes and the antenna. Instead of 12 dBic antenna gain, only 6dBic is required, considerably reducing antenna size, weight and complexity. In Inmarsat parlance, this version of the aeronautical system is referred to as Aero-I, where the "I" denotes the use of the intermediate gain 6 dBic antenna.

Overall, the cost of a ship-set of Aero-I avionics and antenna is between one third and one half of that for an equivalent capability Aero-H AES, typically US$ 85,000.00 - 100,000.00.

The Aero-I system retains all the high-integrity features of the Aero-H/L system design but the opportunity has been taken of advances in voice coding technology made during the last ten years to reduce the encoding bit-rate from 9.6 Kbps to 4.8 Kbps with no loss of voice quality. This, together with a reduction of forward error correction from rate one half to rate two thirds, enables the bit-rate for an Aero-I C-channel to be set at 8.4 Kbps. The forward error correction reduction is possible in part due to the presence of error correction features built into the 4.8 Kbps voice codec which were not available in the 9.6 Kbps codec and in part due to experience with the Aero-H system having shown that the bit error performance of the original C-channel is consistently better than originally predicted.

The combination of lower satellite power and reduced bandwidth per channel results in space segment resource charges for Aero-I circuit-mode services which are significantly lower than for Aero-H.

Aero-I packet-mode data services are provided through the satellites' global beams and are available at bit rates of 600 and 1200 bps. Higher bit rates up to 4.8 Kbps may be provided in the future if required.

Lower avionics cost and lower service charges enable Aero-I to extend the benefits of satellite communications to a range of smaller aircraft types for which the more expensive Aero-H system may not be justifiable. At service launch in May 1998, around a dozen airlines had committed to install Aero-I avionics on more than 300 mostly short- and medium-haul aircraft, including some turbo-props. Both these numbers are expected to grow rapidly as the service becomes established.

The characteristics of the Aero-I version of the system which differ from those of Aero-H and Aero-L, primarily the 6 dBic antenna, the 4.8 Kbps voice codec and the associated C-channel specification, have been submitted to ICAO for adoption as amendments to the AMSS SARPs. This process, which involves technical review and validation of the proposed SARPs changes by ICAO member States' experts, is expected to be complete early in 1999. SARPs recognition will provide the basis for Aero-I to be accepted for safety service use alongside Aero-H and Aero-L.

3.2 Upgrades for Aero-H

Users of the Aero-H system will also be able to enjoy the benefits of the reduced rate voice codec and its associated C-channel as these are to be introduced into the Aero-H system as an optional evolutionary upgrade. The number of simultaneous voice channels handled by an Aero-H AES may be doubled in this way through modifications to the equipment's circuit-mode channel units. Manufacturers may deliver new Aero-H AESs from the factory with this capability incorporated if customers so specify. Use of the Aero-H 12 dBic antenna will further reduce the satellite power needed to close the satellite-to-aircraft link, resulting in space segment resource charges even lower than for Aero-I. The upgraded Aero-H system, which is to be known as "Aero-H+", will offer this service in both global and spot beams .

A further evolutionary upgrade now being studied for commercial viability is the provision of high-speed (32-64 Kbps) packet-mode and circuit-mode data services. Such services are technically feasible though the use of advanced modulation and coding techniques. Providing them would require the definition of additional channel types for the system and the implementation of the associated channel unit hardware and software in GESs and AESs. Market analysis has shown a number of potential applications for high-speed data in the non-safety aircraft passenger service environment. One possible application in the safety-service sector could be the dissemination to aircraft of the monthly navigation system data base updates. This is currently a labour-intensive activity, involving the physical carriage of data loaders onto aircraft and the execution of manual downloading procedures. Needs may also exist for downlinking bulk data from aircraft, e.g., for airline maintenance organisations to acquire in near-real-time aircraft systems and component condition information. If the on-going business studies, which are due to be completed in July 1998, show a positive outcome, high-speed data services could be available in about two years.

3.3 Augmentation Services for GPS & GLONASS

A feature of each Inmarsat-3 satellites not related to its communications services functions is a transponder dedicated to services designed to enhance the performance of the US GPS and the Russian GLONASS satellite navigation systems. Due to their military origins, these systems have certain shortcomings for civil use which can be overcome by the use of such services, which are referred to by ICAO as satellite-based augmentation services (SBAS). Foremost among these services is near real-time GPS/GLONASS system integrity assurance, followed by accuracy and availability enhancement. Integrity assurance and accuracy enhancement rely on the monitoring of the navigation systems satellites by networks of terrestrial stations located at accurately surveyed positions. Anomalous performance due to navigation satellite problems is noted by one or more of these stations and information concerning it encoded into a data message sent to the network's master station. Also encoded into the message is data concerning the instantaneous accuracy of the navigation system which is obtained in part by comparing the position information derived from the navigation systems and the surveyed position of each monitoring station. The network master station collates the data received from the monitoring stations and constructs a data message for uplinking to the navigation transponder on an Inmarsat-3 satellite. This process is shown schematically in Figure 5.

The signal radiated by the satellite is, in fact, a GPS signal having a unique gold code but otherwise identical to the signals transmitted by GPS satellites on the L1 frequency. It is transmitted at a power level which produces a signal flux density at users' receivers similar to that produced by GPS satellite transmissions and carries as modulation the data derived from the monitoring network. To users' GPS receivers, therefore, an Inmarsat-3 satellite appears to be another source of GPS ranging information from whose signals it can also derive near real-time health/status information concerning each GPS and GLONASS satellite, near real-time correction data for short-term accuracy disturbances and correction data for accuracy degradation caused by longer-term ionospheric signal delays. Receivers do require the software and processing capability to deal with this information but are otherwise standard GPS units.

There are two ground monitoring networks currently under construction which will use the SBAS capabilities of Inmarsat's satellites. These are the Wide Area Augmentation System (WAAS) in the United States, which will employ the navigation transponders of the Atlantic Ocean Region-West and Pacific Ocean Region satellites, and the European Geostationary Overlay System (EGNOS), which will use the navigation transponders of the Atlantic Ocean Region-East and Indian Ocean Region satellites. Note that Inmarsat provides only the space-based capability required for these systems and certain designated signatories provide the ground earth stations through which signals are passed to the satellites' navigation transponders. Provision of the terrestrial monitoring networks and responsibility for the augmentation services is undertaken by the civil aviation authorities in whose airspace the services are used.

4 NEXT-GENERATION SATELLITE SYSTEMS

4.1 Low- and Medium Earth Orbit Satcoms Systems

The imminent entry into service of personal communications systems in which satellites in low- or medium-earth orbits provide services directly to hand-held mobile telephones has encouraged at least one ICAO member State to request that organisation to prepare SARPs for such a system to be employed for aeronautical safety services. This request stems from the fact that the wide differences between geostationary and non-geostationary satellite system designs make impractical the use of much of the technical specification material in the existing SARPs. The protagonists of non-geostationary systems contend that the costs of both the avionics and the communications services will be significantly lower than for present systems and that this is a compelling reason for moving towards their use. The AESs for such systems would have to be considerably more sophisticated than a hand-held telephone, of course, and an antenna, although possibly not a steered beam antenna, mounted on the outside of the aircraft's fuselage would be needed. Packet-mode as well as circuit-mode services would be required and the system would have to meet the overall operational requirements already established by ICAO for AMSS.

ICAO's response to this request was to task a working group of the ICAO Aeronautical Mobile Communications Panel (AMCP) to conduct a study of the feasibility of applying non-geostationary satellite system technology to aeronautical safety communications. The study, based largely on information supplied by one personal communications non-geostationary system provider from the ICAO member State which originated the request, showed that this use of the technology is technically feasible. However, it left unanswered the significant question of cost benefit and failed to justify any need to depart from continued adherence to the existing SARPs. Having considered the working group's report, the AMCP concluded at its April, 1998 meeting that the preparation of new SARPs should be based on acceptability criteria for the use of non-geostationary systems to be developed by ICAO.

The Panel's decision not to embrace non-geostationary satellite system technology for aeronautical safety communications before the real need for it has been demonstrated and the costs justified is in keeping with the philosophy adopted by the FANS Committee for introducing satellite CNS into civil aviation in the first place. The committee recognised that in addition to meeting the operational requirements for future air traffic management, the CNS solutions they selected must remain valid for a long-enough time-frame for the needed investment in systems implementation to be justifiable. Thanks to the care taken by ICAO in the definition of these operational requirements and the choice of technical solutions, the satellite systems now in service are capable of meeting those needs for the foreseeable future. Doing this even more cost-effectively in the future will arise from the introduction of evolutionary system upgrades. The first such major upgrade to the Inmarsat aeronautical system was the introduction of Aero-I, which became possible as the consequence of the launch of a constellation of advanced technology satellites. It is quite probable that other system enhancements will follow, not necessarily based on a new space segment but exploiting other areas of evolving technology to provide enhanced services at lower cost. The transition to any new satellite technology should only be considered when this process is no longer capable of providing solutions which meet the operational needs.

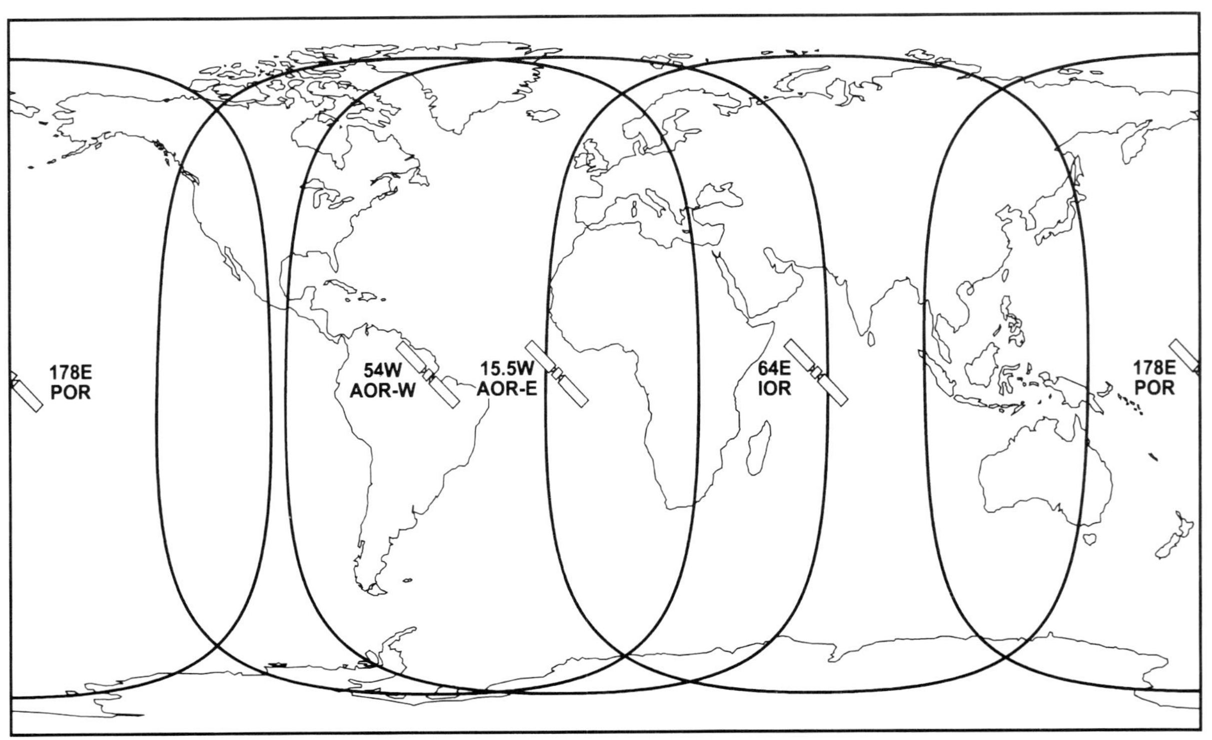

Figure 1: Apart from the poles the Inmarsat-3 satellite constellation covers the globe

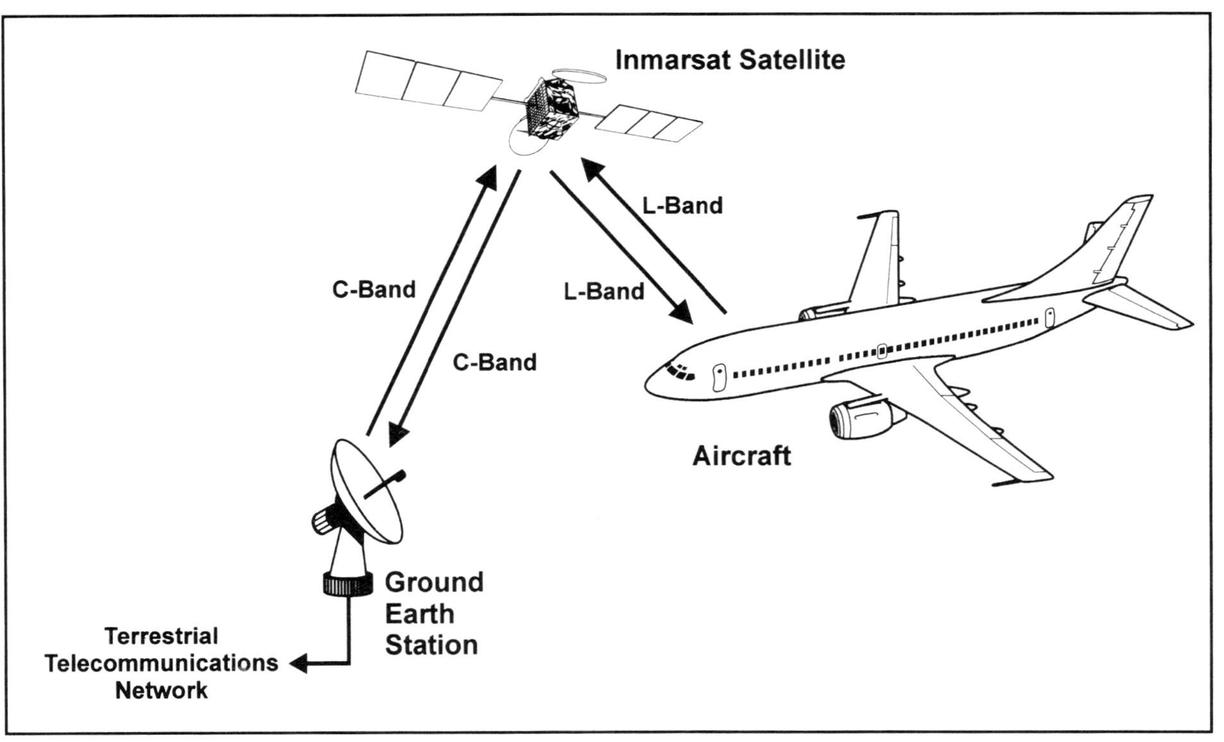

Figure 2: Basic satcoms system configuration

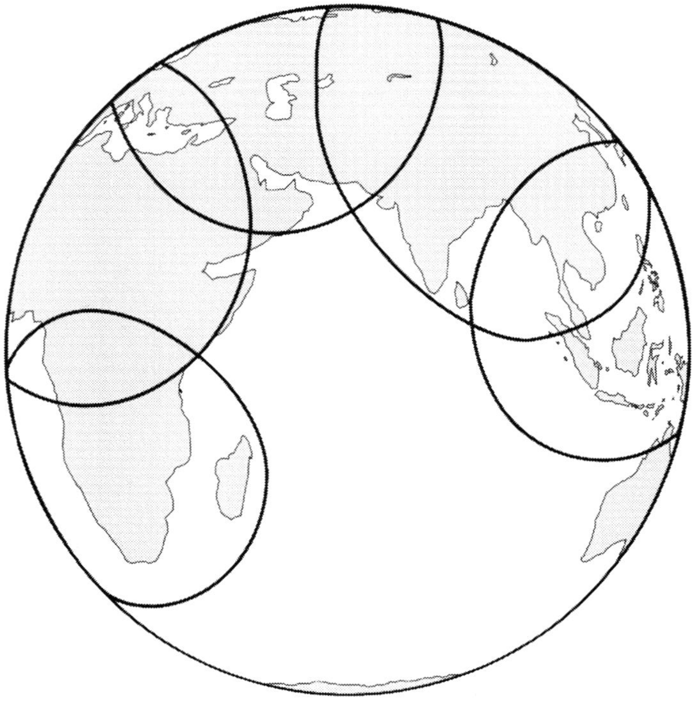

Figure 3: Inmarsat spot beam coverage of the Indian Ocean Region

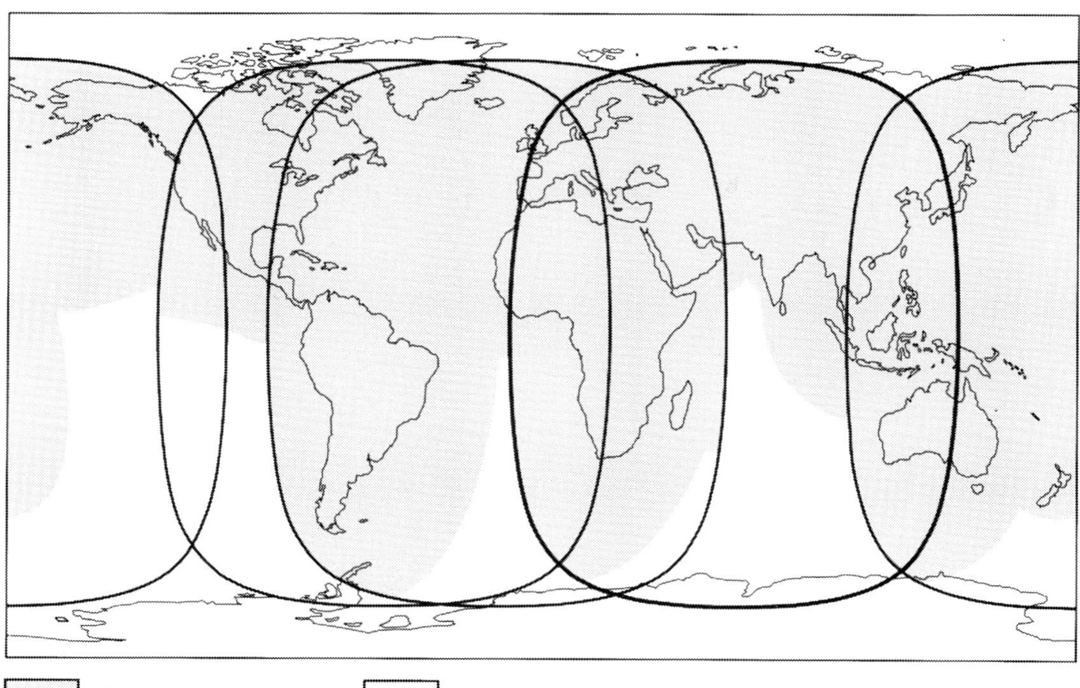

Figure 4: Only a few areas of the world lack spot beam coverage

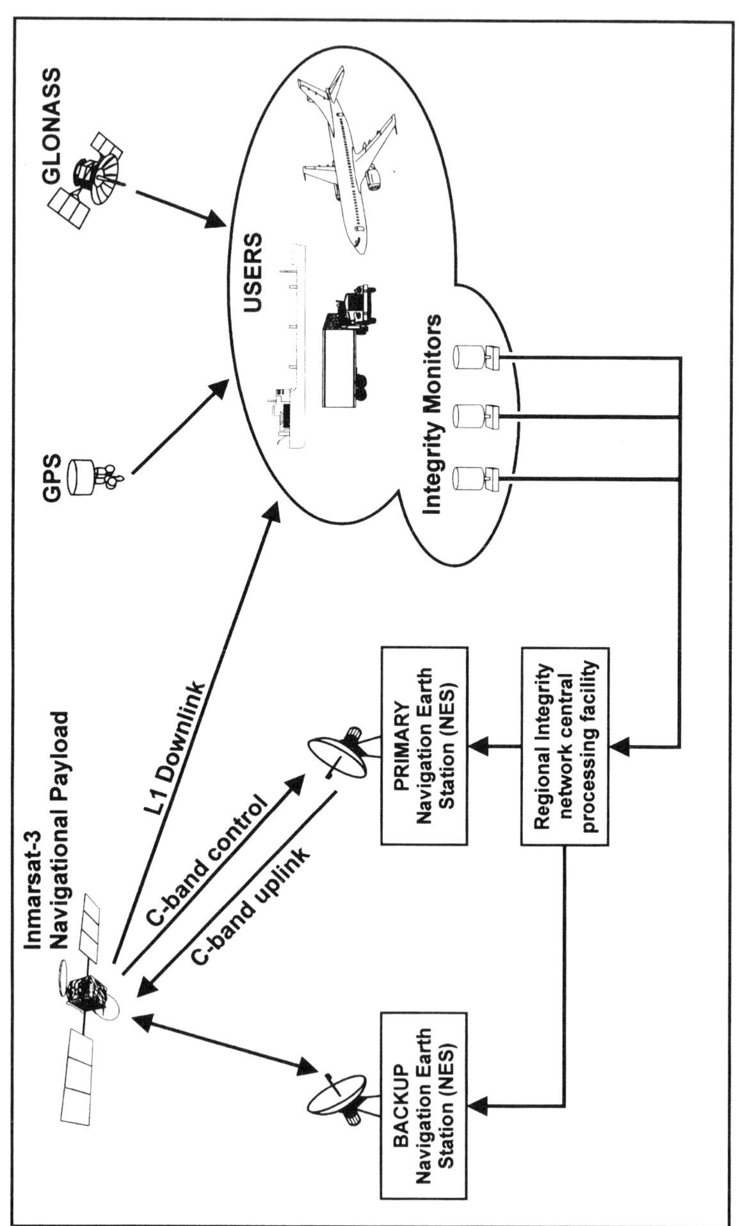

Figure 5: Satellite based augmentation system concept

C545/047/98

RPN RNAV terminal areas – procedures design and approval for Juneau, Alaska

M CRAMER MS
Smiths Industries, USA

While new technology resulting in improved navigation accuracy and integrity provides the potential for shorter and more efficient departure and arrival routes and lower minima, there are a great many corollary considerations that are necessary to make revised operations safe and efficient. At Juneau, given conventional navigation capability, the terrain allows only one route into and out of the airport. However, the integration of the multi-sensor navigation of the Smiths Industries FMS with GPS position sensing and RNP navigation and alerting capability created the potential for using the narrow Gastineau Channel as a sea-level departure and arrival route.

This paper presents an overview of the procedure design and focuses on the operational cockpit integration aspects of how the procedures are to be flown. Details of the equipment and external factors that determine procedure availability are presented along with crew training requirements. Methods for safe extraction of the aircraft in the event of equipment failure are explained, and their evaluation during the demonstration flights is presented.

1. CONVENTIONAL PRACTICE

Instrument departures and arrivals at Juneau are made to and from the west, since the easterly direction must pass through the narrow Gastineau Channel, which is too narrow for safe passage using conventional navigation aids. The normal arrival is a localizer type directional aid (LDA) approach to runway 08 (RW08), which arrives through a saddle in high terrain on the Mendenhall Peninsula which is to the west of the airport. The runway is slightly misaligned with the approach path and terrain restrictions to the east force the missed approach to be a 180 degree turn initiated three miles prior to the runway with an MDA of 1000' and two miles visibility required.(Fig 1)

The usual departure is runway 26 (RW26), which requires a climb over the saddle in the Mendenhall Peninsula. This requires a limitation on maximum takeoff weight to ensure that the aircraft can clear the terrain even with loss of one engine at V_1. A second departure is occasionally used, however, it is considerably more difficult. It is a departure from RW08 which must be done in visual conditions as it demands a 180 degree turn after takeoff to assure the that the aircraft can fly a lateral path which safely clears the terrain.

2. POTENTIAL BENEFIT

Use of the Gastineau Channel as a "back door" route for missed approach, departure and arrival has the potential for significantly improving the restrictive conditions associated with the conventional practice. Conventional navigation does not supply sufficient navigation accuracy and integrity to allow the channel to be safely negotiated in instrument conditions, however advanced FMS/GPS RNP based navigation has the accuracy and integrity needed. Since the Gastineau Channel is essentially a sea level route, its use can lower the minima and alleviate the weight restrictions associated with the current procedures. It also would provide for shorter departure and arrival routes to and from the southeast (Seattle), saving five minutes inbound and six minutes outbound per trip. This distance savings could save 245 flight hours and 740,000 pounds of fuel per year based on current numbers of trips flown.

3. PROCEDURE DEVELOPMENT

The design of these approach and departure procedures required a careful assessment of not only the navigation performance in terms of the accuracy and integrity of position, but an evaluation of the path construction and path following ability of the system as well. This lead to an iterative method of developing the procedures to achieve the most advantageous minima and takeoff weight. In general, the approach procedures were developed by assuming maximum landing gross weight, selecting a desired Required Navigation Performance (RNP) level and performing the analysis to determine a DA(H) based on a lateral containment region of twice the RNP (2 x RNP) with a nominal obstacle clearance of 500' beneath the containment region. In some cases, a minimum of 250' terrain clearance underneath the lateral containment area was allowed with the addition of crew procedures to bound the barometric altitude error.

The second aspect of the procedure development was an analysis of the aircraft's ability to safely negotiate the procedure. This meant an analysis of the avionics performance and integration, since the path construction and path following ability of the aircraft is determined by the integration of the flight control systems and the flight management system. For this application, the integration placed some limitations on the achievable performance gains. Each of these instances is discussed below in the context of the development for each procedure.

3.1 Approach Procedures (Fig 2)

Three approach procedures were initially developed. The first was an approach to RW08 which essentially overlaid the current LDA approach, with the difference being that the missed approach utilized the channel, eliminating the need for a 180 degree turn on the missed approach. This would allow reduced minimums because of the sea level missed approach path, and the removal of the 180 degree turn. The second and third proposed approaches were to RW26 via the channel. The primary goal was again reduced minima. Since some of the considerations for the RW08 missed approach are the same as for the two channel departure procedures, discussion of that design will be covered in the section on departure procedures.(Fig 3)

The objective for development of the first approach to RW26 using the channel was to achieve significant reduction in landing minima at maximum landing weight. Using a nominal 3.5 degree approach gradient, and an RNP of 0.3 nmi, a lateral path was devised which follows the channel past Juneau and ends with a turn of approximately 35 degrees to align with the runway 1.3 miles from the threshold. The missed approach requires a left turn crossing the runway threshold to avoid high terrain, just west of the airport. The terrain near the final turn, and the need to maintain at least 500' beneath the lateral containment area, (2 x RNP on each side of the lateral track) determined with engine out performance at the maximum landing weight placed the DA(H) for this approach at 700'. Comparing this altitude to the vertical path placed the go-around decision point prior to the final turn in the lateral path. This is within the navigation accuracy and integrity provided by the FMS/GPS/IRS navigation system of the aircraft. However, simulations for flight procedure verification, following development, revealed additional flight guidance factors that needed to be addressed.

In simulation, the approach was flown normally, and when a missed approach was initiated at the DA(H), using the Take-off / Go around switch, LNAV and VNAV automatically disengaged and the FCC reverted to flying a constant track. This is normal operation for the flight control computer on the aircraft. However, since the missed approach point is prior to the final turn before the runway (at LEMNN), the disengagement of LNAV meant that, until the crew could re-engage the FMS modes, the aircraft could possibly fly off the desired lateral track in an area of high terrain. This was judged to be unacceptable and the procedure discarded.

Since the difficulty with the preceding proposal was the fact that the DA(H) was reached before the turn at LEMNN, necessitating extra pilot action to perform the missed approach, it was revised by tightening the RNP value for the approach. The controlling factor in the DA(H) was the need to clear high terrain west of the airport in the event of an engine out missed approach. Reducing the RNP value to match the terrain clearance needed on the engine out climb path, permitted use of a lower DA(H). The new procedure was designed to an RNP of 0.2 nmi which reduced the DA(H) to 368'(350'), moving it down the lateral track to a point after the final turn to align with the runway. This better addressed the safety consideration related to initiating a go-around (e.g. initiating TOGA) prior to a turn near high terrain, allowing the procedure to be safely flown. While this is a lower value of RNP than has currently been demonstrated for incorporation in the AFM for Alaska Airline's B737-400 navigation system, anticipated AFM adjustments reflecting an improved level of RNP are expected in the near future. Accordingly, use of RNP 0.2 is planned only in visual conditions until the necessary AFM provisions are completed.

3.2 Departure Procedures

As in the case of the approach design, the desire was to attain lower minima higher gross weight for the takeoff using the sea level route through the channel. A tentative lateral path down the channel was laid out using an initial minimum climb gradient and attempting to place the lateral path such the containment boundaries remained equidistant from the terrain on both sides. It begins with a turn to change from runway heading through a 35 degree right turn to align with the channel and continues through the remaining 12 miles to open water. The design depends on the aircraft's ability to fly this first turn, as reflected in the amount of area needed to assure that the lateral containment boundaries around the lateral path in the turn cannot impact the terrain. This lead to two further architectural considerations that needed to be addressed on an interim basis, until FMS and flight control system designs are eventually updated.

First, with both flight directors on, as is normal practice for takeoff, the TOGA mode in the of the Flight Control Computers (FCCs) are engaged which prevents LNAV from being selected for Flight Director (FD) guidance below 400' AGL. Second, the aircraft bank angle is limited to 8 degrees by the FCC, reducing turn performance at altitudes below 400'. On a normal, full power takeoff, the aircraft climb performance is such that 400' is reached well before the turn to follow the channel needs to be started, however, the procedure must be designed to be flyable with a loss of one engine at V_1.

With the restrictions on turning flight below 400' noted above, it was necessary to consider changes to the normal cockpit procedure so that the engine out takeoff can be flown safely. Instead of having both flight directors on at TOGA, the pilot not flying will leave his flight director off. This prevents the FCC from entering its TOGA mode allowing LNAV to be engaged immediately on takeoff. This gives the pilot flying positive course guidance on the flight director for the initial turn even if an engine is lost and the aircraft does not reach 400' before the turn must be initiated. The 8 degree bank limit is still in effect however, and this means that the turn will be wider than normal for the engine out condition. Specific obstructions encountered on this departure required the RNP to be specified at 0.17 nmi in order to attain the highest allowable takeoff gross weight. Since full guidance is provided for departure the minimums obtained are 1/4 mile visibility and zero ceiling.

An analysis of the availability of GPS signals and the associated HIL required to support an RNP of 0.17 nmi in the FMS navigation showed that it would be unavailable some percentage of the time, so a higher RNP departure was also designed. This provides nearly 100% availability of the procedure as satellite coverage will almost always be sufficient to maintain the required accuracy. This departure still provides the lower minima for takeoff, no ceiling specified and 1/4 mile visibility, however since the RNP was increased to 0.3 nmi, a large containment area results, which includes higher obstacles to be considered. This resulted in a requirement for an earlier turn and reduction in the allowable takeoff gross weight because of the climb gradient required.

The missed approach for the RW08 GPS approach follows the same path as the second of the departures discussed above, which provides an improvement over the existing LDA approach (which it overlays) because it lowers the minima.

4. FLIGHT CREW CONSIDERATIONS

The flight crew retains its normal responsibility for determining whether a particular procedure is still usable just prior to beginning it. This includes checking that the minimum airborne equipment needed to fly the procedure is available and that the procedure required aircraft performance parameters are within the required range. Dispatchers will predict the availability of appropriate GPS coverage and the weather expected relative to the minimums. The flight crew will still need to ascertain that all required equipment is functional and all other conditions are sufficient to actually commit to the procedure. The following will highlight the new aspects of these RNP based procedures in terms of crew decision making.

4.1 Equipment Availability

Aircraft Systems : The navigation equipment will be used in a manner which provides for normal operations, rare normal operations (e.g. winds and turbulence, etc.), and non-normal operations (e.g. navigation or aircraft failure conditions). No additional backup means of navigation is necessary, appropriate, or available, to meet the RNP for the procedures. The reliability of the FMS, GPS, EFIS and IRS components is such that dual redundancy is sufficient to assure that once the procedure is begun, it can be safely completed with any probable single equipment failure. Operational dual FMS, dual EFIS, dual GPS and Dual IRS components are considered to be the minimum required for the navigation systems. This of course is in conjunction with the normal requirement for FCC's and flight directors.

Flight checking of the procedures also showed that the weather radar could be used to "paint" the terrain in the channel. This provided an exceptional aid to crew situational awareness, adding to the integrity of the navigation, so the weather radar was added to the minimum equipment list for using these procedures.

External Systems (GPS): While checking airborne equipment availability is a flight crew responsibility, the availability of the supporting external systems is first predicted by dispatch to allow filing a flight plan to use one of the procedures. This process models the GPS constellation visible over Juneau for the half hour around the scheduled arrival or departure time to assess the RNP level available for the procedure. If the predicted RNP is sufficient, the desired procedure can be planned for use and the aircraft dispatched normally. For the approaches, this must be updated in flight for any changes which would preclude the use of the planned procedure.

4.2 Approach Procedures

There are specific flight crew checks to be made prior to beginning either the approach to RW08 with the channel missed approach or the approach to RW26 through the channel. The minimum equipment must be operational, and the RNP for the procedure must be entered into the FMS. The first waypoint of the departure must be cross checked from the chart, and course and distance verified to all subsequent waypoints. The actual navigation performance (ANP) of the system must be equal to or less than the RNP at throughout the procedure. The FMS provides the proper level of alerting relative to navigation performance by comparing ANP to RNP and warning the crew when ANP exceeds RNP. For the approach to RW26 or the missed approach for RW08, flying through the channel, the weather radar may be utilized to provide increased terrain awareness as the approach is flown.

4.3 Departure Procedures

The departures require a crew check of actual navigation performance (ANP) relative to a specified limit when at the runway threshold. If the ANP is less than the limit, at that time, the pilot may commit to the departure. This is based on the fact that for actual performance better than the limit, even a loss of GPS at takeoff will not cause an unacceptable loss of navigation performance prior to reaching the end of the departure procedure. The ground calibration of the IRS through the use of GPS into the FMS up to the departure is sufficient.

As an interim procedure, the crew must only engage one flight director for the Pilot Flying, leaving the other one off to prevent the LNAV "lockout" due to TOGA mode in the FCC. They must also verify the latitude and longitude of the first departure waypoint (at the turn), and then verify the course and distance between each of the subsequent waypoints of the departure route. The waypoint coordinates printed on the departure chart may not be directly derived from the navigation database source data so the check of the FMS waypoint location is truly independent.

4.4 Training Requirements

The planned procedures were flown in the simulator to form the basis for a training program. It was determined that a ground school time of 4 hours, followed by 4 hours in the simulator would be adequate to prepare crews to fly RNP, use the new FMS features (e.g. GPS, etc.) and safely fly the new procedures. At certain critical terrain airports (e.g. those specified through FAR 121.445) the airline will require actually flying the procedures under supervision before crews are qualified to use them in revenue service.

5. FLIGHT DEMONSTRATION

During June of 1995, demonstration flights were conducted for the FAA to operationally verify navigation system performance, aircraft performance, flight procedures, failure condition safety, and crew procedures. . These flights were intended to expose any problems with the procedure design as well as to demonstrate their flyability. Each procedure was flown normally first, then various combinations of equipment "failures" were introduced. Each of the abnormal conditions is discussed below, with the resulting actions or restrictions to use of the procedures noted.

5.1 Loss of GPS

Since the FMS/GPS/IRS combination is used as primary navigation for these procedures and dual systems are required, loss of a single GPS, IRS or FMS function does not affect the completion of the procedure. However, GPS signal-in-space could be lost, which would remove all GPS updating capability from the navigation. This condition was studied analytically to assure appropriate levels of ANP and derive the departure ANP minimum threshold. Since the FMS is calibrating the IRS using the GPS inputs pre-departure, and the ANP is calculated from the IRS model and the measurement inputs, the study projected IRS drifts following calibration and assured that the RNP could be met for the whole departure. In the departure case, the ANP must be less or equal to the predetermined threshold value, and in the approach case, it must be less than or equal to the RNP for the approach throughout the procedure. The check prior to departure assures that total loss of GPS updating as early as the start of the procedure will not affect the ability to complete it safely and the continuous comparison of ANP to RNP during the approach procedure assures timely warning to allow the crew to perform a missed approach if the ANP is too large for the approach.

5.2 Loss of Coupled Navigation

The approaches are predicated on the aircraft being flown with autopilot engaged in LNAV and VNAV modes. However, a backup mode is available based on use of the flight director using LNAV or VNAV or manual flying using the EFIS map mode (e.g. the pilot manually follows the map display magenta line). Each of these was demonstrated for all the procedures to allow safe extraction of the aircraft in the event of dual failures which could prevent full autopilot coupled LNAV/VNAV guidance.

5.3 Loss of Generators

Noting that the navigation system is used as a "primary means of navigation", an assessment had to be made that there were no probable system failure modes which could cause total loss of navigation. This was verified. Nonetheless, in the very rare instance that such an event were to occur, a method to address that possibility was identified. To assure adequate electrical power, the FMS and CDUs are connected to the main generators and, at present, neither side can be run from the standby power system. To prevent a total loss of navigation in the event that all generators fail or they cannot be re-connected in a timely manner, an interim APU electrical supply procedure is used. Until an aircraft change can be made to place one FMS/MDCU on the standby bus, the APU must be running to provide a backup generator for each of these procedures.

5.4 Loss of Engine

There were several instances where the loss of an engine needed to be considered. It had to be demonstrated that the departures could be safely flown with loss of an engine occurring at the most critical point during takeoff, and that both a rejected landing and the planned missed approach could be performed in the equivalent condition. Each of these cases was demonstrated by reducing thrust of one engine to idle at the appropriate critical time and completing the scenario with the remaining operating engine thrust level and providing performance at a value equivalent to that of limit weight takeoff or approach. Each of the procedures was demonstrated considering the most critical failures and no revisions were necessary based on the trials.

6. CONCLUSIONS

The experience of designing procedures to solve difficult operational scenarios such as mountainous airport terrain clearance problems, while improving takeoff and landing weather minimums and improving allowable gross weight associated with a particular runway leads one to some interesting conclusions. First and foremost, the enabling technical achievement was the integration of FMS/IRS RNP based navigation with EFIS map displays, supported by the accuracy of GPS. The integrity of the navigation derives from the advanced FMS/IRS/GPS combination, the RNP / ANP alerting capability and the GPS accuracy which speaks for itself.

However, with the available FMS solution to the navigation integrity and accuracy problems to levels necessary to implement these procedures, only the first part of the procedure development and utilization problem has been solved. More cooperative attempts to develop beneficial procedures using the current aircraft equipment are needed to reveal technical integration problems which can reduce the effectiveness of the added accuracy and integrity.

The procedure development at Juneau shortened departure and arrival routes significantly, reducing the time of flight by 5-6 minutes each. It also reduced weather minima from an MDA(H) of 1000' for RW08 to a DA(H) of 750' and provided an advantageous new approach from the southeast with a DA(H) of 368'(350')' and provided takeoffs to the east with minima of a zero ceiling and 1/4 mile visibility. These are extremely important to the reduction of cancellations and diversions for the airline, translating directly into cost avoidance.

Fortunately, the architectural characteristics of the avionics systems, which initially were keyed to routine operations using conventional scenarios at typical airports, did not preclude a significant reduction in mimima at a special airport such as Juneau. However, as the currently available accuracy and integrity of FMS/IRS/GPS navigation and EFIS map displays are increasingly utilized to address problems of airport access or efficiency, such as those at Juneau, the solution of architectural issues described above take on increasing importance and should be addressed.

The ability to perform an LNAV takeoff, with appropriate lateral control authority increasing as altitude is initially gained should become standard. Coupled LNAV based guidance for Go-Around should be normal for FMS navigation systems, as should fully coupled VNAV. Once changes like these are made, and as other issues are identified and dealt with during further development of similar procedures, the full potential of existing RNP capable FMS/EFIS/GPS aircraft can be realized in practice.

These capabilities are important not only for critical terrain airports such as Juneau, Alaska, but they are also key to many future safety, efficiency, and capacity enhancements throughout the NAS and globally. This is true because many airports will require and benefit from similar RNP based critical arrival and departure procedures for traffic capacity increases, as well as for terrain clearance considerations.

Licensed to SMITHS INDUSTRIES LTD, . Printed 3/3/98.
Notice: After 3/12/98 0901Z, this chart may no longer be valid. Check JeppView.
JEPPESEN
19 SEP 97
11-1
JUNEAU, ALASKA
JUNEAU INTL
LDA-1 Rwy 8
LDA 109.9 IJDL
Apt. Elev 19'
*ATIS 126.4
ANCHORAGE Center 133.9
*JUNEAU Tower 118.7
*Ground 121.9
JUNEAU Radio (LAA) CTAF 118.7 when Twr inop.
Alt Set: INCHES
MSA
CGL NDB
6300'
8200'
5700'
Use IJDL DME when on LOC course. For night circling, if runway lights not visible over the MAP, execute missed approach.
COGHLAN I
212 CGL
D2.0 IJDL
DIBOL
D18.5 IJDL
(IAF)
LYNNS
D14.0
IJDL
BARLO
D8.0 IJDL
062°
242°
to LYNNS
242° 12.0
7000
(IAF)
SISTERS I
114.0 SSR
LDA DME
062° 109.9 IJDL
BERGS
39.0
7000 093°
(IAF)
ELEPHANT
391 EEF
to LYNNS
357° 12.8
5200 NoPT
When DME is not used, continuous ADF monitoring of CGL NDB is required to identify the MAP. Any go-around commenced after passing the MAP will not provide standard obstruction clearance.
DIBOL
D18.5 IJDL
LYNNS
D14.0 IJDL
CGL NDB
D2.0 IJDL
BARLO
D8.0 IJDL
10 NM from LYNNS
6100'
(6081')
5200'
(5181')
3400'
(3381')
063°
TDZE 19'
4.5
6.0
6.0
3.2
MISSED APPROACH: Immediate climbing RIGHT turn via IJDL LDA SOUTHWEST course to cross BARLO INT/D8.0 IJDL LDA at or above 3000'. Continue climb to 5200' direct to SSR VOR or EEF NDB and hold.
STRAIGHT-IN LANDING RWY 8
CIRCLE-TO-LAND
Not Authorized North of Rwy 8-26
DAY
NIGHT
MDA(H)
Max Kts
A 2120' (2101')-4 90 2120' (2101')-4 3000' (2981')-5
B 2700' (2681')-4 120 2700' (2681')-4
C 2880' (2861')-4 140 2880' (2861')-4 3000' (2981')-10
D 3200' (3181')-4 165 3200' (3181')-4 3200' (3181')-10
MAP at CGL NDB/ D2.0 IJDL
CHANGES: See other side.
© JEPPESEN SANDERSON, INC., 1987, 1995. ALL RIGHTS RESERVED.

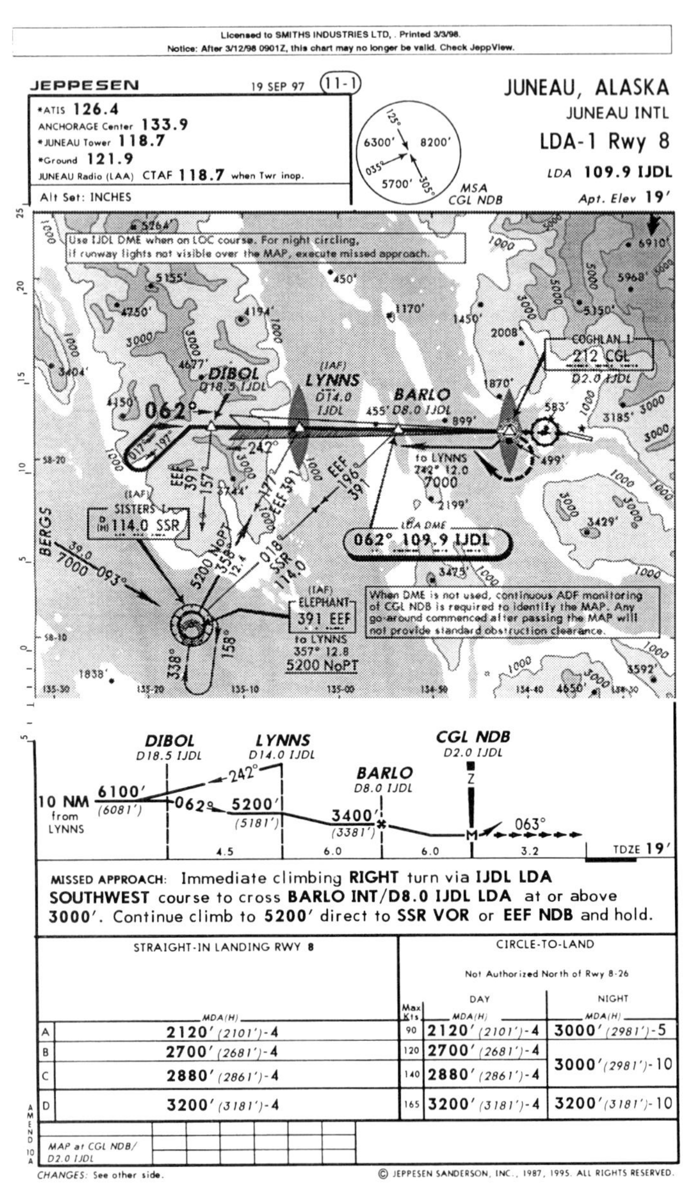

Figure 1

Alaska Airlines
PROOF
24 APR 98
12-7-1
JUNEAU, ALASKA
JUNEAU INTL
RNP RNAV Rwy 8
Special Aircrew
Training Required
RNV Ø8
Apt. Elev 19'
*ATIS 126.4
ANCHORAGE Center 133.9
*JUNEAU Tower 118.7
*Ground 121.9
JUNEAU Radio (LAA) CTAF 118.7 when Twr inop.
Alt Set: INCHES
6500'
8300'
5800'
MSA
RW08
RNP 0.3 RW26 to BARLO, RNP 0.5 BARLO to FUNTR.
COGHLAN I
212 CGL
BARLO
LEMNN
SMGLR
RW26
SALMN
TRDWL
MARMN
FUNTR
MISSED APCH FIX
NOT TO SCALE
210 Kt
230 Kt authorized 6001'-14000'
RNV26
For use by SLANT E (/E), dual GPS equipped, RNP certified aircraft only.
See chart 10-2 for RNP RNAV Rwy 8 approach transitions.
DIBOL
LYNNS
BARLO
CGL NDB
MND NDB
RW08
10 NM from LYNNS
6100' (6081')
5200' (5181')
3540' (3521')
1270' (1251')
CIRCLE
TCH 49'
TDZE 19'
MISSED APPROACH: Climb to 7000' via the RNP RNAV missed approach track to to FUNTR and hold. Cross ADMRL at or above 7000'. Between TRDWL and MARMN right turn direct ADMRL at or above 5600' authorized.
STRAIGHT-IN LANDING RWY 8
RNP 0.15 DA(H) 724'(705')
RNP 0.2 DA(H) 782'(763')
RNP 0.3 DA(H) 822'(803')
LDIN inop
CIRCLE-TO-LAND
Not Authorized North of Rwy 8-26
DAY
NIGHT
MDA(H)
140 2300'(2281')-3
165 2820'(2801')-3
3000'(2981')-5
Gnd speed-Kts 70 90 100 120 140 160
VNAV 3.50° 439 565 628 753 879 1004
Vis - 4 if LDIN inop.
CHANGES: New procedure. PROOF COPY ONLY.
© JEPPESEN SANDERSON, INC., 1998. ALL RIGHTS RESERVED.

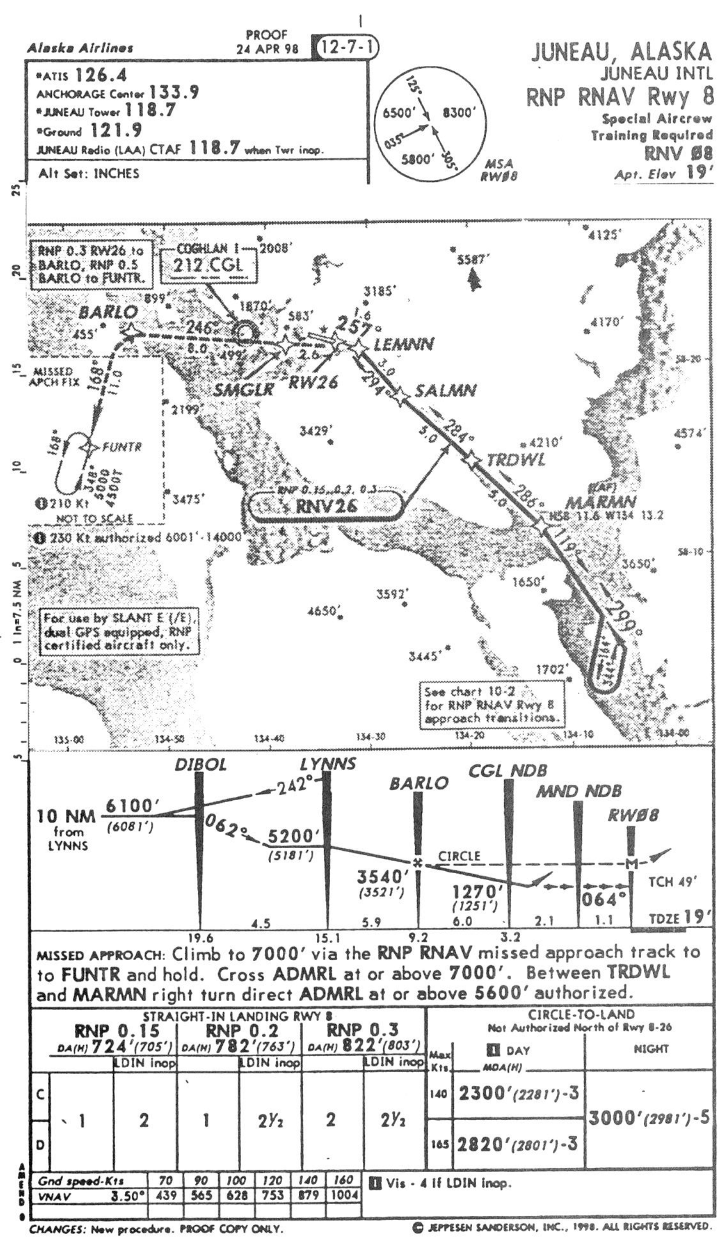

Figure 2

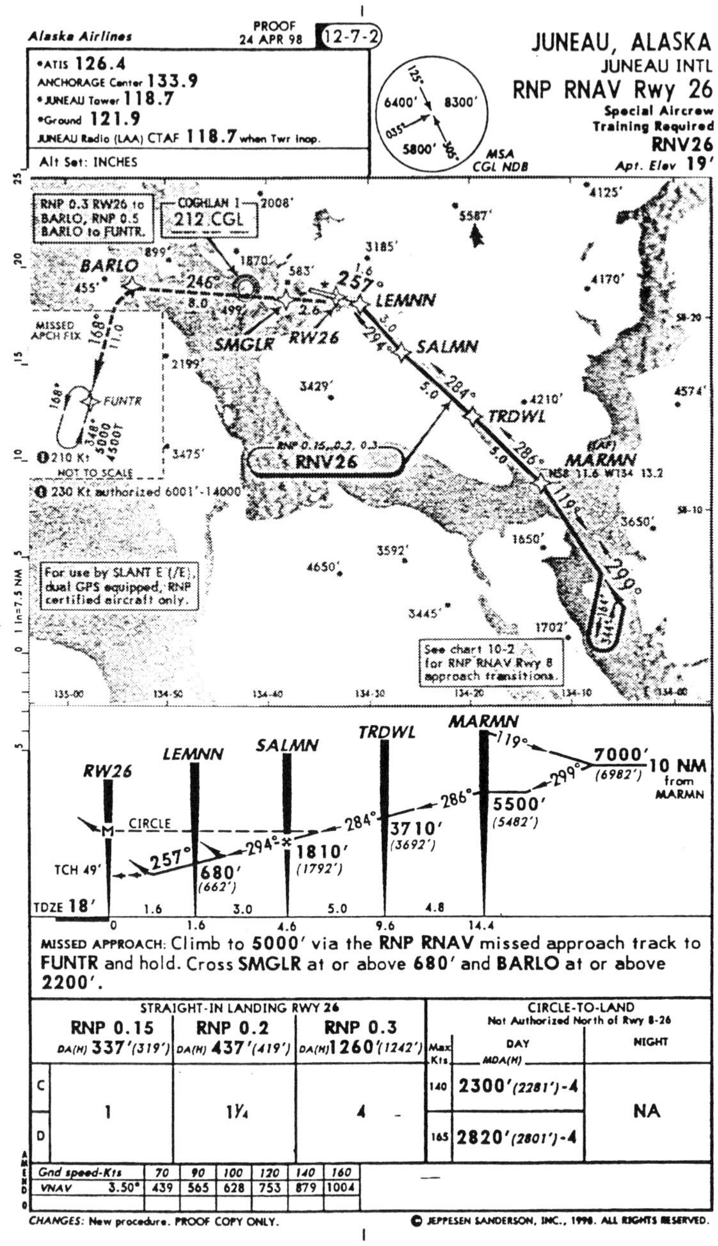

MISSED APPROACH: Climb to **5000'** via the **RNP RNAV** missed approach track to **FUNTR** and hold. Cross **SMGLR** at or above **680'** and **BARLO** at or above **2200'**.

	STRAIGHT-IN LANDING RWY 26			CIRCLE-TO-LAND Not Authorized North of Rwy 8-26		
	RNP 0.15 DA(H) 337'(319')	RNP 0.2 DA(H) 437'(419')	RNP 0.3 DA(H) 1260'(1242')	Max Kts	DAY MDA(H)	NIGHT
C	1	1¼	4	140	2300'(2281')-4	NA
D				165	2820'(2801')-4	

Gnd speed-Kts	70	90	100	120	140	160
VNAV 3.50°	439	565	628	753	879	1004

CHANGES: New procedure. PROOF COPY ONLY.

Figure 3

Authors' Index